Application of Deep Learning in Underwater Image Processing

Application of Deep Learning in Underwater Image Processing

Guest Editors

Chia-Hung Yeh
Chua-Chin Wang
Guo-Shiang Lin

MDPI

Basel • Beijing • Wuhan • Barcelona • Belgrade • Novi Sad • Cluj • Manchester

Guest Editors
Chia-Hung Yeh
Department of Electrical Engineering
National Taiwan Normal University
Taipei
Taiwan

Chua-Chin Wang
Department of Electrical Engineering
National Sun Yat-Sen University
Kaohsiung
Taiwan

Guo-Shiang Lin
Department of Computer Science and Information Engineering
National Chin-Yi University of Technology
Taichung
Taiwan

Editorial Office
MDPI AG
Grosspeteranlage 5
4052 Basel, Switzerland

This is a reprint of the Special Issue, published open access by the journal *Journal of Marine Science and Engineering* (ISSN 2077-1312), freely accessible at: https://www.mdpi.com/journal/jmse/special_issues/NABN08KHD7.

For citation purposes, cite each article independently as indicated on the article page online and as indicated below:

Lastname, A.A.; Lastname, B.B. Article Title. *Journal Name* **Year**, *Volume Number*, Page Range.

ISBN 978-3-7258-5585-8 (Hbk)
ISBN 978-3-7258-5586-5 (PDF)
https://doi.org/10.3390/books978-3-7258-5586-5

Contents

About the Editors

Chia-Hung Yeh

Chia-Hung Yeh received B.S. and Ph.D. degrees from the Department of Electrical Engineering, National Chung Cheng University, Chiayi, Taiwan, in 1997 and 2002, respectively. From August 2002 to December 2004, he was a Postdoctoral Fellow with the Department of Electrical Engineering Systems, University of Southern California, Los Angeles, CA, USA. He was an assistant professor from 2007 to 2010, an associate professor from 2010 to 2013, and a professor from 2013 to 2017, with the Department of Electrical Engineering, National Sun Yat-sen University, Kaohsiung, Taiwan. He is currently a distinguished professor with National Taiwan Normal University, Taipei, Taiwan. He has coauthored more than 300 technical international conferences and journal papers and held more than 50 patents in the USA, Taiwan, and China. His research interests include deep learning, video coding, and image/video processing. He was the Associate Editor for the *Journal of Visual Communication and Image Representation, EURASIP Journal on Advances in Signal Processing, ICT Express, and APSIPA Transactions on Signal and Information Processing*. He was the recipient of the 2013 Distinguished Young Researcher Award of NSYSU, the 2013 IEEE MMSP Top 10% Paper Award, the 2014 IEEE GCCE Outstanding Poster Award, the 2015 APSIPA Distinguished Lecturer, the 2017 Distinguished Professor Award of NTNU, the IEEE Outstanding Technical Achievement Award (IEEE Tainan Section), and IET Fellow.

Chua-Chin Wang

Chua-Chin Wang is a Full Professor with Department of Electrical Engineering, National Sun Yat-Sen University (NSYSU), Kaohsiung, Taiwan. His research interests include memory and logic circuit design, communication circuit design, biomedical circuits, and interfacing I/O circuits. He was bestowed with the Distinguished Engineering Professor designation by the Chinese Institute of Engineers and the Fellow honored by IET, and he received the Outstanding Research Award from NSYSU in 2012. His accomplishments earned him the ASE Chair Professor designation in 2013. He was designated as the Distinguished Lecturer at the IEEE CASS from 2019 to 2021. He was Dean of Engineering College and VP of Research and Development Office in NSYSU, respectively, in 2014–2017 and 2021–2024. He is now Director of the CHIP4H Program funded by the National Science and Technology Council (NSTC) of Taiwan since 2024.

Guo-Shiang Lin

Guo-Shiang Lin is a full professor at the Department of Computer Science and Information Engineering, National Chin-Yi University of Technology, Taichung, Taiwan. His research interests include multimedia signal processing, computer vision, machine learning (deep learning), and multimedia applications. He served as the guest editor of *Multimedia Tools and Applications, Journal of Information Science and Engineering*, Special Session co-Chair of IWAIT2026, Workshop co-chair of AVSS2025, Publication chair of IS3C2025, Special session organization chair of ICCE-TW2024, Tutorial co-chair of APSIPA ASC 2023, and Program co-chair of CVGIP2022. He was a visiting scholar at the Department of Mathematics and Computer Science, University of Münster, Germany, and Muroran Institute of Technology, Japan.

Journal of
*Marine Science
and Engineering*

MDPI

Editorial

Application of Deep Learning in Underwater Image Processing

Chia-Hung Yeh [1,2,*], Chua-Chin Wang [2] and Guo-Shiang Lin [3]

1 Department of Electrical Engineering, National Taiwan Normal University, Taipei 106, Taiwan
2 Department of Electrical Engineering, National Sun Yat-sen University, Kaohsiung 80424, Taiwan; ccwang@ee.nsysu.edu.tw
3 Department of Computer Science and Information Engineering, National Chin-Yi University of Technology, Taichung 402, Taiwan; gslin@ncut.edu.tw
* Correspondence: chyeh@ntnu.edu.tw

1. Introduction

The ocean covers a significant portion of the Earth's surface and harbors abundant natural resources, meaning that underwater exploration is of significant importance. Various underwater technologies, including autonomous underwater vehicles (AUVs), rely heavily on visual data for navigation, mapping, and environmental analysis, making underwater imaging a critical component of ocean exploration. As a result, underwater image processing has become a vital technology across multiple domains, including marine biology, oceanography, and underwater robotics. However, acquiring and processing high-quality underwater images is exceptionally challenging due to the complex and uncontrollable nature of the underwater environment. Common issues, often caused by artificial lighting, such as light attenuation, color distortion, low contrast, blurred details, and noise, significantly degrade image quality. These problems not only hinder human visual perception but also reduce the effectiveness of automated analysis. Moreover, the unique optical properties of underwater imaging, including selective light absorption and scattering, mean that traditional in-air imaging methods or conventional enhancement algorithms are insufficient. Specially designed algorithms for underwater image enhancement, correction, and analysis are essential in addressing these challenges and advancing underwater exploration.

Classic image enhancement methods in underwater settings rely on the physical modeling of light propagation to approximate and reverse image degradation [1] or employ priors such as dark-channel estimation [2] and color information [3] to improve visual quality. Beyond enhancement, conventional approaches to underwater object detection, segmentation, and tracking have relied primarily on handcrafted features and heuristic rules [4,5]. Although these approaches can improve underwater image quality, they typically depend on fixed assumptions that may not generalize well to the diverse and dynamic conditions encountered in underwater environments. In recent years, deep learning-based methods have emerged as powerful alternatives for underwater image processing. These methods leverage large amounts of data and neural network architectures to automatically learn representations and achieve impressive performance. For example, several studies have applied convolutional neural networks, transformers, and state-space models to enhance and restore underwater images [6–10]. In addition, deep learning techniques have been widely adopted for underwater object detection and segmentation [11–15], demonstrating strong performance in various underwater environments. Compared to traditional methods, deep learning-based approaches offer improved robustness, leading to enhanced image detail, higher accuracy, and better adaptability to complex underwater conditions.

This book, *Application of Deep Learning in Underwater Image Processing*, presents nine innovative approaches that leverage deep learning techniques to address the key challenges

 https://doi.org/10.3390/jmse13081591

in underwater image processing. In addition to a quantitative study on underwater image enhancement, this publication includes a comprehensive introduction to related topics, including image restoration, underwater object detection and segmentation, and sonar image analysis. Researchers have conducted in-depth analyses and integrated advanced techniques, including Generative Adversarial Networks (GANs) and the You Only Look Once (YOLO) series, to further advance deep learning applications in underwater image processing. Furthermore, we provide a thorough comparative analysis of conventional and deep learning-based methods. In summary, this book serves as a comprehensive resource exploring recent breakthroughs in deep learning for underwater imaging, offering both practical tools and conceptual frameworks for professionals engaged in marine science, engineering, and computer vision research.

2. An Overview of Published Articles

Awan et al. (contribution 1) address the challenging problem of underwater image degradation caused by color distortion and contrast loss due to light attenuation and scattering. Existing methods typically use linear transformations for color compensation followed by image enhancement. However, the authors observed that linear transformations for color compensation may fail to enhance images across a variety of underwater scenes. To address this problem, the authors propose a dual-pathway framework that uses a classifier to categorize underwater images as Type I or Type II based on their color characteristics. Type I images benefit from linear transformation, whereas Type II images decline in quality when a linear transform is applied. Depending on the classification, images are then processed using either the Deep Line Model or the Deep Curve Model, which perform linear or nonlinear transformations, respectively. The framework demonstrates superior performance in restoring underwater images on benchmark datasets. Future research directions include refining the classifier to handle ambiguous cases, extending the model to more nuanced image categories, and integrating richer performance metrics for deeper insights into image quality restoration.

Fu et al. (contribution 2) tackle the challenges of underwater image degradation caused by absorption, scattering, and complex lighting conditions. Existing GAN-based methods often suffer from instability during training, resulting in suboptimal enhancement. To address these issues, the authors propose the Multi-scale Evolutionary Generative Adversarial Network (MEvo-GAN), which integrates genetic algorithms into GANs to enhance underwater images. The MEvoGAN framework employs a multi-path generator architecture to extract features at different spatial scales. This design improves the network's ability to recover fine textures and global structures from degraded underwater images. In addition, the authors integrated an evolutionary algorithm composed of variation, evaluation, and selection modules. These components guide generator training by simulating natural selection, generating multiple offspring models, and choosing the best-performing candidates based on metrics. The experimental results show that MEvoGAN outperforms existing methods in restoring underwater images in benchmark datasets.

Gayá-Vilar et al. (contribution 3) address the challenge of efficiently monitoring cold-water coral habitats in deep-sea environments. These habitats are difficult to assess due to limited visibility, light attenuation, and complex backgrounds. Traditional object detection methods are limited in terms of computational efficiency and real-time performance. To overcome these limitations, the authors utilized the YOLOv8l-seg to detect and segment coral species in underwater images. The experimental results demonstrated that YOLOv8l-seg is effective in monitoring cold-water coral species. Furthermore, the study revealed that the coral distribution is highly uneven and is greatly influenced by environmental conditions.

Jiang et al. (contribution 4) propose an improved YOLOv8 for identifying and classifying marine debris. Marine debris has caused significant environmental damage, emphasizing the necessity of cleanup. However, the identification and classification of marine debris are hindered by low underwater visibility, which slows down cleanup operations. To overcome these limitations, the authors integrated the clo block transformer module into the YOLOv8 backbone network. This enhancement improves the extraction of both high- and low-frequency features from underwater debris images, thereby enhancing the perception of crucial image information, particularly for small, indistinct targets. Furthermore, the authors introduced the coarse-to-fine spatial and channel reconstruction module to reduce spatial and channel redundancy and enhance feature representation to handle confusion caused by suspended matter and varying light intensities underwater. The experimental results show that the improved model outperforms the original YOLOv8 in marine debris detection tasks.

Zheng et al. (contribution 5) propose Multi-gradient Feature Fusion YOLOv7 (MFF-YOLOv7) to address the challenges of target detection in underwater sonar images. These challenges include the complex underwater environment, low-quality sonar image data, and limited sample sizes. MFF-YOLOv7 involves several key modifications to the YOLOv7 model, including replacing its spatial pyramid pooling channel shuffling and pixel-level convolution with a multi-scale information fusion module to enhance multi-scale feature processing and reduce missed detections for various target sizes. In addition, the authors introduced recurrent feature aggregation convolution to improve feature extraction and adaptability to noisy sonar images. This allows the model to better learn and represent target features. Furthermore, a spatial and channel synergistic attention mechanism was integrated to help the model focus on crucial features, thereby boosting recognition accuracy and robustness in challenging underwater environments. The experimental results show that MFF-YOLOv7 achieves higher accuracy than other object detection approaches.

Yang et al. (contribution 6) address the challenges of target recognition in side-scan sonar (SSS) images, which suffer from distortion and noise, leading to blurred details and feature loss. Existing models often have limitations when deployed on edge devices due to their high computational complexity and resource consumption. To address these issues, the authors propose SS-YOLO, a lightweight deep learning model focused on SSS target detection that aims to achieve both a lightweight design and enhanced accuracy. The SS-YOLO framework improves the YOLOv8 model by replacing the complex convolutional layer in the coarse-to-fine module with a combination of partial convolution and pointwise convolution to reduce redundant computations and memory access. Additionally, they integrated an adaptive scale spatial fusion module using 3D convolution to combine multi-scale feature maps, maximizing the extraction of invariant features and addressing information loss. The authors also included an improved multi-head self-attention mechanism in the detection head, which enhanced the model's ability to focus on important features with low computational load. Furthermore, the authors propose a new side-scan sonar dataset, created by combining self-collected and public data and expanding it through augmentation to overcome limited sample sizes. The experimental results demonstrate that SS-YOLO outperforms the original YOLOv8 model in terms of accuracy while maintaining lower model complexity.

Lu et al. (contribution 7) address the challenges of underwater object detection caused by the limitations of sonar imaging, such as noise, low resolution, and the lack of texture and color information. To overcome these challenges and improve detection accuracy, the authors developed AquaYOLO, an enhanced version of YOLOv8. YOLOv8 consists of a backbone module, neck module, and prediction module. The backbone module performs initial feature extraction, the neck module refines and captures detailed features, and the prediction head identifies and localizes objects. The authors propose a residual block to replace traditional convolutional layers in the backbone module for improved feature extraction. In addition, they propose a dynamic selection aggregation module in the neck module to dynamically fuse multi-layer features and enhance feature correlation. The experimental results show that AquaYOLO achieves superior performance on a custom sonar image dataset.

Lin et al. (contribution 8) address the challenges of underwater image degradation stemming from color attenuation, scattering, and noise from artificial illumination. In contrast to traditional GAN-based restoration models, which often require paired data or suffer from color inconsistencies, the authors propose a Dual-CycleGAN framework with dynamic guidance for robust underwater image restoration. Their framework comprises two collaboratively trained CycleGANs: a Light Field CycleGAN, which generates enhanced light field guidance images, and a Restoration CycleGAN, which performs the actual restoration process. Integrating light field information into the model guides the restoration process, significantly improving color fidelity and structural detail. A comprehensive set of loss functions supports the training process, including adversarial, cycle consistency, perceptual, identity, patch-based contrast quality index, intermediate output, and color balance losses. These functions ensure enhanced training stability and perceptual quality. The Dual-CycleGAN achieves state-of-the-art performance with reduced computational complexity.

Ma et al. (contribution 9) conducted a comparative study on a traditional physical model and four deep learning-based approaches for underwater image enhancement. The traditional method employed techniques such as color balance correction, LAB spatial decomposition, adaptive histogram, bilateral filter, and Laplace pyramid decomposition. The deep learning-based methods included water-net, UWCNN, UWCycleGAN, and U-shape Transformer. Based on evaluations using the UIEB dataset, the authors conclude that traditional methods are more effective in shallow and stable waters environments, while deep learning-based methods are better suited to diverse and dynamic underwater conditions. For future work, the authors suggest incorporating physical priors into deep learning architectures, developing lightweight models, and exploring adaptive enhancement strategies.

Conflicts of Interest: The authors declare no conflicts of interest.

List of Contributions:

1. Awan, H.S.A.; Mahmood, M.T. Deep Dynamic Weights for Underwater Image Restoration. *J. Mar. Sci. Eng.* **2024**, *12*, 1208. https://doi.org/10.3390/jmse12071208.
2. Fu, F.; Liu, P.; Shao, Z.; Xu, J.; Fang, M. MEvo-GAN: A Multi-Scale Evolutionary Generative Adversarial Network for Underwater Image Enhancement. *J. Mar. Sci. Eng.* **2024**, *12*, 1210. https://doi.org/10.3390/jmse12071210.
3. Gayá-Vilar, A.; Abad-Uribarren, A.; Rodríguez-Basalo, A.; Ríos, P.; Cristobo, J.; Prado, E. Deep Learning Based Characterization of Cold-Water Coral Habitat at Central Cantabrian Natura 2000 Sites Using YOLOv8. *J. Mar. Sci. Eng.* **2024**, *12*, 1617. https://doi.org/10.3390/jmse12091617.
4. Jiang, W.; Yang, L.; Bu, Y. Research on the Identification and Classification of Marine Debris Based on Improved YOLOv8. *J. Mar. Sci. Eng.* **2024**, *12*, 1748. https://doi.org/10.3390/jmse12101748.
5. Zheng, K.; Liang, H.; Zhao, H.; Chen, Z.; Xie, G.; Li, L.; Lu, J.; Long, Z. Application and Analysis of the MFF-YOLOv7 Model in Underwater Sonar Image Target Detection. *J. Mar. Sci. Eng.* **2024**, *12*, 2326. https://doi.org/10.3390/jmse12122326.
6. Yang, N.; Li, G.; Wang, S.; Wei, Z.; Ren, H.; Zhang, X.; Pei, Y. SS-YOLO: A Lightweight Deep Learning Model Focused on Side-Scan Sonar Target Detection. *J. Mar. Sci. Eng.* **2025**, *13*, 66. https://doi.org/10.3390/jmse13010066.

7. Lu, Y.; Zhang, J.; Chen, Q.; Xu, C.; Irfan, M.; Chen, Z. AquaYOLO: Enhancing YOLOv8 for Accurate Underwater Object Detection for Sonar Images. *J. Mar. Sci. Eng.* **2025**, *13*, 73. https://doi.org/10.3390/jmse13010073.
8. Lin, Y.-Y.; Huang, W.-J.; Yeh, C.-H. Dual-CycleGANs with Dynamic Guidance for Robust Underwater Image Restoration. *J. Mar. Sci. Eng.* **2025**, *13*, 231. https://doi.org/10.3390/jmse13020231.
9. Ma, Y.; Cheng, Y.; Zhang, D. Comparative Analysis of Traditional and Deep Learning Approaches for Underwater Remote Sensing Image Enhancement: A Quantitative Study. *J. Mar. Sci. Eng.* **2025**, *13*, 899. https://doi.org/10.3390/jmse13050899.

References

1. Trucco, E.; Olmos-Antillon, A.T. Self-tuning underwater image restoration. *IEEE J. Ocean. Eng.* **2006**, *31*, 511–519. [CrossRef]
2. Chiang, Y.-W.; Chen, Y.-C. Underwater image enhancement by wavelength compensation and dehazing. *IEEE Trans. Image Process.* **2012**, *21*, 1765–1769. [CrossRef] [PubMed]
3. Drews, P.L.J.; Nascimento, E.R.; Botelho, S.S.C.; Campos, M.F.M. Underwater depth estimation and image restoration based on single images. *IEEE Comput. Graph. Appl.* **2016**, *36*, 24–35. [CrossRef] [PubMed]
4. Lee, D.; Kim, G.; Kim, D.; Myung, H.; Choi, H.-T. Vision-based object detection and tracking for autonomous navigation of underwater robots. *Ocean Eng.* **2012**, *48*, 59–68. [CrossRef]
5. Walter, D.; Edgingtion, D.R.; Koch, C. Detection and tracking of objects in underwater video. In Proceedings of the IEEE Computer Society Conference on Computer Vision and Pattern Recognition, Washington, DC, USA, 27 June–2 July 2004.
6. Wang, Y.; Zhang, J.; Cao, Y.; Wang, Z. A deep CNN method for underwater image enhancement. In Proceedings of the IEEE International Conference on Image Processing, Beijing, China, 17–20 September 2017.
7. Li, C.; Anwar, S.; Porikli, F. Underwater scene prior inspired deep underwater image and video enhancement. *Pattern Recogit.* **2020**, *98*, 107038. [CrossRef]
8. Peng, L.; Zhu, C.; Bian, L. U-shape transformer for underwater image enhancement. *IEEE Trans. Image Process.* **2023**, *32*, 3066–3079. [CrossRef] [PubMed]
9. Peng, Y.-T.; Chen, Y.-R.; Chen, G.-R.; Liao, C.-J. Histoformer: Histogram-based transformer for efficient underwater image enhancement. *IEEE J. Ocean. Eng.* **2025**, *50*, 164–177. [CrossRef]
10. Lin, W.-T.; Lin, Y.-X.; Chen, J.-W.; Hua, K.-L. PixMamba: Leveraging state space models in a dual-level architecture for underwater image enhancement. In Proceedings of the Asian Conference on Computer Vision, Hanoi, Vietnam, 8–12 December 2024.
11. Gao, J.; Zhang, Y.; Geng, X.; Tang, H.; Bhatti, U.A. PE-Transformer: Path enhancement transformer for improving underwater object detection. *Expert Syst. Appl.* **2024**, *246*, 123253. [CrossRef]
12. Dai, L.; Liu, H.; Song, P.; Liu, M. Composited FishNet: Fish detection and species recognition from low-quality underwater videos. *IEEE Trans. Image Process.* **2021**, *30*, 4719–4734.
13. Zhou, J.; He, Z.; Lam, K.-M.; Wang, Y.; Zhang, W.; Guo, C.; Li, C. AMSP-UOD: When vortex convolution and stochastic perturbation meet underwater object detection. In Proceedings of the AAAI Conference on Artificial Intelligence, Vancouver, BC, Canada, 26–27 February 2024.
14. Lian, S.; Li, H.; Cong, R.; Li, S.; Zhang, W.; Sam, K. WaterMask: Instance segmentation for underwater imagery. In Proceedings of the IEEE/CVF International Conference on Computer Vision, Paris, France, 4–6 October 2023.
15. Yao, M.; Tam, K.M.; Wang, M.; Li, L.; Kawakami, R. Language-guided reasoning segmentation for underwater images. *Inf. Fusion* **2025**, *122*, 103177. [CrossRef]

Journal of
Marine Science and Engineering

MDPI

Article

Deep Dynamic Weights for Underwater Image Restoration

Hafiz Shakeel Ahmad Awan and Muhammad Tariq Mahmood *

Future Convergence Engineering, School of Computer Science and Engineering, Korea University of Technology and Education, 1600 Chungjeolro, Byeongcheonmyeon, Cheonan 31253, Republic of Korea
* Correspondence: tariq@koreatech.ac.kr; Tel.: +82-041-560-1483

Abstract: Underwater imaging presents unique challenges, notably color distortions and reduced contrast due to light attenuation and scattering. Most underwater image enhancement methods first use linear transformations for color compensation and then enhance the image. We observed that linear transformation for color compensation is not suitable for certain images. For such images, non-linear mapping is a better choice. This paper introduces a unique underwater image restoration approach leveraging a streamlined convolutional neural network (CNN) for dynamic weight learning for linear and non-linear mapping. In the first phase, a classifier is applied that classifies the input images as Type I or Type II. In the second phase, we use the Deep Line Model (DLM) for Type-I images and the Deep Curve Model (DCM) for Type-II images. For mapping an input image to an output image, the DLM creatively combines color compensation and contrast adjustment in a single step and uses deep lines for transformation, whereas the DCM employs higher-order curves. Both models utilize lightweight neural networks that learn per-pixel dynamic weights based on the input image's characteristics. Comprehensive evaluations on benchmark datasets using metrics like peak signal-to-noise ratio (PSNR) and root mean square error (RMSE) affirm our method's effectiveness in accurately restoring underwater images, outperforming existing techniques.

Keywords: underwater images; underwater image restoration; underwater image enhancement; color restoration; lightweight network; deep learning

1. Introduction

Underwater imaging plays an important role in ocean observation and marine engineering applications. However, underwater images suffer from several artifacts. While capturing underwater images, a considerable portion of the light is absorbed during its propagation in the water, resulting in color distortion [1]. Moreover, backward–forward light scattering severely affects the contrast and details of images, which further deteriorates the performance of underwater industrial applications [2]. Therefore, underwater image enhancement—addressing color restoration, enhancing contrast, and improving details—is an essential task in marine engineering and observation applications.

In the literature, many methods have been proposed for improving underwater image quality. These methods can be broadly categorized into prior-based, imaging-based, and machine-deep learning-based techniques [2–5]. Generally, prior-based methods heavily depend on hand-crafted priors and excel in dehazing outdoor images. However, their performance is less than satisfactory for underwater images, and they struggle to correctly manage color shifts. Although imaging-based approaches significantly improve the color and contrast of underwater images, they often overlook the specificities of underwater imaging models. This oversight can result in over-enhanced or over-saturated final images. Machine/deep learning methods provide better results; however, they usually suffer from generalization problems. We observed that most of the methods from all three categories behave differently for different input images. It means that their performance depends on the characteristics of input images. They may work well on some images, while they may not provide good results for other images. Therefore, it is important to study the characteristics of the input images.

In this study, we propose a method for underwater image restoration that employs linear or non linear mapping depending on the type of the input image. First, an input image is classified as Type I or Type II. Then, Type-I images are enhanced using the Deep Line Model (DLM), while the Deep Curve Model (DCM) is employed for Type-II images. The DLM effectively integrates color compensation and contrast adjustment in a unified process, utilizing deep lines for transformation, whereas the DCM is focused on applying higher-order curves for image enhancement. Both models utilize lightweight neural networks that learn per-pixel dynamic weights based on the input image's characteristics. The main contributions of the paper are summarized below:

- We observed that color components of the degraded underwater images have linear and non-linear relationships among them. So, images are classified as Type I or Type II. Different treatment is suggested for different types of images, and it yields better results.
- The Deep Line Model (DLM) is proposed for input images having linear relationships among their color components. As the color components have a linear relationship, pixels can be improved using a linear (line) model, whereas the DLM learns the parameters of the line for each pixel.
- The Deep Curve Model (DCM) is proposed for images having non-linear relationships among their color components. As the color components have a non-linear relationship, pixels may not be improved using a line model. In this case, a curve is more appropriate and effective in improving the color components. Thus, the DCM learns the parameters of the curve for each pixel.

The efficacy of the proposed solution is measured by conducting experiments on benchmark datasets and using quantitative metrics: the peak signal-to-noise ratio (PSNR) and root mean square error (RMSE). The comparative analysis affirms our method's effectiveness in accurately restoring underwater images, outperforming existing techniques.

2. Related Work

2.1. Underwater Physical Imaging Model

The underwater image formation model (IFM) also known as the atmospheric scattering model (ASM) is depicted in Figure 1. It can be seen that three types of lights are received by the image-capturing device: (1) the reflected light that comes to the camera directly after striking the object, (2) the forward-scattered light that deviates from the original direction after striking the object, (3) the back-scattering light that comes to the camera after encountering particles. The attenuation of light depends both on the distance of the device to the object and the light's wavelengths and is affected by seasonal, geographic, and climate variations. These factors, in turn, severely affect the quality of the captured images, and the image restoration become a challenging task. There are various image formation methods that describe the formation of images in scattering media, but we adopt the model proposed in Schechner and Kopeika [6] in this work. According to this model, the intensity of the image in each color channel $c \in \{R, G, B\}$ at each pixel is composed of two components: the attenuated signal, which represents the amount of light absorbed by the underwater medium, and the veiling light, which represents the light scattered by the medium.

$$I_c(\mathbf{x}) = J_c(\mathbf{x})t_c(\mathbf{x}) + (1 - t_c(\mathbf{x})) \cdot A_c\,, \tag{1}$$

where bold denotes vectors, x is the pixel coordinate, $I_c(x)$ is the acquired image value in color channel c, $t_c(x)$ is the transmission of that color channel, and $J_c(x)$ is the object radiance. The global veiling-light component A_c is the scene value in areas with no objects ($t_c = 0,\ \forall c \in \{R, G, B\}$). In homogeneous media, the transmission map (TM) can be described by $t_c(x) = e^{-\beta_c z(x)}$, where β is the medium attenuation coefficient and $z(x)$ is the depth. The primary objective of underwater image restoration methods is to restore the original image $J(x)$ with corrected colors from the observed and degraded image $I_c(x)$.

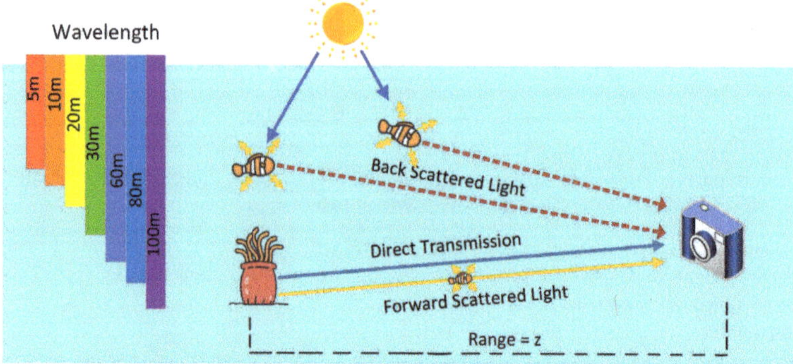

Figure 1. Formation of underwater images. The direct transmission of light contains valuable information about the scene, while the backscattered light degrades the image by reflecting off the suspended particles in the water column. The distance between the camera and the object, denoted by z, affects the clarity of the image. Red light is absorbed more quickly than other wavelengths, making it less effective for underwater imaging. Additionally, forward scattered light can blur the scene, further reducing image quality.

2.2. Underwater Restoration Techniques

Underwater image restoration techniques can broadly be categorized into prior-based, imaging-based, and machine-deep learning-based techniques. Prior-based methods utilize the underwater image formation model (IFM) and draw priors from the degraded images. Initially, a transmission map (TM) is derived from priors such as the dark-channel prior (DCP) [3], red-channel prior (RCP) [7], medium-channel prior (MDP) [8], and haze-line prior (NLD) [9]. Subsequently, the image is restored using the IFM, which is equipped with the TM and atmospheric light. In [10], a red-channel prior (RCP)-guided variational framework is introduced to enhance the TM, and the image is restored utilizing the IFM. In contrast to the prior-based methods, imaging-based methods do not utilize the IFM. Instead, they rely on foundational image enhancement techniques such as contrast enhancement, histogram equalization, image fusion, and depth estimation. Peng et al. [2] proposed a depth estimation technique for underwater scenes that relies on image blurriness and light absorption. This depth information is fed into the IFM to restore and enhance the underwater visuals. In another study by Ancuti et al. [11], a combined approach of color compensation and white balancing is applied to the original degraded image to restore its clarity. Zhang et al. [12] introduced a strategy guided by the minimum color loss principle and maximum attenuation map to adjust for color shifts. In another recent work by Zhang et al. [13], a Retinex-inspired color correction mechanism is employed to eliminate color cast. The research further incorporates both local and global contrast-enhanced versions of the image to refine the color output.

On the other hand, deep learning methods are mainly divided into ASM-based and non-ASM-based techniques. ASM-based methods use the atmospheric scattering model (ASM) to clear up hazy images. For instance, DehazeNet [14] by Cai et al. applies a deep architectural approach to estimate transmission maps, generating clear images. Similarly, MSCNN [15] by Ren et al. uses a multi-scale network to learn the mapping between hazy images and their corresponding transmission maps. AOD-Net [16] by Li et al. directly creates clear images with a lightweight CNN, and DCPDN [17] by Zhang et al. leverages an innovative network architecture, focusing on multi-level pyramid pooling to optimize the dehazing performance. In contrast, non-ASM-based methods rely on various network designs to transform hazy images directly into clear ones through various structures like the encoder–decoder, GAN-based, attention-based, knowledge transfer, and transformer-based networks. Encoder–decoder structures like the gated fusion network (GFN) by Ren et al. [18] and Gated Context Aggregation Network (GCANet) by Chen et al. [19] utilize multiple inputs and dilated convolutions to effectively reduce halo effects and enhance feature extraction. GAN-based networks such as Cycle-Dehaze by Engin et al. [20] and BPPNet by Singh et al. [21] offer unpaired training processes and are capable of learning multiple complexities, thereby yielding high-quality dehazing results even with minimal training datasets. Attention-based networks like GridDehazeNet by Liu et al. [22] and FFA-Net by Qin et al. [23] implement adaptive and attention-based techniques, providing more flexibility and efficiently dealing with non-homogeneous haze. Knowledge transfer methods like KTDN by Wu et al. [5] leverage teacher–student networks, enhancing performance in non-homogeneous haze conditions by transferring the robust knowledge acquired by the teacher network. In [24], to tackle the problems of low contrast, color distortion and poor visual appearance, a sequence of operations such as white balancing, gamma correction, sharpening, and manipulating weight maps are performed on the input image. In [25], a CycleGAN-based network is proposed that uses the U-Net structure in the generator part, as the long skip connection of U-Net will obtain more detailed information. The pixel-level attention block is appended in the network for detail structure modeling. Transformer-based networks like DehazeFormer by Song et al. [26] make significant modifications in traditional structures and employ innovative techniques like SoftReLU and RescaleNorm, presenting better performance in dehazing tasks with efficient computational cost and parameter utilization. In a more recent deep learning-based method [27], a style transfer network is used to synthesize underwater images from clear images. Then, an underwater image enhancement network with a U-shaped convolutional variational autoencoder is constructed for underwater image restoration. In another work [28], a physical imaging-based model is proposed that includes a multi-scale progressive enhancement module to enrich the image details and a chromatic aberration correction mechanism for color balance. In [29], an underwater image enhancement scheme is proposed that incorporates domain adaptation. Firstly, an underwater dataset fitting model (UDFM) is developed for merging degraded datasets. Then, an underwater image enhancement model (UIEM) is suggested for image enhancement. In a more recent work [30], we propose a multi-task fusion where fusion weights are obtained from the similarity measures. Fusion based on such weights provides better image enhancement and restoration capabilities.

3. Motivation

We conducted an experiment using 11,950 images with ground truths (GTs) from two well-known datasets: EUVP [31] and UIEBD [32], as well as various methods from the literature: ACT [33], AOD [16], DNet [14], FGAN [31], MMLE [12], NLD [9], RLP [34], SCNet [35], UDCP [36], and UNTV [10]. First, we computed the root mean square error (RMSE) and peak signal-to-noise ratio (PSNR) using the input images and their GTs. Then, these methods were applied to the 11,950 images, and the metrics RMSE and PSNR were computed using the restored images and their GTs. We then compared the metric values before and after applying the methods to the images. After counting the number of images with improved and declined metrics, the results were unexpected. Figure 2 shows the number of images that have improved and declined metrics. It can be observed that a larger number of images, after applying the restoration methods, have not improved the metrics RMSE and PSNR. This indicates that the performance of the methods depends on the characteristics of the input images. They may work well on some images, while they may not provide expected results for others. Therefore, it is important to study the characteristics of the input images. For this purpose, we investigate the relationships between the color components of the input images.

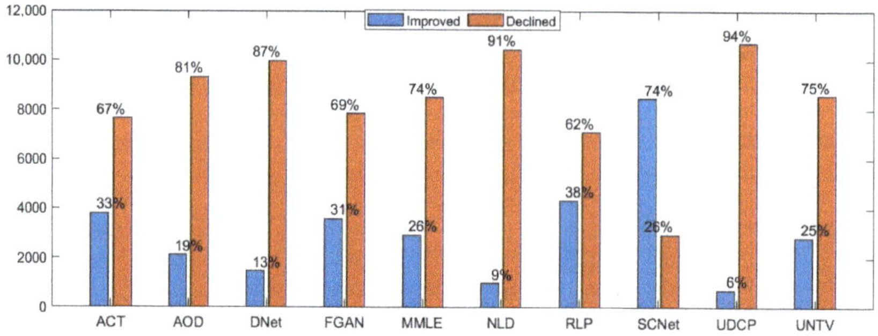

Figure 2. Methods including ACT [33], AOD [16], DNet [14], FGAN [31], MMLE [12], NLD [9], RLP [34], SCNet [35], UDCP [36], and UNTV [10] are applied on 11,950 images. The numbers and their percentages are shown for the number of images where the metrics RMSE and PSNR are improved and declined, respectively.

Let $I_c(\boldsymbol{x})$ represent a degraded underwater color image, where $\boldsymbol{x} = (x, y)$ denotes the coordinates of the image pixels and $c \in \{r, g, b\}$ signifies the red, green, and blue color channels, respectively. The color components of the image can thus be denoted as $\{I_r(\boldsymbol{x}), I_g(\boldsymbol{x}), I_b(\boldsymbol{x})\}$. In underwater imaging, differential color attenuation across wavelengths frequently leads to compromised visual fidelity, predominantly impacting the red channel while leaving the green comparatively unaltered [11]. Conventional restoration techniques typically adopt a sequential approach: initial color correction to balance channel disparities, followed by linear enhancement methods such as contrast stretching to mitigate the attenuation effects.

In the literature, many methods use the mean values from each channel for color compensation [11,37–39]. This approach is grounded in the Gray World assumption, which suggests that all channels should exhibit equal mean intensities in an undistorted image [40], leading to a straightforward approach for color compensation:

$$\begin{cases} I_r(\boldsymbol{x}) = I_r(\boldsymbol{x}) + (\bar{I}_g(\boldsymbol{x}) - \bar{I}_r(\boldsymbol{x})) \cdot (1 - I_r(\boldsymbol{x}) \cdot I_g(\boldsymbol{x})), \\ I_g(\boldsymbol{x}) = I_g(\boldsymbol{x}), \\ I_b(\boldsymbol{x}) = I_b(\boldsymbol{x}) + (\bar{I}_g(\boldsymbol{x}) - \bar{I}_b(\boldsymbol{x})) \cdot (1 - I_b(\boldsymbol{x}) \cdot I_g(\boldsymbol{x})). \end{cases} \tag{2}$$

where $\bar{I}_r(\boldsymbol{x})$, $\bar{I}_g(\boldsymbol{x})$, and $\bar{I}_b(\boldsymbol{x})$ denote the mean values of the degraded color components of the underwater image $I_c(\boldsymbol{x})$.

Although additive adjustments can compensate for color distortions in red and blue channels, our study reveals that this compensation may worsen the color composition in many cases, leading to inferior quality in restored images. As demonstrated in Figure 3, two distinct outcomes are observed: Type-I images benefit from color correction, with spectral intensities approaching the ground truth, enhancing visual quality. Conversely, Type-II images experience worsened color discrepancies, resulting in suboptimal restoration. This necessitates a dual restoration approach. Our method uses a classifier to categorize images, which is followed by the application of the DLM for Type-I images and the DCM for Type-II images. This strategy ensures precise, adaptive restoration aligned with the specific requirements of each image category.

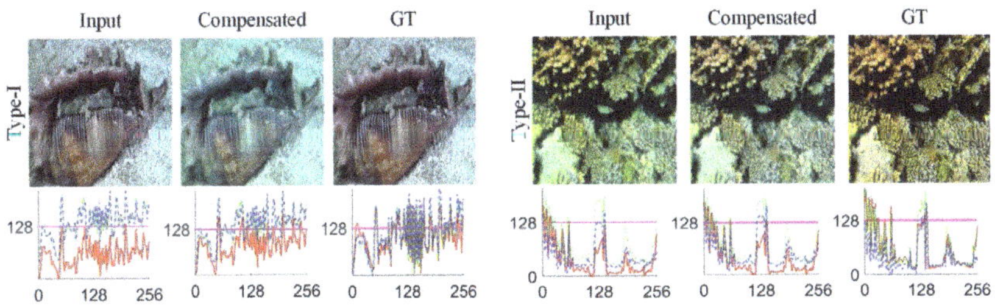

Figure 3. The effect of color compensation operation on two different types of images. For Type I, color compensation aligns colors closer to the ground truth (GT), enhancing visual fidelity. In contrast, for Type II, the compensation results in color distortion when compared to GT.

4. Proposed Method

The proposed methodology restores images through a two-phase process. Initially, an image classifier categorizes each image as either Type I or Type II. Subsequently, Type-I images are processed using the Deep Line Model (DLM), while Type-II images undergo enhancement through the Deep Curve Model (DCM). The complete framework depicted in Figure 4 showcases the complete process, from initial classification to the final output, highlighting the effectiveness and adaptability of both the DLM and DCM in underwater image restoration.

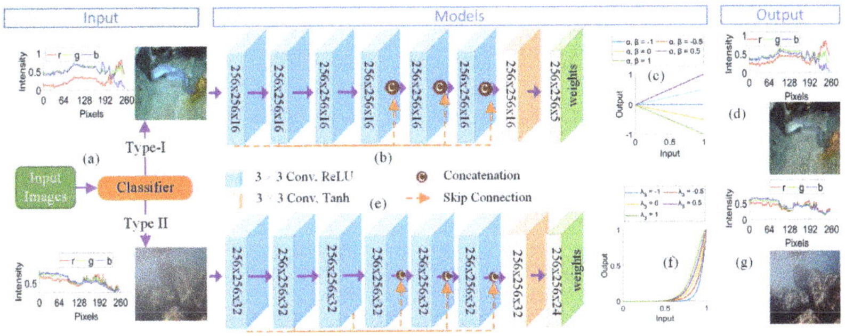

Figure 4. Overview of the proposed framework illustrating adaptive mapping capabilities of the models: (**a**) input images with intensity profiles, (**b**) architecture of the Deep Line Model (DLM), (**c**) examples of lines varying with parameters α, β, (**d**) output images processed by the DLM, (**e**) architecture of the Deep Curve Model (DCM), (**f**) examples of curves demonstrating higher-order adjustments, (**g**) output images processed by the DCM.

4.1. Image Classifier

Based on the observations in images' profile intensity and metrics, images have been categorized into two distinct types. Type-I images are those that retain or improve in quality following linear transformation for color compensation, while Type-II images are characterized by a decline in quality after the same transformation. Following this classification, we applied linear transformations and computed metrics such as RMSE and PSNR and labeled the images accordingly. For the classification task, a small neural network is designed that takes the features obtained through DenseNet121 [41] and analyzes the intrinsic properties of an input image and guides it toward the most appropriate restoration pathway—the DLM for Type-I images and the DCM for Type-II images—thereby enhancing feature extraction and image restoration. The subsequent sections delve into the detailed frameworks of each model.

4.2. Deep Line Model

Mostly, underwater image enhancement methods are executed in two linear operations. Initially, a color-corrected image $\acute{I}_c(x)$ is obtained by compensating the channels through additive adjustment factors [11,42]. The generalized form of this operation is as follows:

$$\acute{I}_c(x) = I_c(x) + \eta_c(x), \tag{3}$$

where η_c represents the additive adjustment factors for each channel $c \in \{r, g, b\}$. In the second step, the restored image $\hat{I}_c(x)$ is obtained by improving contrast. It is achieved by applying another linear operation that stretches the pixel values of the color-corrected image.

$$\hat{I}_c(x) = \alpha_c \cdot \acute{I}_c(x) + \beta_c, \tag{4}$$

where α_c and β_c, are constants utilized to represent weights for each channel that are mostly applied globally.

Instead of performing two separate linear operations and using global weights, we suggest a Deep Line Model that combines two steps and uses per-pixel weights. The proposed model is expressed as

$$\hat{I}_c(x) = \alpha_c(x) \cdot I_c(x) + \beta_c(x) \cdot \eta_c(x), \tag{5}$$

where $\alpha_c(x)$ and $\beta_c(x)$ are weight matrices which are learned through the deep network and η_c represents the color compensation factors for each channel $c \in \{r, g, b\}$. The color compensation factors η_c are computed by using the mean guided compensations [11] through the following expressions.

$$\begin{cases} \eta_r = (\bar{I}_g(\mathbf{x}) - \bar{I}_r(\mathbf{x}))(1 - I_r(\mathbf{x})I_g(\mathbf{x})), \\ \eta_g = 0, \\ \eta_b = (\bar{I}_g(\mathbf{x}) - \bar{I}_b(\mathbf{x}))(1 - I_b(\mathbf{x})I_g(\mathbf{x})). \end{cases} \tag{6}$$

where $\bar{I}_r(\mathbf{x})$, $\bar{I}_g(\mathbf{x})$, and $\bar{I}_b(\mathbf{x})$ denote the mean values of degraded color components of underwater image $I_c(x)$.

Now, the computational challenge exists in the derivation of dynamic weight matrices. To overcome this, we employ a lightweight deep neural network, as illustrated in Figure 4b. The network's architecture encompasses seven convolutional layers; the first six are equipped with 16 filters each, kernel 3×3 and utilizing ReLU activation functions, while the seventh layer adopts a *Tanh* activation to produce the required weight matrices. This setup is specifically designed to facilitate localized adjustments, empowering the deep line model delineated in Equation (5) to competently address the complex characteristics of underwater images and effectuate precise, adaptive enhancements. Figure 4c exemplifies this capability, depicting the generic behavior of deep lines with varying random parameters α, β, spanning a range from -1 to 1. Our model is not only effective but also efficient, comprising a mere 18,390 trainable parameters and requiring only 54.07 MB of memory, rendering it an optimal solution for resource-constrained systems.

4.3. Deep Curve Model

The images that are not restored through linear operations require non-linear transformations. Inspired by the work [43] for low light image enhancement, we propose a Deep Curve Model for underwater image restoration. A second-order polynomial is a simple non-linear mapping between an input image $I_c(x)$ and the output image $\hat{I}_c(x)$, which is also differential.

$$\hat{I}_c(x) = \gamma_{c,1}(x) \cdot (I_c(x))^2 + \gamma_{c,2}(x) \cdot I_c(x) + \gamma_{c,3}(x), \tag{7}$$

where $\gamma_{c,1}(x)$, $\gamma_{c,2}(x)$, and $\gamma_{c,3}(x)$ are pixel-wise coefficients for each channel $c \in \{r, g, b\}$. By setting $\gamma_{c,1}(x) = \gamma_{c,2}(x)$, and $\gamma_{c,3}(x) = zeros(x)$, the non-linear mapping can be simplified and re-written as

$$\hat{I}_c(x) = I_c(x) + \gamma_c(x) \cdot I_c(x) \cdot (1 - I_c(x)) \tag{8}$$

where $\gamma_c(x)$ represents the weight matrices for the non-linear mapping (curves) for each channel $c \in \{r, g, b\}$. It means that the curves are applied separately to each of the three RGB channels, allowing for better restoration by preserving the inherent color and by reducing the risk of over-saturation. Furthermore, the image is restored by applying mapping inside the network, so it should be differentiable for forward and backward propagation. While the second-order curves can provide satisfactory restoration results, they can further be improved by applying higher-order curves. One simple way to achieve higher-order mapping is to apply a second-order mapping iterative fashion. The iterative version of the deep curve model can be expressed as

$$\hat{I}_c^{(n)}(x) = \hat{I}_c^{(n-1)}(x) + \gamma_c^{(n-1)}(x) \cdot \hat{I}_c^{(n-1)}(x) \cdot (1 - \hat{I}_c^{(n-1)}(x)), \tag{9}$$

where n indicates the iteration number and $\gamma_c^{(n-1)}(x)$ represents the weight matrices for the $(n-1)th$ iteration. For $n = 1$, $\hat{I}_c^{(1)}(x)$ is computed through (7). Furthermore, for n iterations, $n \times 3$ weight matrices are required. In this work, we have set the eight, i.e., $n \in \{1, 2, \ldots, 8\}$, so $(8 \times 3 = 24)$ dynamic weight matrices need to be learned. Now, the problem is how to compute dynamic weight matrices.

To learn the weight matrices (curve parameter maps), we adopted a technique similar to that used in the deep line model discussed in the previous section. A lightweight deep neural network, as shown in Figure 4e, is employed to compute these dynamic weight matrices. The network takes the input image $I_c(x)$ and learns a set of pixel-wise curve parameter maps corresponding to higher-order curves. The behavior of such curves is illustrated in Figure 4f, for instance, λ_3, with λ_1 and λ_2 set to -1 and the number of iterations n equal to 3, showcasing the advanced adjustment capabilities with these curves. The network's architecture comprises seven convolutional layers with the first six layers each containing 32 kernels of size 3×3 with a stride of 1, which are followed by a ReLU activation function to introduce non-linearity into the model. The final layer consists of 24 convolutional kernels of the same size and stride but employs a *Tanh* activation function, ensuring that the output values are constrained within the range of -1 to 1. This layer produces a total of 24 curve parameter maps (dynamic weight matrices) across eight iterations with each iteration providing three curve parameter maps for each channel $c \in \{r, g, b\}$.

5. Results and Discussion

5.1. Datasets

In this study, we utilized two primary datasets: EUVP [31] and UIEBD [32]. Both datasets comprise subsets containing paired and unpaired images. We aggregated paired images from subsets of EUVP, including underwater_dark, underwater_scenes, and underwater_imagenet, resulting in a combined total of 11,950 images. These images were used for both training and testing within our proposed method. Similarly, from the UIEBD dataset, we selected 890 images from the raw-890 subset for the same purpose. For performance evaluation, we used the test_samples subsets from both EUVP and UIEBD, which consist of 515 and 240 images, respectively.

5.2. Evaluation Metrics

In assessing the quality of restored images, we adopted widely recognized metrics such as RMSE and PSNR [44]. RMSE was calculated as follows:

$$RMSE = \sqrt{\frac{1}{MN}\sum_{x=1}^{M}\sum_{y=1}^{N}(I_c'(x,y) - \hat{I}_c(x,y))^2} \tag{10}$$

where M and N represent the dimensions of the images, $I_c'(x,y)$ is the color-corrected input image, and $I_c'(x,y)$ is the output image. A lower value of RMSE indicates better results. In addition to RMSE, the peak signal-to-noise ratio (PSNR) was used as an evaluation metric during the validation phase. PSNR is a standard measure for assessing the quality of reconstructed images in comparison with the original ones. PSNR was calculated using the formula:

$$PSNR = 20 \cdot \log_{10}\left(\frac{MAX_I}{\sqrt{MSE}}\right) \tag{11}$$

where MAX_I is the maximum possible pixel value of the image, and MSE represents the mean squared error, which is calculated as the squared difference between $I_c'(x,y)$ and $\hat{I}_c(x,y)$ averaged over all of the pixels. A higher value of PSNR represented better quality.

5.3. Implementation

In our implementation, we constructed the framework using PyTorch and executed on an NVIDIA GeForce RTX3090 GPU. The proposed method integrates three core models: (1) Image Classifier, (2) DLM, and (3) DCM. Image Classifier differentiates between Type-I and Type-II images. For the training of Image Classifier, first, we applied color compensation through (2) on 11,434 images and labeled them with 1 if the image is improved with respect to the input image; otherwise, its label was set to 0. From a total of 11,434 images, 9000 images were used for training and 2434 images were utilized for testing. The training process also utilized a learning rate of 1×10^{-4} and a batch size of 2 and 100 epochs. The training accuracy obtained was 99.01%, whereas accuracy on test data was noted as 84.18%. The DLM targets the enhancement of Type-I images with configuration settings such as batch sizes of 64 and 80 epochs for EUVP and batch sizes of 2 and 200 epochs for UIEBD; the DCM focuses on the restoration of Type-II images. This model adopts optimizer configurations similar to the DLM but with a specific batch size of 2 and 100 epochs for EUVP and solely 200 epochs for UIEBD. Throughout their training phases, both the DLM and DCM models employ the RMSE (L2) as a loss function. Using the assembled dataset, a classifier was trained to categorize the images into two types: Type I and Type II. This classification was carried out on 11,435 images from EUVP and 890 images from UIEBD. As a result, 758 images from EUVP and 240 images from UIEBD were identified as Type II, while the remaining images were designated as Type I. The DLM was then trained on 10,677 Type-I images from EUVP and 500 images from UIEBD. In contrast, the DCM was trained using 758 Type-II images from EUVP and 140 Type-II images from UIEBD. All models utilized the Adam optimizer to ensure efficient convergence. For both training and evaluation, we processed images with dimensions 256 × 256. Importantly, our settings incorporated gradient clipping, normalized to 0.1, to prevent gradient explosions, along with a weight decay of 0.0001 to provide regularization.

5.4. Comparative Analysis

In order to rigorously assess the efficacy of our proposed approach, we contrasted it against leading-edge methods in the domain. This encompasses non-learning techniques such as NLD [9], RLP [34], MMLE [12], and UNTV [10], as well as learning-driven paradigms including ACT [33], FGAN [31], DNet [14], AOD [16], and SCNet [35]. First, input images are restored through the above-mentioned methods and the proposed method, and then restoration accuracy is compared by utilizing the widely accepted quantitative metrics, root mean square error (RMSE) and peak signal-to-noise ratio (PSNR). A lower value for RMSE and a higher value for PSNR indicate better results. Table 1 presents the quantitative results for test samples across both datasets. The performance of the proposed networks for the EUVP dataset is denoted as "Ours". Focusing on the EUVP dataset, a quick examination of Table 1 reveals that the DLM exhibits superior performance in Type-I images, registering an RMSE of 0.08 and a PSNR of 22.30 dB, which outperforms state-of-the-art (SOTA) methods. Similarly, for Type-II images, the DCM surpasses competing methods by achieving 0.06 RMSE and 25.00 dB PSNR. Notably, when contrasted with SCNet and ACT—the models boasting the second-best performance for Type-I and Type-II images, respectively—the DLM excels with a differential of 0.02 in RMSE and 2.3 dB in PSNR, whereas the DCM showcases a marked improvement with 0.87 RMSE and 4.4 dB PSNR. Transitioning to the UIEBD dataset, SCNet leads in performance for Type-I images, registering a 0.07 RMSE and 23.60 dB PSNR. This is contrasted with the DLM, which holds the position for the second-best performance, achieving 0.11 RMSE and 20.10 dB PSNR. Conversely, for Type-II images, the DCM stands out by attaining 0.08 RMSE and 22.20 dB PSNR. When compared with SCNet—the runner-up in performance—our DCM establishes a superior benchmark with a difference of 0.02 in RMSE and an elevation of 1.5 dB in PSNR.

Table 1. Comparison of average RMSE and PSNR for non-learning and learning-based methods on EUVP and UIEBD test images. The best results are in **bold**, and the second-best results are underlined.

				Non-Learning-Based Methods				Learning-Based Methods					
Measure	Dataset	Type	Input	NLD	RLP	MMLE	UNTV	ACT	DNet	AOD	FGAN	SCNet	Ours
RMSE	EUVP	Type-I	0.11	0.15	0.15	0.18	0.14	0.84	0.12	0.78	0.73	0.10	**0.08**
		Type-II	0.08	0.12	0.15	0.17	0.11	0.93	0.10	0.87	0.84	0.11	**0.06**
	UIEBD	Type-I	0.15	0.18	0.14	0.21	0.16	0.81	0.17	0.75	0.70	**0.07**	0.11
		Type-II	0.14	0.12	0.11	0.24	0.17	0.92	0.14	0.86	0.53	0.10	**0.08**
PSNR	EUVP	Type-I	20.00	17.00	17.30	15.30	17.40	17.50	19.10	15.10	13.70	20.00	**22.30**
		Type-II	22.00	18.40	16.90	15.60	19.20	20.60	20.10	13.30	15.50	19.70	**25.00**
	UIEBD	Type-I	17.60	15.30	17.50	14.30	16.20	15.50	15.90	15.10	12.70	**23.60**	20.10
		Type-II	18.60	18.70	19.20	12.60	15.60	19.70	17.80	13.10	9.99	20.70	**22.20**

To assess the robustness of the proposed method for qualitative evaluation, we selected six images from each dataset of Type-I and Type-II categories and the visual outcomes from established approaches and the proposed method shown in Figure 5. It can be observed from the figure that methods such as MMLE tend to over-darken certain areas in their dehazing results. Additionally, techniques like NLD, RLP, and UNTV exhibit notable color distortions and texture degradation. Meanwhile, ACT, DNet, and AOD-Net struggle with the remaining haze effects. FGAN's outcomes lean excessively toward reddish tones, and while SCNet shows an improvement over previous methods, especially for Type-I images, it still presents slight color distortions and tends to produce overly bright images. Remarkably, ACT performs commendably on Type-II images when compared to other competitors. In contrast, our proposed method excels by restoring finer details and achieves more visually appealing restorations, outperforming both traditional and learning-based counterparts.

Similarly, to further verify the effectiveness of our approach, we extended our comparison to the six randomly selected images from the UIEBD dataset, representing both Type-I and Type-II categories. The restored images for the comparative methods along with the proposed method are presented in Figure 6. It can be observed from the figure that prevalent methods such as UDCP, ACT, D-Net, and AOD-Net struggle with lasting haze. MMLE and UNTV, in particular, introduce significant color distortions, to preserve texture details and edge sharpness, whereas F-GAN tends to bias the image restoration toward a reddish tint. RLP and SCNet offer superior visual clarity compared to their counterparts, yet their dehazed images exhibit excessive brightness when compared to our model. In contrast, our method not only restores natural colors and sharp edges but also excels in processing Type-II images, consistently surpassing both conventional and learning-based algorithms.

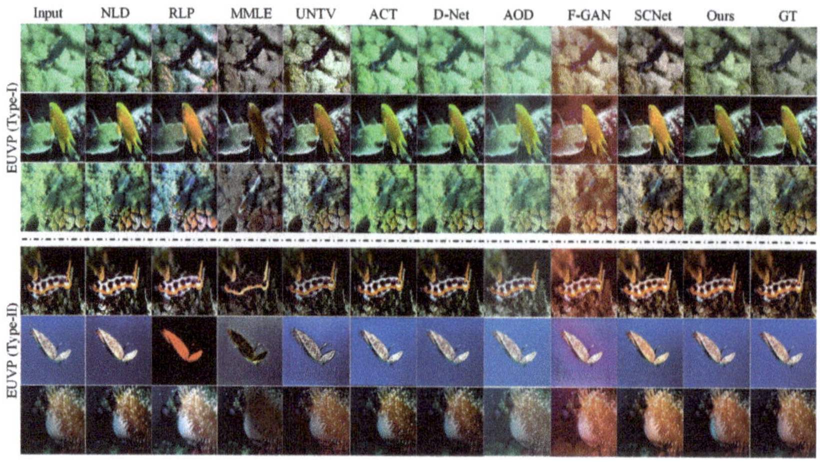

Figure 5. Visual enhancements of the EUVP dataset utilizing non-learning and learning techniques, inclusive of our proposed method. Refer to Tables 2 and 3 for associated RMSE and PSNR values.

Table 2. Comparison of average PSNR for non-learning and learning-based methods on EUVP and UIEBD test images. The best results are in **bold**, and the second-best results are underlined.

			Non-Learning-Based Methods				Learning-Based Methods					
Dataset	**Image**	**Input**	**NLD**	**RLP**	**MMLE**	**UNTV**	**ACT**	**D-Net**	**AOD**	**F-GAN**	**SCNet**	**Ours**
EUVP	test_p84_	19.30	15.70	15.10	17.33	17.52	17.00	18.90	15.16	15.50	15.62	**26.70**
Type-I	test_p404_	17.20	15.00	16.00	18.79	15.96	14.99	17.10	15.24	14.60	13.96	**28.30**
	test_p510_	22.70	21.10	22.20	17.48	20.77	20.11	20.40	13.73	13.60	17.55	**24.50**
EUVP	test_p171_	23.70	20.70	19.60	15.61	20.38	22.04	19.90	12.13	15.20	20.22	**27.70**
Type-II	test_p255_	26.50	18.90	8.45	10.45	18.61	24.88	26.40	15.73	13.50	24.27	**29.90**
	test_p327_	23.70	20.20	16.20	15.68	20.24	21.18	20.50	13.44	15.00	18.99	**26.10**
UIEBD	375_img_	20.80	19.50	16.80	16.54	20.10	19.40	19.30	15.01	15.40	15.57	**22.90**
Type-I	495_img_	18.10	15.40	14.50	15.11	17.51	14.62	17.20	13.94	14.00	13.80	**27.00**
	619_img_	22.70	21.10	22.20	17.49	20.78	20.12	20.41	13.74	13.61	17.56	**24.51**
UIEBD	746_img_	23.70	20.70	19.60	15.62	20.39	22.05	19.91	12.14	15.21	20.23	**27.71**
Type-II	845_img_	26.50	18.90	8.46	10.46	18.62	24.89	26.41	15.74	13.51	24.28	**29.91**
	967_img_	23.70	20.20	16.21	15.69	20.25	21.19	20.51	13.45	15.01	19.00	**26.11**

Table 3. Comparison of average RMSE for non-learning and learning-based methods on EUVP and UIEBD test images. The best results are in **bold**, and the second-best results are underlined.

			Non-Learning-Based Methods				Learning-Based Methods					
Dataset	**Image**	**Input**	**NLD**	**RLP**	**MMLE**	**UNTV**	**ACT**	**DNet**	**AOD**	**FGAN**	**SCNet**	**Ours**
EUVP	test_p84_	0.11	0.16	0.20	0.14	0.13	0.14	0.11	0.18	0.17	0.17	**0.05**
Type-I	test_p404_	0.14	0.18	0.20	0.11	0.16	0.18	0.14	0.17	0.19	0.20	**0.04**
	test_p510_	0.07	0.09	0.10	0.13	0.09	0.10	0.10	0.21	0.21	0.13	**0.06**
EUVP	test_p171_	0.07	0.09	0.10	0.17	0.10	0.08	0.10	0.25	0.17	0.10	**0.04**
Type-II	test_p255_	0.05	0.11	0.40	0.30	0.12	0.06	0.05	0.16	0.21	0.06	**0.03**
	test_p327_	0.07	0.10	0.20	0.16	0.10	0.09	0.09	0.21	0.18	0.11	**0.05**
UIEBD	375_img_	0.09	0.11	0.14	0.15	0.10	0.11	0.11	0.18	0.17	0.17	**0.07**
Type-I	495_img_	0.14	0.20	0.20	0.18	0.14	0.19	0.14	0.20	0.20	0.20	**0.04**
	619_img_	0.07	0.09	0.10	0.13	0.09	0.10	0.10	0.21	0.21	0.13	**0.06**
UIEBD	746_img_	0.07	0.09	0.10	0.17	0.10	0.08	0.10	0.25	0.17	0.10	**0.04**
Type-II	845_img_	0.05	0.11	0.40	0.30	0.12	0.06	0.05	0.16	0.21	0.06	**0.03**
	967_img_	0.07	0.10	0.20	0.16	0.10	0.09	0.09	0.21	0.18	0.11	**0.05**

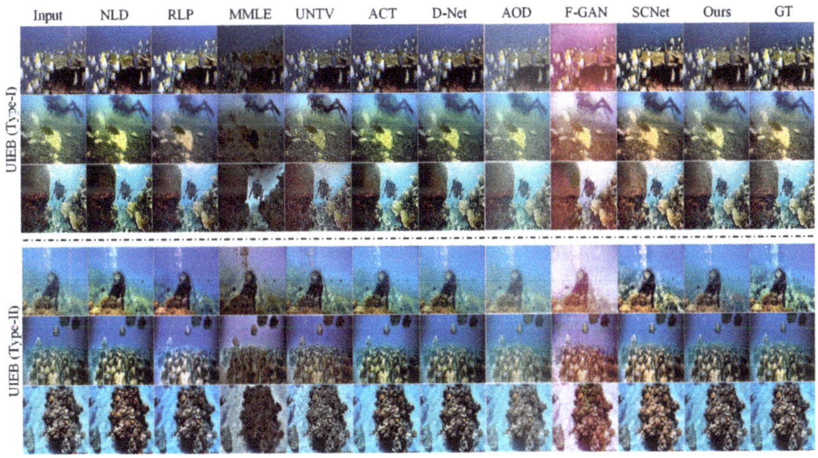

Figure 6. Visual enhancements on the UIEBD dataset utilizing non-learning and learning techniques, inclusive of our proposed method. Refer to Tables 2 and 3 for associated RMSE and PSNR values.

Furthermore, Tables 2 and 3 enumerate the quantitative results for images depicted in Figures 5 and 6. For both datasets, the performance of the proposed method is indicated as "Ours". In examining the EUVP dataset for Type-I images, the DLM consistently showcases superior performance across all images, registering a minimum RMSE of 0.04 and a PSNR of 28.30dB. Similarly, for Type-II images, the DCM emerges as the top performer across all images, notching a minimum RMSE of 0.03 and a PSNR of 29.90 dB. Transitioning to the UIEBD dataset, for Type-I images, the DLM stands out in performance for two out of three images, achieving a minimum RMSE of 0.05 and a PSNR of 26.90 dB. For Type-II images, the DCM exhibits top-tier performance in 2 out of 3 images, recording a minimum RMSE of 0.04 and a PSNR of 29 dB. In a comprehensive analysis, both the DLM and DCM prove their efficacy across both datasets, outperforming in 10 out of 12 images. In comparison to other methods, with RLP being the second-best performer, it excelled in 4 out of 12 images.

5.5. Ablation Study

The proposed method was tested using 515 images from the EUVP dataset and 240 images from the UIEBD dataset. The proposed classifier designated 486 to Type I and 29 images to Type II out of 515 images from the EUVP dataset, setting them up as the testing sets for the DLM and DCM, respectively. Similarly, from the UIEBD dataset, test images classified 213 images as Type I and 27 images as Type II. All datasets were provided to each of the proposed DLM and DCM for the restoration. RMSE and PSNR metrics were computed for each image in the datasets, and the average values of the PSNR and RMSE are shown in Table 4. The DLM is designed for refining Type-I images and the DCM is developed for the restoration of Type-II images. From the table, it can be observed that when models are applied on Type-I and Type-II images, respectively, improved PSNR and RMSE measures are obtained. Whereas, declined or marginally improved metrics are obtained when models are applied on Type-II and Type-I images, respectively. Hence, the DLM and DCM are effective for the restoration of Type-I and Type-II images, respectively.

In addition, for qualitative analysis of the DLM and DCM, we selected two images from each type, and their restored versions are shown in Figure 7. From the analysis of the figure, it is evident that the DLM performs better for Type-I images in both datasets, yielding restored images that are closer to the ground truth (GT). However, when the DCM is applied to Type-I images, although certain areas appear clearer, there is a tendency for colors to become denser. For instance, the blue color intensifies, resulting in a more bluish appearance of the image. Conversely, Type-II images restored using the DCM for both datasets exhibit a cleaner look and are more closely aligned with the GT, whereas the DLM shows suboptimal performance, either causing color distortion or producing blurry images.

Table 4. Quantitative results of ablation study comparing the DLM and DCM. Average RMSE/PSNR for Type-I and Type-II images. (The best values are highlighted).

Datasets	Group	Input	DLM	DCM
EUVP	Type-I	0.11/20.00	0.93/22.35	0.89/20.63
	Type-II	0.08/22.00	0.97/24.18	0.97/25.03
UIEBD	Type-I	0.15/17.60	0.93/20.08	0.85/16.37
	Type-II	0.14/18.60	0.94/17.81	0.95/22.20

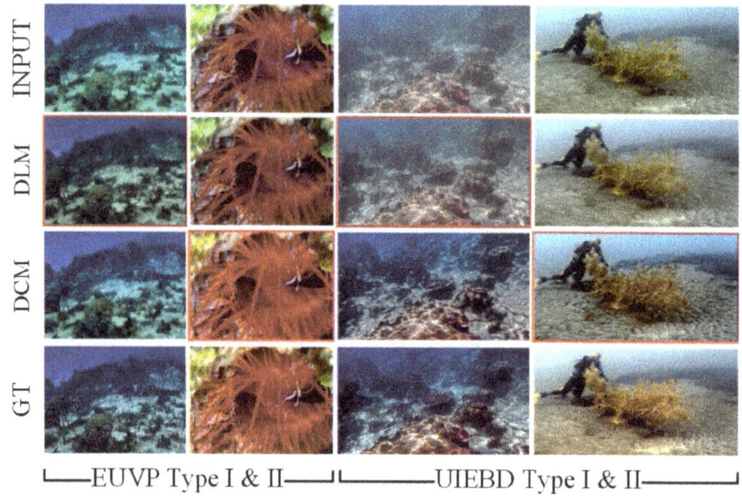

Figure 7. Ablation study evaluating the DLM and DCM for underwater image restoration. For Type-I images, the DLM achieves RMSE/PSNR values of 0.05/26.73, while the DCM yields 0.05/26.14 for Type-II images, and 0.06/24.80 in the UIEBD dataset, compared to the DLM's 0.16/16.10. Red boxes indicate qualitative differences between the models.

In our detailed examination of the Deep Curve Model (DCM) performance, we observed a consistent enhancement in the restoration quality with increasing iterations. Utilizing two representative images from the UIEBD dataset, as illustrated in Figure 8, we record the progression of quality improvements. For instance, as shown in Figure 8b, the restoration quality at the fourth iteration (I4) manifests a considerable enhancement from the input, which is evidenced by the RMSE/PSNR values of 0.09/20.95. This trend of enhancement persists through iterations I4 and I6, as highlighted by the corresponding RMSE/PSNR figures, which signify the improved clarity and overall quality of the dehazed images. The peak of visual clarity is achieved at iteration I8, registering the lowest RMSE of 0.06 and the highest PSNR of 24.66 for the top image, with the bottom image exhibiting similarly positive metrics. Notably, limiting the model to merely one or two iterations does not invariably lead to inferior dehazing outcomes; in some cases, the image quality may remain stable or even slightly enhance, suggesting that the optimal number of iterations for image enhancement is variable. The correlation between the iteration count and the quality of dehazing is evident with higher iterations yielding superior visual and quantitative results.

Figure 8. Ablation study of the effect of different iterations. The RMSE/PSNR for the input and the corresponding iterations are written beneath each subfigure.

5.6. Complexity of Models

In order to estimate the time taken by different models, we run all three model classifiers, DML, and DCM turn by turn on the system X64 bit-based PC having an 11th generation Intel Core i9-11900K processor with 32 GB RAM for 2700 images and calculated the processing time. Table 5 shows the numerical values for time in seconds for all three models. As expected, the classifier is based on DenseNet121, which has much more trainable parameters. For each Type-1 image, approximately 0.0155 s were consumed, whereas for the processing of a Type-II image, 0.0161 s time was taken. So, it can also be observed that the classifier is taking a larger portion of the time. The DML and DCM are light-weight models, so they consume a tiny portion of the total time per image.

Table 5. Time taken by different models used in the proposed solution (in seconds).

Number of Images	Classifier	DLM	DCM
2700	37.82	4.11	5.84
1	0.0140	0.0015	0.0021

5.7. Limitations and Future Work

Although the proposed framework typically yields improved restoration outcomes, its efficacy can be compromised by the misclassification of images. Such wrong classifications are more likely when images possess mixed characteristics or when the changes in their characteristics are slight. These occurrences underline the need to refine the classifier's accuracy to ensure dependable performance in real-world applications. An improved classifier can be designed by cultivating better image labeling procedures and exploiting deep image features. Moreover, in this study, we divided the images into two types, which may also not be optimal. It is anticipated that dividing images into various categories and developing models according to the characteristics of the images will further improve the results. In addition, a thorough study about the other performance metrics for input and restored images is required for further investigation. Such a study not only will help in better understanding the problem of underwater image restoration problem but also will be helpful in designing better solutions.

6. Conclusions

In this paper, we presented an underwater image restoration solution that initially, categorizes input images into Type I or Type II. Afterward, based on the classification, the DLM is applied to restore Type-I images, while the DCM is used for the restoration of Type-II images. Both models utilize lightweight neural networks for learning per-pixel weight matrices based on the input image's characteristics. The efficacy of the proposed solution is measured by conducting experiments on benchmark datasets and using quantitative metrics PSNR and RMSE. Experimental results and comparative analysis demonstrate the efficacy of the proposed method.

Author Contributions: Conceptualization, H.S.A.A. and M.T.M.; methodology, M.T.M.; software, H.S.A.A.; validation, H.S.A.A. and M.T.M.; formal analysis, H.S.A.A.; investigation, H.S.A.A.; resources, M.T.M.; data curation, H.S.A.A.; writing—original draft preparation, H.S.A.A.; writing—review and editing, M.T.M.; visualization, H.S.A.A.; supervision, M.T.M.; project administration, M.T.M.; funding acquisition, M.T.M. All authors have read and agreed to the published version of the manuscript.

Funding: This work was supported by the Education and Research Promotion Program of KOREATECH (2024).

Institutional Review Board Statement: Not applicable.

Informed Consent Statement: Not applicable.

Data Availability Statement: The datasets used for this study are publicly available through the link https://irvlab.cs.umn.edu/resources/euvp-dataset (accessed on 16 July 2024).

Conflicts of Interest: The authors declare no conflicts of interest.

Abbreviations

The following abbreviations are used in this manuscript:

ASM	atmospheric scattering model
CNN	convolutional neural network
DCP	dark-channel prior
DCM	Deep Curve Model
DLM	Deep Line Model
GFN	gated fusion network
IFM	image formation model
MCP	medium-channel prior
PSNR	peak signal-to-noise ratio
RCP	red-channel prior
RMSE	root mean square error
TM	transmission map
UIEM	underwater image enhancement model

References

1. Akkaynak, D.; Treibitz, T.; Shlesinger, T.; Loya, Y.; Tamir, R.; Iluz, D. What is the space of attenuation coefficients in underwater computer vision? In Proceedings of the IEEE Conference on Computer Vision and Pattern Recognition, Honolulu, HI, USA, 21–26 July 2017; pp. 4931–4940.
2. Peng, Y.T.; Cosman, P.C. Underwater image restoration based on image blurriness and light absorption. *IEEE Trans. Image Process.* **2017**, *26*, 1579–1594. [CrossRef] [PubMed]
3. He, K.; Sun, J.; Tang, X. Single image haze removal using dark channel prior. *IEEE Trans. Pattern Anal. Mach. Intell.* **2010**, *33*, 2341–2353. [PubMed]
4. Raveendran, S.; Patil, M.D.; Birajdar, G.K. Underwater image enhancement: A comprehensive review, recent trends, challenges and applications. *Artif. Intell. Rev.* **2021**, *54*, 5413–5467. [CrossRef]
5. Wu, H.; Liu, J.; Xie, Y.; Qu, Y.; Ma, L. Knowledge transfer dehazing network for nonhomogeneous dehazing. In Proceedings of the IEEE/CVF Conference on Computer Vision and Pattern Recognition Workshops, Virtual, 14–19 June 2020; pp. 478–479.
6. Schechner, Y.Y.; Karpel, N. Recovery of underwater visibility and structure by polarization analysis. *IEEE J. Ocean. Eng.* **2005**, *30*, 570–587. [CrossRef]
7. Galdran, A.; Pardo, D.; Picón, A.; Alvarez-Gila, A. Automatic Red-Channel underwater image restoration. *J. Vis. Commun. Image Represent.* **2015**, *26*, 132–145. [CrossRef]
8. Gibson, K.B.; Vo, D.T.; Nguyen, T.Q. An investigation of dehazing effects on image and video coding. *IEEE Trans. Image Process.* **2011**, *21*, 662–673. [CrossRef]

9. Berman, D.; Levy, D.; Avidan, S.; Treibitz, T. Underwater single image color restoration using haze-lines and a new quantitative dataset. *IEEE Trans. Pattern Anal. Mach. Intell.* **2020**, *43*, 2822–2837. [CrossRef]
10. Xie, J.; Hou, G.; Wang, G.; Pan, Z. A Variational Framework for Underwater Image Dehazing and Deblurring. *IEEE Trans. Circuits Syst. Video Technol.* **2022**, *32*, 3514–3526. [CrossRef]
11. Ancuti, C.O.; Ancuti, C.; De Vleeschouwer, C.; Bekaert, P. Color balance and fusion for underwater image enhancement. *IEEE Trans. Image Process.* **2017**, *27*, 379–393. [CrossRef]
12. Zhang, W.; Zhuang, P.; Sun, H.H.; Li, G.; Kwong, S.; Li, C. Underwater image enhancement via minimal color loss and locally adaptive contrast enhancement. *IEEE Trans. Image Process.* **2022**, *31*, 3997–4010. [CrossRef]
13. Zhang, W.; Dong, L.; Xu, W. Retinex-inspired color correction and detail preserved fusion for underwater image enhancement. *Comput. Electron. Agric.* **2022**, *192*, 106585. [CrossRef]
14. Cai, B.; Xu, X.; Jia, K.; Qing, C.; Tao, D. Dehazenet: An End-to-End System for Single Image Haze Removal. *IEEE Trans. Image Process.* **2016**, *25*, 5187–5198. [CrossRef]
15. Ren, W.; Liu, S.; Zhang, H.; Pan, J.; Cao, X.; Yang, M.H. Single image dehazing via multi-scale convolutional neural networks. In Proceedings of the Computer Vision–ECCV 2016: 14th European Conference, Amsterdam, The Netherlands, 11–14 October 2016; Proceedings, Part II 14; Springer: Cham, Switzerland, 2016; pp. 154–169.
16. Li, B.; Peng, X.; Wang, Z.; Xu, J.; Feng, D. Aod-Net: All-in-One Dehazing Network. In Proceedings of the IEEE International Conference on Computer Vision, Venice, Italy, 22–29 October 2017.
17. Zhang, H.; Patel, V.M. Densely connected pyramid dehazing network. In Proceedings of the IEEE Conference on Computer Vision and Pattern Recognition, Salt Lake City, UT, USA, 18–23 June 2018; pp. 3194–3203.
18. Ren, W.; Ma, L.; Zhang, J.; Pan, J.; Cao, X.; Liu, W.; Yang, M.H. Gated fusion network for single image dehazing. In Proceedings of the IEEE Conference on Computer Vision and Pattern Recognition, Salt Lake City, UT, USA, 18–23 June 2018; pp. 3253–3261.
19. Chen, D.; He, M.; Fan, Q.; Liao, J.; Zhang, L.; Hou, D.; Yuan, L.; Hua, G. Gated context aggregation network for image dehazing and deraining. In Proceedings of the 2019 IEEE Winter Conference on Applications of Computer Vision (WACV), Waikoloa Village, HI, USA, 7–11 January 2019; IEEE: Piscataway, NJ, USA, 2019; pp. 1375–1383.
20. Engin, D.; Genç, A.; Kemal Ekenel, H. Cycle-dehaze: Enhanced cyclegan for single image dehazing. In Proceedings of the IEEE Conference on Computer Vision and Pattern Recognition Workshops, Salt Lake City, UT, USA, 18–22 June 2018; pp. 825–833.
21. Singh, A.; Bhave, A.; Prasad, D.K. Single image dehazing for a variety of haze scenarios using back projected pyramid network. In Proceedings of the Computer Vision—ECCV 2020 Workshops, Glasgow, UK, 23–28 August 2020; Proceedings, Part IV 16; Springer: Cham, Switzerland, 2020; pp. 166–181.
22. Liu, X.; Ma, Y.; Shi, Z.; Chen, J. Griddehazenet: Attention-based multi-scale network for image dehazing. In Proceedings of the IEEE/CVF International Conference on Computer Vision, Seoul, Republic of Korea, 27 October–2 November 2019; pp. 7314–7323.
23. Qin, X.; Wang, Z.; Bai, Y.; Xie, X.; Jia, H. FFA-Net: Feature fusion attention network for single image dehazing. In Proceedings of the AAAI conference on Artificial Intelligence, New York, NY, USA, 7–12 February 2020; Volume 34, pp. 11908–11915.
24. Mohan, S.; Simon, P. Underwater image enhancement based on histogram manipulation and multiscale fusion. *Procedia Comput. Sci.* **2020**, *171*, 941–950. [CrossRef]
25. Wang, Z.; Liu, W.; Wang, Y.; Liu, B. Agcyclegan: Attention-guided cyclegan for single underwater image restoration. In Proceedings of the ICASSP 2022—2022 IEEE International Conference on Acoustics, Speech and Signal Processing (ICASSP), Singapore, 23–27 May 2022; IEEE: Piscataway, NJ, USA, 2022; pp. 2779–2783.
26. Song, Y.; He, Z.; Qian, H.; Du, X. Vision transformers for single image dehazing. *IEEE Trans. Image Process.* **2023**, *32*, 1927–1941. [CrossRef] [PubMed]
27. Wang, Z.; Zhang, K.; Yang, Z.; Da, Z.; Huang, S.; Wang, P. Underwater Image Enhancement Based on Improved U-Net Convolutional Neural Network. In Proceedings of the 2023 IEEE 18th Conference on Industrial Electronics and Applications (ICIEA), Ningbo, China, 18–22 August 2023; pp. 1902–1908.
28. Yang, J.; Li, C.; Li, X. Underwater image restoration with light-aware progressive network. In Proceedings of the ICASSP 2023–2023 IEEE International Conference on Acoustics, Speech and Signal Processing (ICASSP), Rhodes Island, Greece, 4–10 June 2023; pp. 1–5.
29. Deng, X.; Liu, T.; He, S.; Xiao, X.; Li, P.; Gu, Y. An underwater image enhancement model for domain adaptation. *Front. Mar. Sci.* **2023**, *10*, 1138013. [CrossRef]
30. Liao, K.; Peng, X. Underwater image enhancement using multi-task fusion. *PLoS ONE* **2024**, *19*, e0299110. [CrossRef] [PubMed]
31. Islam, M.J., Xia, Y., Sattar, J. Fast Underwater Image Enhancement for Improved Visual Perception. *IEEE Robot. Autom. Lett.* **2020**, *5*, 3227–3234. [CrossRef]
32. Li, C.; Guo, C.; Ren, W.; Cong, R.; Hou, J.; Kwong, S.; Tao, D. An Underwater Image Enhancement Benchmark Dataset and Beyond. *IEEE Trans. Image Process.* **2020**, *29*, 4376–4389. [CrossRef]
33. Yang, H.H.; Fu, Y. Wavelet U-Net and the Chromatic Adaptation Transform for Single Image Dehazing. In Proceedings of the 2019 IEEE International Conference on Image Processing (ICIP), Taipei, Taiwan, 22–25 September 2019.
34. Ju, M.; Ding, C.; Guo, C.A.; Ren, W.; Tao, D. IDRLP: Image dehazing using region line prior. *IEEE Trans. Image Process.* **2021**, *30*, 9043–9057. [CrossRef] [PubMed]

35. Fu, Z.; Lin, X.; Wang, W.; Huang, Y.; Ding, X. Underwater image enhancement via learning water type desensitized representations. In Proceedings of the ICASSP 2022—2022 IEEE International Conference on Acoustics, Speech and Signal Processing (ICASSP), Singapore, 23–27 May 2022; pp. 2764–2768.
36. Drews, P.; Nascimento, E.; Moraes, F.; Botelho, S.; Campos, M. Transmission estimation in underwater single images. In Proceedings of the IEEE International Conference on Computer Vision Workshops, Sydney, Australia, 2–8 December 2013; pp. 825–830.
37. Ancuti, C.O.; Ancuti, C.; De Vleeschouwer, C.; Sbetr, M. Color channel transfer for image dehazing. *IEEE Signal Process. Lett.* **2019**, *26*, 1413–1417. [CrossRef]
38. Liu, C.; Shu, X.; Pan, L.; Shi, J.; Han, B. MultiScale Underwater Image Enhancement in RGB and HSV Color Spaces. *IEEE Trans. Instrum. Meas.* **2023**, 72, 5021814. [CrossRef]
39. Liang, Z.; Zhang, W.; Ruan, R.; Zhuang, P.; Xie, X.; Li, C. Underwater Image Quality Improvement via Color, Detail, and Contrast Restoration. *IEEE Trans. Circuits Syst. Video Technol.* **2023**, *34*, 1726–1742 [CrossRef]
40. Buchsbaum, G. A spatial processor model for object colour perception. *J. Frankl. Inst.* **1980**, *310*, 1–26. [CrossRef]
41. Huang, G.; Liu, Z.; Van Der Maaten, L.; Weinberger, K.Q. Densely connected convolutional networks. In Proceedings of the IEEE Conference on Computer Vision and Pattern Recognition, Honolulu, HI, USA, 21–26 July 2017; pp. 4700–4708.
42. Ebner, M. *Color Constancy*; John Wiley & Sons: Hoboken, NJ, USA, 2007; Volume 7.
43. Guo, C.; Li, C.; Guo, J.; Loy, C.C.; Hou, J.; Kwong, S.; Cong, R. Zero-reference deep curve estimation for low-light image enhancement. In Proceedings of the IEEE/CVF Conference on Computer Vision and Pattern Recognition, Seattle, WA, USA, 14–19 June 2020; pp. 1780–1789.
44. Hore, A.; Ziou, D. Image quality metrics: PSNR vs. SSIM. In Proceedings of the 2010 20th International Conference on Pattern Recognition, Istanbul, Turkey, 23–26 August 2010; IEEE: Piscataway, NJ, USA, 2010; pp. 2366–2369.

Journal of
*Marine Science
and Engineering*

MDPI

Article

MEvo-GAN: A Multi-Scale Evolutionary Generative Adversarial Network for Underwater Image Enhancement

Feiran Fu [1,2], Peng Liu [3,*], Zhen Shao [3], Jing Xu [2,3] and Ming Fang [1,2]

[1] College of Artificial Intelligence, Changchun University of Science and Technology, Changchun 130022, China; fufeiran@cust.edu.cn (F.F.); fangming@cust.edu.cn (M.F.)
[2] Zhongshan Institute of Changchun University of Science and Technology, Zhongshan 528400, China; xujing@cust.edu.cn
[3] College of Computer Science and Technology, Changchun University of Science and Technology, Changchun 130022, China; custsz@cust.edu.cn
* Correspondence: 2022101039@mails.cust.edu.cn

Abstract: In underwater imaging, achieving high-quality imagery is essential but challenging due to factors such as wavelength-dependent absorption and complex lighting dynamics. This paper introduces MEvo-GAN, a novel methodology designed to address these challenges by combining generative adversarial networks with genetic algorithms. The key innovation lies in the integration of genetic algorithm principles with multi-scale generator and discriminator structures in Generative Adversarial Networks (GANs). This approach enhances image details and structural integrity while significantly improving training stability. This combination enables more effective exploration and optimization of the solution space, leading to reduced oscillation, mitigated mode collapse, and smoother convergence to high-quality generative outcomes. By analyzing various public datasets in a quantitative and qualitative manner, the results confirm the effectiveness of MEvo-GAN in improving the clarity, color fidelity, and detail accuracy of underwater images. The results of the experiments on the UIEB dataset are remarkable, with MEvo-GAN attaining a Peak Signal-to-Noise Ratio (PSNR) of 21.2758, Structural Similarity Index (SSIM) of 0.8662, and Underwater Color Image Quality Evaluation (UCIQE) of 0.6597.

Keywords: underwater image enhancement; color transfer; genetic algorithms

1. Introduction

The field of underwater imaging technology plays a vital role in numerous applications, including marine resource exploitation, marine ecological protection, and biodiversity monitoring. It is a fundamental component of marine information collection. Nevertheless, the underwater environment presents a number of significant challenges, including strong scattering, absorption, and background noise, which can degrade image quality by affecting contrast, sharpness, and color [1]. These challenges present significant barriers to the effective application of underwater imaging techniques and require advances in imaging technology to overcome them.

Deep learning-based methods are more effective than traditional methods at capturing useful information in underwater images and providing more accurate and adaptive enhancements. This is achieved by utilizing deep neural networks to learn complex features and map functions of the image. Generative adversarial networks, which are powerful deep learning models, have been successfully applied to underwater image enhancement tasks with remarkable results. These methods use end-to-end mapping without relying on any underwater imaging models and prior knowledge, are widely applicable, and achieve better results than traditional methods.

However, GAN training for underwater image enhancement is challenging. The typical low contrast, blurriness, and color distortion in underwater images lead to unstable training processes, resulting in images with low fidelity; insufficient texture detail; and, often, color bias.

Addressing these issues, this paper introduces a novel underwater image enhancement algorithm called Multi-scale Evolutionary Generative Adversarial Networks (MEvo-GAN). MEvo-GAN enhances the traditional GAN framework by improving the network's loss function, integrating deep residual shrinkage network blocks, and employing multi-scale generative networks. This method effectively learns the mapping relationship between degraded and clear underwater images, capturing diverse scale features and details more comprehensively. It significantly improves image clarity, addressing low contrast, blurriness, and color distortion more efficiently. Additionally, the incorporation of a genetic algorithm stabilizes the training process by selecting the most fit offspring.

The contributions of MEvo-GAN are twofold. First, MEvo-GAN employs a multi-path approach in its generator and discriminator, a strategy crucial for capturing a broader range of features at different scales. This multi-scale processing enables the network to more effectively extract complex features inherent in underwater imagery, such as varying light patterns and obscure textures, thereby substantially improving the restoration of image details and structure. Secondly, to address the specificity of underwater images, underwater image quality metrics are also taken into account when evaluating the offspring. This makes the genetic algorithm integrated in MEvo-GAN play a key role in optimizing the generator parameters. Targeting this approach reflects the evolutionary process, selectively retaining and combining effective features passed on from generation to generation, thus improving the diversity and quality of the generated underwater images. Such optimization ensures a more nuanced adaptation to the unique challenges of underwater environments, enhancing the realism of the restored images.

In summary, these advances make MEvo-GAN a significant advancement in the field, providing powerful solutions to the complex challenges of underwater imaging and opening up new avenues for ocean exploration and research.

2. Related Work

2.1. Underwater Image Enhancement

With the rapid development in the fields of computer vision and image processing, researchers have continued to explore and improve methods for underwater image enhancement. The field encompasses a range of approaches, from traditional physics-based methods to contemporary deep learning techniques.

In the field of physics-based methods, researchers often employ physical models to simulate underwater light propagation, coupled with complicated mathematical operations for image restoration. One prominent example is the Dark Channel Prior (DCP) algorithm by He et al., which ingeniously utilizes the darkest points in hazy images to restore them, integrating physical models of image propagation [2]. These algorithms typically require the formulation of underwater imaging models, including the estimation of scattering light components and attenuation coefficients. But because of the complex and variable nature of the underwater environment, it is difficult to establish a precise model and estimate robust parameters. Building on the DCP framework, Chiang et al. introduced a novel method that amalgamates DCP with wavelength-dependent compensation, adeptly restoring color balance in underwater imagery [3]. Similarly, Galdran et al. developed an enhanced underwater image restoration algorithm, considering the distinct influences of natural and artificial light sources, by modifying the red channel/dark channel approach [4]. Drews et al. contributed to this field with their Underwater Dark Channel Prior (UDCP) algorithm, focusing on the attenuation characteristics of red light underwater [5]. Peng et al. expanded the DCP concept through their Generalized DCP (GDCP) algorithm, which incorporates adaptive color correction into the restoration model, offering more versatility in underwater image enhancement [6]. Hou et al. designed a variational model with an L0 norm term, constraint term, and gradient term by integrating the proposed ICSP into an extended underwater image formation model [7]. Despite the efficacy of these methods, they rely on heuristic enhancement strategies and specific prior knowledge, which renders them incapable of addressing the intricate and multifaceted degradation issues encountered in real-world underwater scenarios. Consequently, they are subject to inherent limitations.

In contrast, deep learning approaches, based on deep neural networks, have been shown to be particularly effective in the field of image enhancement. These methods are especially effective in capturing vital information from underwater images and providing accurate, adaptive enhancement. In recent years, generative adversarial networks have been widely used in this field. For instance, Yao et al. used a deep learning-based approach to solve the underwater image degradation problem. They constructed Gaussian pyramids of multiple dimensions to extract shallow features. Then, they enhanced the high-dimensional salient features using a VGG16-based progressive enhancement neural network [8]. Zhang et al. proposed an adversarial learning-based approach to enhance underwater images, addressing issues such as color casting. They also utilized pre-processing techniques and improvements in generative adversarial networks and evaluated their approach using public datasets [9]. The ECO-GAN method proposed by Jiang et al. successfully solves the problems of color distortion, low contrast, and motion blur in underwater images by means of an innovative generative adversarial network and a specific decoder design. This demonstrates its significant contribution and potential for extension in the field of underwater image enhancement [10]. Chen et al. introduced a hybrid restoration scheme that combines filtering techniques in the Fourier domain with GAN-based enhancement, demonstrating significant improvements in image quality [11]. In an innovative approach, Li et al. developed WaterGAN, a network that incorporates depth estimation and color correction modules, utilizing unsupervised learning to generate realistic underwater images from aerial image and depth pairings for color correction [12]. Yang's contribution involves a CGAN-based approach using multi-scale generative networks and dual discriminator networks, specifically targeting underwater image distortion [13]. In addition, Li et al. proposed a new approach to improve the traditional loss function of CycleGAN, which provides a two-step learning strategy to enhance the performance of underwater images [14]. Cong et al. designed dual discriminators for the style-content adversarial constraint, promoting the authenticity and visual aesthetics of the results [15]. Wang et al. divided underwater enhanced images into different domains and utilized a feature vector to measure the distance from the raw image domain to each enhanced image domain [16]. In a further development, Li et al. designed a template-free color transfer learning framework for predicting transfer parameters, which are more easily captured and described [17].

GAN-based approaches have been demonstrated to outperform conventional methods in mapping degraded underwater images to visually clear outputs. However, the utilization of GANs in this context is not without challenges. Common issues include training instabilities, model collapse, and the necessity of significant time and computational resources. These problems can result in inaccurate color restoration and unclear images in underwater photography, necessitating further research and optimization in this area.

2.2. Genetic Algorithms with GAN

GAN-based methods still face problems such as training instability, mode collapse, and other problems, which restrict their application in the field of underwater image enhancement, particularly in the early stages. The adversarial process between the generator and the discriminator may result in local optimization. Mode collapse occurs when a generator is trapped in a specific pattern, producing similar samples uniformly, which leads to a lack of diversity and variability.

Evolutionary algorithms optimize GAN generators by simulating biological evolution processes like selection, crossover, and mutation. Combining evolutionary algorithms with GANs can improve stability and enhance the expressive capability of generators. Wang et al. proposed a novel framework named Evolutionary Generative Adversarial Networks (E-GANs), evolving a group of generators to compete against each other [18]. Experiments show that E-GANs can overcome the limitations of a single adversarial training objective and consistently retain well-performing offspring, further advancing GAN success and progress. Chen et al. introduced the CDE-GAN framework, integrating dual evolution of generators and discriminators into a unified evolutionary adversarial framework, utilizing their complementary properties and injecting dual mutation diversity during training, effectively conducting adversarial multi-objective optimization, stably capturing multi-modal estimated densities, and improving generative performance [19]. Mu et al. employed mutation operations from genetic algorithms, retaining well-performing generators for subsequent training, effectively capturing data distributions, and mitigating mode collapse in standard GANs [20]. He et al. proposed a multi-objective evolutionary algorithm driven by GANs, classifying parental solutions as real and fake samples to train GANs, then improving stability and the quality of training [21]. Zhang et al. presented a GAN based on the PSO algorithm to enhance training stability, particularly by improving the inertia weight of particle swarms and assessing generator performance, achieving notable results in face generation [22]. Liu et al. proposed EvoGAN, an evolutionary algorithm (EA)-assisted GAN method for generating various composite expressions, accurately generating target composite expressions [23]. Xue et al. incorporated evolutionary mechanisms into CycleGAN, continuously improving generator weight configurations and enhancing generated image quality and details through a channel attention mechanism [24].

Evolutionary mechanisms in generative adversarial network training can enhance stability, generative effects, and diversity. These improvements lead to increased efficiency and generative capacity in GANs. However, challenges remain, including selecting appropriate loss functions, designing effective network structures, and optimizing the efficiency of the evolutionary algorithms.

3. Proposed Method

Contemporary deep learning methodologies for underwater image enhancement encounter challenges in processing multi-scale underwater images, particularly in addressing the varying physical properties inherent at different scales. This can lead to noise and unwanted artifacts in the generation process, further reducing image clarity and negatively impacting subsequent visual tasks.

To address these issues, we introduce a novel underwater image enhancement method, MEvo-GAN. As depicted in Figure 1, the MEvo-GAN network comprises two generators, namely $G_{X \to Y}$ and $G_{Y \to X}$, alongside two discriminators, namely DX and DY. Generator $G_{X \to Y}$ is tasked with transforming degraded underwater images into clear counterparts, whereas $G_{Y \to X}$ performs the inverse function. The discriminators, DX and DY, ascertain the authenticity of images produced by these generators.

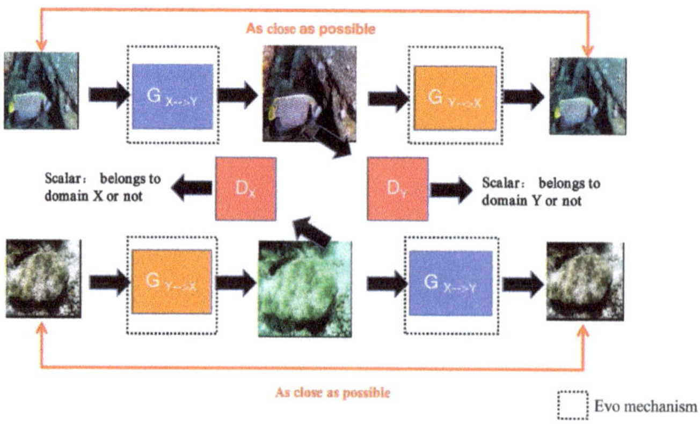

Figure 1. General MEvo-GAN architecture.

In Figure 1, the double-headed arrows labeled "as close as possible" represent the various loss functions. These loss functions are designed to ensure that images transformed between domains maintain maximum similarity to the original version after a round-trip conversion. By minimizing the differences between the original and enhanced images, our approach achieves the ability to enhance images that are both sharp and retain the corresponding detail of the original image.

We propose a multi-scale generator that is more sensitive to image details. This multi-scale approach captures a wider range of features at different scales, enabling the network to more effectively extract complex features inherent in underwater images, such as changing light patterns and blurred underwater textures. The multi-scale generator achieves this by processing the input image through a number of different dilated convolution kernels, allowing it to focus on both fine detail and broader structural features, resulting in greatly improved recovery of image detail and structure.

Furthermore, by incorporating evolutionary algorithms, the parameters of the generator are gradually optimized during the training process. In each iteration, offspring with higher fitness are selected from the generators and used as parents for the next generation, progressively enhancing the network's performance. This evolutionary process mirrors natural selection, improving the overall quality and robustness of the generated images.

3.1. Generators and Discriminators

The primary function of the generator is to convert degraded input images into clear underwater representations. For enhanced detail retrieval, the generator employs a multipath methodology in feature extraction, leveraging convolutional kernels of varied dimensions and sizes. This approach significantly mitigates computational load while concurrently augmenting processing speed. A deep residual contraction network is also used, which includes a deep residual network and a soft thresholding learning network function [25]. Soft thresholding is a nonlinear transformation method whose formula can be expressed as follows:

$$f(x) = \begin{cases} x - \lambda, & x > \lambda \\ x + \lambda, & x < -\lambda \\ 0, & -\lambda \leq x \leq \lambda \end{cases} \tag{1}$$

This function subtracts the absolute value of the signal from the threshold value, and when the absolute value is less than the threshold value, the output is zero. Such an operation effectively attenuates small fluctuations in the signal and retains larger signal changes, thus removing noise or unimportant fluctuations from the signal. Specifically, the soft thresholding function is applied after each residual block to ensure that all small noise fluctuations are effectively filtered out during feature extraction, while larger useful signals are retained. This approach not only helps to reduce noise but also enhances image details. After obtaining a series of thresholds, the soft threshold learning network achieves channel weighting, which reduces redundant information and helps to suppress the effects of noise. The innovation of these three path networks incorporates deep residual contraction networks in order to capture different levels of features and form a multi-scale, high-level semantic feature map.

The purpose of the discriminator is to differentiate whether the input enhanced images are real. The discriminator and generator engage in adversarial learning, prompting the generator to produce more realistic images, thereby improving the quality of the generated images. With a multi-scale discriminator structure, the global structure and local details of images are considered simultaneously, further enhancing the realism and fidelity of the generated images.

In summary, MEvo-GAN is able to extract valuable features from clear underwater images by utilizing multi-scale paths and depth residual shrinkage blocks, adeptly extracting valuable features from clear underwater images. These features are then reinfused into the generated images through a series of encoding, transforming, and decoding steps. Detailed generator and discriminator architecture as shown in Figure 2.

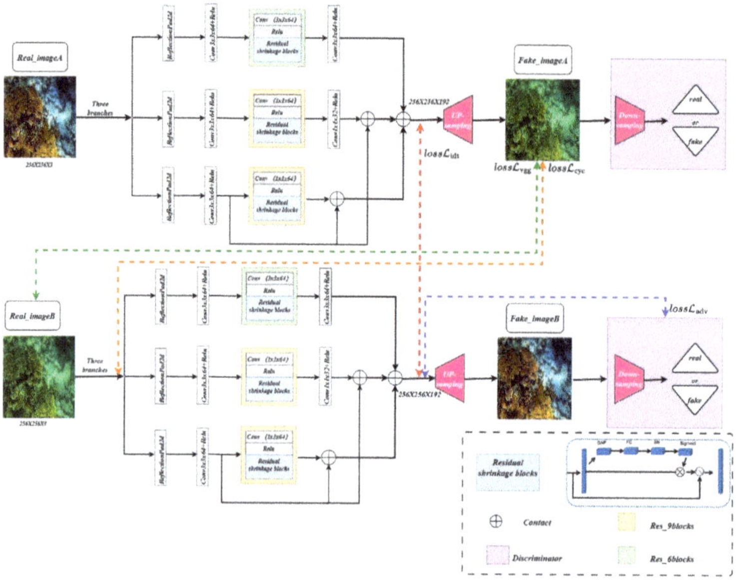

Figure 2. Detailed generator and discriminator architecture.

3.2. Genetic Algorithm

Genetic algorithms mimic the process of biological evolution by simulating the processes of natural selection, crossover, and mutation in order to optimize solutions and improve performance. There are three modules at the core of the genetic algorithm, namely G-Variations (mutation), G-Evaluation (evaluation), and G-Selection (selection). These modules are combined to optimize the parameters of the generator, as shown in Figure 3. With the genetic algorithm module, the network is able to gradually optimize the parameters of the generator during the training process, and the variations module enables the network to converge faster and adapt to different input situations, which enhances the stability and generalization of the network. The evaluation and selection modules combine underwater physical imaging characteristics to generate higher-quality and clearer underwater images.

In the evolutionary process, we randomly initialize a set of generators from an extensive parameter space. These generators constitute the initial set of parent generators. The parameters (θ) of each generator are chosen by random distribution to ensure diversity in the parameter space. The initial parent generators can be represented as follows:

$$\{G_{X\to Y}(\theta^1), G_{X\to Y}(\theta^2), \ldots, G_{X\to Y}(\theta^J)\}, \{G_{Y\to X}(\theta^1), G_{Y\to X}(\theta^2), \ldots, G_{Y\to X}(\theta^J)\},$$

where θ^j represents the parameter set for the (j)-th generator. Then, each parent generator ($G_{X\to Y}(\theta^j)$ and $G_{Y\to X}(\theta^j)$) produces M offspring through variation, resulting in the following:

$$\{G_{X\to Y}(\theta_1^j), G_{X\to Y}(\theta_2^j), \ldots, G_{X\to Y}(\theta_M^J)\}, \{G_{Y\to X}(\theta_1^j), G_{Y\to X}(\theta_2^j), \ldots, G_{Y\to X}(\theta_M^J)\}.$$

where θ_m^j denotes the parameter set of the (m)-th offspring derived from the (j)-th parent. Thus, we generate a total of $J \times M$ offspring generators for each generation. These offspring are then evaluated, and the best-performing ones are selected as the new parents for the next generation as follows:

$$\{G_{X\to Y}(\theta_1^1), G_{X\to Y}(\theta_1^2), \ldots, G_{X\to Y}(\theta_1^J)\}, \{G_{Y\to X}(\theta_1^1), G_{Y\to X}(\theta_1^2), \ldots, G_{Y\to X}(\theta_1^J)\}.$$

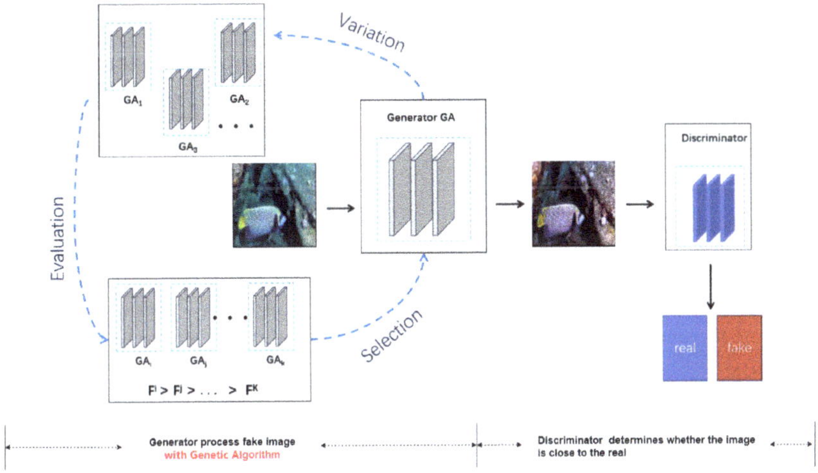

Figure 3. General genetic algorithm architecture.

In the G-Variations module, three different mutation strategies are applied, each corresponding to different minimization objectives of the generator, namely G-minimax mutation, G-heuristic mutation, and G-least-square mutation, corresponding to a traditional GAN, Non-Saturated GAN (NS-GAN), and Least Squares GAN (LSGAN), respectively [26].

G-Minimax Mutation: This mutation strategy aims to minimize the difference between generated and real samples, making the generated samples more realistic. Its objective function is to minimize the Jensen–Shannon divergence between the generated samples and the real samples.

$$\mathcal{M}_G^{\text{minimax}} = \frac{1}{2}\mathbb{E}_{z\sim p_z}[\log(1 - D(G_\theta(z)))]. \quad (2)$$

In Equation (2) and the equations that follow, D is the discriminator, G_θ is the generator, z is the noise sample, and p_z is the noise distribution. It is evident that the G-minimax mutation is the most effective in terms of offspring development during the training process. However, this mutation fails when the discriminator can discriminate well between samples generated by the generator.

G-Heuristic variant: In contrast to the minimax mutation, the G-heuristic mutation is non-saturating when the discriminator effectively rejects the generated samples. This avoids the phenomenon of gradient vanishing.

$$\mathcal{M}_G^{\text{heuristic}} = -\frac{1}{2}\mathbb{E}_{z\sim p_z}[\log(D(G_\theta(z)))]. \quad (3)$$

Nevertheless, G-heuristic mutation may result in instability and fluctuations in generative quality due to the pushing apart of data and model distributions.

G-Least-Square variant: The G-least-square mutation is effective in preventing gradient vanishing when the discriminator easily recognizes the generated samples. In addition, G-least-square mutations do not impose extremely high penalties for generating false samples, nor do they impose extremely low penalties for pattern dropping, thus helping to avoid mode collapse.

$$\mathcal{M}_G^{\text{ls}} = \mathbb{E}_{z\sim p_z}[(D(G_\theta(z)) - 1)^2]. \quad (4)$$

Thus, three different mutations provide multiple training strategies for the generator. The overall goal is to optimize the generator's parameters for effective and high-quality underwater image generation.

Then, each of the initial generators is evaluated using the G-Evaluation module. In the G-Evaluation module, we assess the quality and diversity of individual generators and decide the parents for the next generation. We introduce the Quality Fitness Score (FG_q), Underwater Image Quality Score (FG_u), and Diversity Fitness Score (FG_d), which consider the quality and diversity of the produced samples. The goal is to make informed decisions about the selection of parents for the next generation based on these evaluations.

Quality Fitness Score (FG_q): This score is used to evaluate the quality of generated samples by calculating the cumulative output of the generated samples across multiple discriminators.

$$FG_q = \mathbb{E}_{z \sim pz}[D(G_\theta(z))]. \tag{5}$$

The Quality Fitness Score (FG_q) measures the acceptance level of the generated samples across the discriminators, representing how well the generator's samples conform to the real distribution.

Underwater Image Quality Score (FG_u): The UCIQE (Underwater Color Image Quality Evaluation) is a metric specifically designed to assess the quality of underwater images [27]. It analyzes the color, clarity, and contrast of images to measure image quality. A higher overall UCIQE value indicates clearer images with higher contrast, more details, and better restoration effects.

Diversity Fitness Score (FG_d): This score is primarily used to assess the diversity of generated samples, i.e., the difference between the distributions of generated samples and real samples [28]. The formula for the diversity fitness score is expresses as follows:

$$\begin{aligned} FG_d = &-\log ||\nabla_D - \mathbb{E}_{x \sim p_{data}}[\log D(x)] \\ &- \mathbb{E}_{z \sim p_z}\Big[\log\Big(1 - D\big(G_\theta(z)\big)\Big)\Big]||. \end{aligned} \tag{6}$$

The Comprehensive Fitness Score (FG) combines the Underwater Image Quality Score and the Diversity Fitness Score to holistically evaluate the performance of individual generators as follows:

$$FG = FG_q + \gamma FG_d + \eta FG_u, \tag{7}$$

where γ and η are weight coefficients used to balance the Underwater Image Quality Score and the Diversity Fitness Score. The Comprehensive Fitness Score is used to evaluate the performance of individual generators, determining which generators will be selected as parents for the next generation and continuously optimizing the generator's parameters during the evolutionary process. This approach is in line with the evolutionary principles, guiding the selection of parents for the next generation in a way that enhances both the quality and diversity of the generated underwater images.

In the G-Selection module, the next generation's parents are determined by comparing the fitness scores of individual generators. We use the (μ, λ) selection strategy, which is a variant of the selection process in evolutionary algorithms. This strategy balances exploration (μ: parent population size) and exploitation (λ: offspring population size) by selecting the best individuals from both parent and offspring populations.

After sorting, J individuals possessing the maximum fitness score can survive for the next evolution during adversarial training. This process is formulated as follows:

$$\theta^1, \theta^2, \ldots, \theta^J \leftarrow \theta_1^1, \theta_1^2, \ldots, \theta_1^J. \tag{8}$$

The pseudo-code of genetic algorithm involved in MEvo-GAN is shown in Algorithm 1.

Algorithm 1 The algorithm of MEvo-GAN

Require: The generators $G_{X\to Y}, G_{Y\to X}$; the discriminators D_X, D_Y; the number of iterations T; the number of parents for Generators J; the number of mutations for each generator M; the hyper-parameter γ, η of fitness function of Generators.

Initialize $G_{X\to Y}$'s parameter $\{G_{X\to Y}(\theta^1), G_{X\to Y}(\theta^2), \ldots, G_{X\to Y}(\theta^J)\}$, initialize $G_{Y\to X}$'s parameter $\{G_{Y\to X}(\theta^1), G_{Y\to X}(\theta^2), \ldots, G_{Y\to X}(\theta^J)\}$.
Initialize D_X, D_Y parameters.
for $t = 1$ to T **do**
    Sample a batch of $x_{realX} \sim P_{data}$.
    Sample a batch of $z \sim P_z$, and generate a batch of x_{fake} with Generators.
    Update D_X, D_Y parameters.
end for
for $j = 1$ to J **do**
    Sample a batch of $z \sim P_z$.
    for $m = 1$ to M **do**
        $G_{X\to Y}(\theta^j)$ and $G_{Y\to X}(\theta^j)$ produce offspring $G_{X\to Y}(\theta^j_m)$ and $G_{Y\to X}(\theta^j_m)$ via Equation (2), Equation (3), and Equation (4) respectively.
    end for
end for
Evaluate the $J \times M$ evolved offspring of Generators via Equation (7).
Select the best-performing offspring $\{G_{X\to Y}(\theta^1_1), G_{X\to Y}(\theta^2_1), \ldots, G_{X\to Y}(\theta^J_1)\}$ for $G_{X\to Y}$ and $\{G_{Y\to X}(\theta^1_1), G_{Y\to X}(\theta^2_1), \ldots, G_{Y\to X}(\theta^J_1)\}$ for $G_{Y\to X}$ as the next generation's parents of Generators.

3.3. Loss Functions

The loss functions comprise four main components, each playing different roles in training the generator and discriminator. By balancing these components, the generator is guided to produce the desired transformation results.

(1) Adversarial loss is primarily used to train the generator and discriminator, enabling the generator to create realistic target-domain images and allowing the discriminator to distinguish between generated and real images. The loss function for the generator G is expressed as follows:

$$\mathcal{L}_{\text{adv}}(G, D_Y) = \mathbb{E}_{x\sim p_{\text{data}}(x)}[\log D_Y(G(x))]. \tag{9}$$

The loss function of the discriminator (D) usually consists of the following two parts: the loss for generated images and the loss for real images. The goal of these losses is to enable the discriminator to accurately distinguish between generated and real images. The loss function is expressed as follows:

$$\begin{aligned}\mathcal{L}_{\text{adv}}(D_Y, G) = &-\mathbb{E}_{y\sim p_{\text{data}}(y)}[\log D_Y(y)] \\ &- \mathbb{E}_{x\sim p_{\text{data}}(x)}[\log(1 - D_Y(G(x)))].\end{aligned} \tag{10}$$

This indicates that the discriminator (D) aims to maximize this loss function while the generator (G) seeks to minimize it. By alternating optimization of the generator and discriminator's losses during training, the generator gradually produces more realistic images, and the discriminator improves its discrimination capability. This adversarial training process leads to the generation of high-quality images.

(2) Cycle consistency loss ensures that an image, after being transformed by the generator then reversed back, maintains its original form. This helps the generator learn the mapping between the source and target domains and prevents mode collapse. Cycle consistency loss consists of two parts—for transformations from the source to target domain and vice versa.

$$\begin{aligned}\mathcal{L}_{cyc}(G, F) = &E_{X\sim P_{dats}(x)}[\|F(G(x)) - x\|_1] \\ &+ E_{y\sim P_{data}(y)}[\|G(F(y)) - y\|_1].\end{aligned} \tag{11}$$

(3) Identity consistency loss ensures that the input image retains its own characteristics after being transformed by the generator, i.e., the input and generated images are similar to a certain extent. This helps reduce information loss during image transformation.

$$\mathcal{L}_{\text{idt}}(G,F) = \mathbb{E}_{y\sim p_{\text{data}}(y)}[\| G(y) - y \|_1] + \mathbb{E}_{x\sim p_{\text{data}}(x)}[\| F(x) - x \|_1]. \tag{12}$$

(4) To further improve image quality, perceptual loss is introduced to reduce detail loss, improve image blur, and make enhanced images more realistic. The VGG network is trained on large-scale datasets such as ImageNet, making it visually perceptive for feature extraction. The use of VGG loss ensures that the generated images are visually perceived to be consistent with the real images, thus enhancing the subjective quality of the images.

$$\mathcal{L}_{\text{VGG}} = \sum_{l=1}^{L} w_l \| \phi_l(G(x)) - \phi_l(y) \|_1 . \tag{13}$$

Both the generated image (G(x)) and the target image (y) are input into the VGG network to extract feature mappings at various layers. Then, the L1 distance between these mappings is calculated as the VGG loss. By minimizing this loss, results closer to the real image in terms of perception are obtained.

$$\mathcal{L}_{\text{G}} = \mathcal{L}_{\text{adv}}(G, D_Y) + \lambda_{\text{cyc}}\mathcal{L}_{\text{cyc}}(G,F) + \lambda_{\text{idt}}\mathcal{L}_{\text{idt}}(G,F) + \lambda_{\text{vgg}}\mathcal{L}_{\text{vgg}}(G,F). \tag{14}$$

Here, λ_{cyc}, λ_{idt}, and λ_{vgg} are hyperparameters controlling the weights of cycle consistency, identity consistency, and perceptual losses, respectively. The overall generator loss balances these various parts, guiding the generator to learn the necessary transformations.

4. Experimental Results and Analysis

4.1. Datasets

We used publicly available underwater image enhancement datasets EUVP [29], UIEB [30], and UFO-120 [31]. These datasets were carefully chosen for their diverse characteristics, allowing us to comprehensively train and test MEvo-GAN across a range of underwater imaging conditions.

The EUVP dataset includes a wide range of underwater images, both paired and unpaired. These images were taken with seven different cameras and cover various scenarios, such as marine exploration and human–robot cooperation. The dataset is diverse in terms of visibility conditions and locations, making it a realistic representation of underwater environments. It also includes images from public YouTube videos, showcasing different water types and lighting conditions. The EUVP dataset is divided into the following three subsets: synthesized underwater dark-scene images, degraded underwater images generated using ImageNet, and authentic underwater-scene images. We randomly selected 80% of the dataset for training and kept the remaining 20% for testing.

The UIEB dataset encompasses 890 pairs of underwater images, each vividly illustrating various underwater scene degradations, such as insufficient lighting and blurriness. Unique to this dataset is that each image pair includes an original, unenhanced image alongside a high-quality reference image. These reference images were carefully curated and enhanced using various algorithms, providing a valuable benchmark for image quality. In line with our data handling protocol, 80% of the UIEB dataset was randomly selected for the training of MEvo-GAN, with the remaining 20% allocated for testing.

Lastly, the UFO-120 dataset, comprising 120 underwater light fields, primarily consists of images captured across different marine environments and time periods. This dataset highlights the complexity and diversity inherent in underwater environments, making it an ideal tool for testing the adaptability of MEvo-GAN. Unlike the other datasets, UFO-120 is primarily utilized for testing, providing a robust platform to evaluate the effectiveness of MEvo-GAN in real-world scenarios. By training and testing MEvo-GAN with these diverse datasets, we gained a comprehensive understanding of the algorithm's performance and its potential for practical application in the processing of images of real-world marine environments.

4.2. Training Details

In our network training setup, we categorized the images into the following two sets: degraded underwater images in the TrainA folder and clearer counterparts in the TrainB folder. This organization streamlined the training process without separating generator and discriminator training phases. To optimize the training for both speed and memory efficiency, we adjusted the input sample size to a resolution of 256 × 256 pixels. Moreover, we set the batch size to 1 and defined the training duration as 200 epochs, balancing computational demands with performance. The implemented evolutionary algorithm included a specification of three offspring per generation, coupled with the adaptiveness parameters, set at values of 1 and 0.1, respectively. These settings were chosen to effectively balance exploration and exploitation in the learning process. For visual analysis and progress tracking, we employed the Visidom tool, which enabled us to periodically save and visualize the reconstruction results every five iterations. This approach provided a more intuitive monitoring of the network's learning trajectory. During the testing phase of the network model, we designed the system to allow for flexible adjustment of the input sample size to accommodate various testing scenarios. The initialization of parameters was conducted using the Kaiming algorithm, a method known for its effectiveness in neural network initialization. In all our experiments, we utilized the Adam optimizer, a widely used optimization algorithm in machine learning, setting the initial learning rates for the generator and discriminator at 1×10^{-3} and 2×10^{-3}, respectively, to achieve a balanced optimization. Training parameters λ_{vgg}, λ_{cyc}, and λ_{idt} were meticulously set at 1, 12, and 0.6, respectively, after careful consideration of their impact on the network's performance in terms of feature extraction, cycle consistency, and identity mapping.

4.3. Comparison of Visual Quality of Enhancement

In this section, detailed experimental results of MEvo-GAN are provided and compared with existing underwater enhancement algorithms. Tests included the EUVP, UIEB, and UFO-120 datasets, demonstrating MEvo-GAN's performance in various underwater environments.

The color chart recovery test evaluated MEvo-GAN's effectiveness in underwater image color correction using color chart recovery. Based on a distortion-free color chart that undergoes color degradation due to complex underwater imaging environments, the processing of degraded images validated the method's color restoration effectiveness. Color Fidelity Error (CFE) was used to quantify results, measuring the color difference between enhanced and original color chart images. A lower CFE value indicates better color restoration. The results of a comparison of MEvo-GAN with classical methods are shown in Figure 4.

In the color chart recovery test, UDnet [32] generally resulted in darker images. CWR [33] and FunieGAN [34] caused intermingling of color tones, with some overexposure effects affecting actual perception. Shallow-UWnet [35] and URSCT [36] produced images with a general grayish tone, with shallow color information recovery. UGAN [37] and WaterNet restored the color chart image more naturally but with lower distinction in the same color series. RAUnet [38] showed natural color restoration but with uneven color in some blocks. In contrast, MEvo-GAN displayed bright colors with clear distinction among various color series in the color chart images, such as higher contrast in dark blocks, closely resembling the real color chart, offering a good visual effect, and having the smallest CFE index, making it closer to the standard color chart image. Comparative images before and after enhancement are shown in Figure 5. FunieGAN, CWR, and UDnet increased the brightness of enhanced images but were not effective in removing color bias, as especially noticeable in some images with areas of over-enhancement leading to color distortion. Shallow-UWnet and WaterNet effectively removed color bias but were not as effective in removing blurriness. UGAN and URSCT were effective in removing color bias in underwater images, making the colors more natural and maintaining brightness well. However, compared to MEvo-GAN, their restored colors were still not as realistic and vivid. In some areas, URSCT-treated images still had slight blurriness or obstructions, not achieving complete clarity. RAUnet appeared to restore colors naturally without significant color distortion. Clarity and contrast were moderately improved, but there was still room for improvement in some areas. In contrast, MEvo-GAN not only successfully removed color bias and blurriness but also excellently restored image brightness and details. The color restoration appeared both natural and vivid, especially in red and blue recovery, making underwater images truer to life and clear.

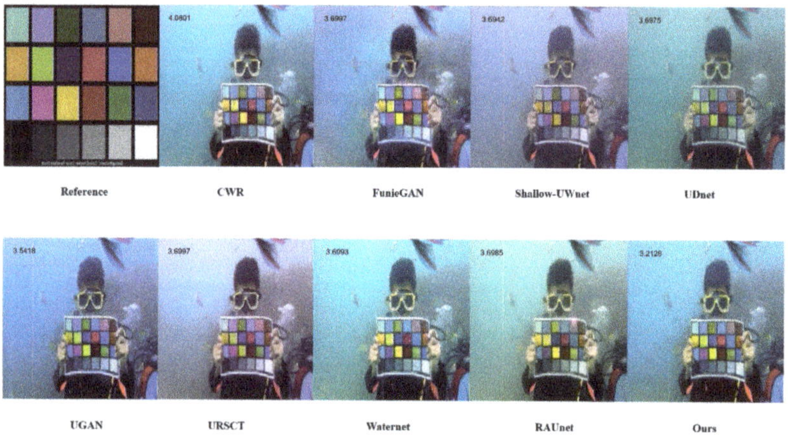

Figure 4. Results of the 9-method color-card recovery experiment (CFE indicator in the upper left of the image).

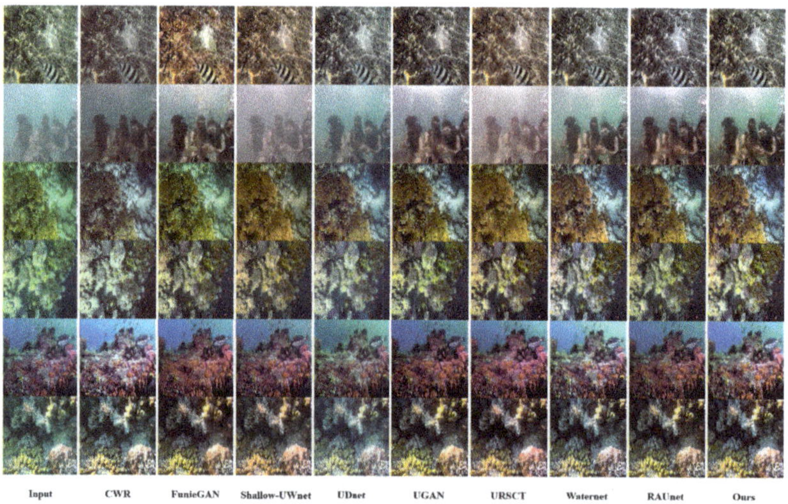

Figure 5. Visual comparison of enhancements of images from the EUVP, UIEB, and UFO-120 datasets.

Then, we compared MEvo-GAN with other mainstream underwater image enhancement methods in terms of various evaluation metrics. As shown in Table 1, MEvo-GAN demonstrated superior performance in metrics like PSNR, SSIM, and UCIQE, especially excelling in the UCIQE index, indicating its significant advantage in improving the overall quality of underwater images that other methods did not have.

Table 1. Comparison on UIEBD, EUVP, and UFO-120 datasets. The best and second-best scores are indicated in red and blue, respectively.

Method	UIEBD			EUVP			UFO-120		
Metric	PSNR	SSIM	UCIQE	PSNR	SSIM	UCIQE	PSNR	SSIM	UCIQE
UGAN	19.4947	0.7496	0.6476	19.6102	0.8131	0.6169	23.1764	0.7959	0.6487
WaterNet	21.1659	0.8290	0.6414	19.4398	0.8492	0.6628	19.6768	0.7704	0.6373
FunieGAN	16.3028	0.7045	0.6434	20.3005	0.7721	0.6451	23.4593	0.7959	0.6487
CWR	16.8157	0.7451	0.5334	16.2670	0.6820	0.6230	16.3482	0.6120	0.6346
Shallow-UWnet	16.9228	0.6857	0.5457	18.9380	0.8288	0.5367	22.2391	0.7796	0.5682
UDnet	18.3965	0.7959	0.5509	20.0486	0.8251	0.5594	19.4468	0.7560	0.6206
URSCT	17.8031	0.6609	0.5432	17.1730	0.8114	0.4231	21.3893	0.7930	0.4314
RAUnet	22.9179	0.8148	0.6467	19.9144	0.8092	0.5809	24.0392	0.8224	0.5961
Ours	21.2758	0.8662	0.6597	20.0502	0.8255	0.6727	19.4011	0.7989	0.7001

4.4. Multi-Scale Visualization

MEvo-GAN implements a multi-branch architecture that integrates convolutional kernels of various sizes and layers of different depths to capture features on multiple scales, thus enhancing image detail. Specifically, the model contains several sub-models, like conv1, conv2, conv3, and conv4, which apply 3×3, 1×1, and 5×5 dilated convolutional kernels, capturing local details and comprehensive contextual information from larger areas. Additionally, the inclusion of a deep residual shrinkage network helps further improve feature extraction efficiency.

Figure 6 illustrates the feature maps generated by these sub-models. A detailed observation of these maps reveals the specialized functions of each sub-model. Conv1 is adept at extracting texture and structural information, playing a pivotal role in restoring details that are often lost in underwater haziness, particularly around object edges and textures. Conv2 is tailored to the extraction of local features, thereby sharpening image detail and enhancing contrast. This enhancement is crucial in making underwater objects more discernible and visually striking. Conv3 produces a binarized effect, concentrating mainly on prominent contours and shapes within the image. This functionality is key to improving the distinction between foreground and background elements, thus highlighting the subject matter more effectively. Conv4, on the other hand, is primarily responsible for capturing color information and luminance levels. This capability is vital for reinstating the original colors of underwater images and for optimizing their dynamic range. The 'result' feature map is a synthesis of the outputs from these four sub-models. This collective integration harnesses their individual strengths, leading to a marked enhancement in image details, contrast, saturation, and color fidelity. When compared to the final 'output', it is evident that this structured, multi-scale approach significantly enriches the quality and realism of the output images.

4.5. Ablation Study

To rigorously evaluate the effectiveness of MEvo-GAN and the contributions of its individual components, we executed an ablation study. This process involved the sequential removal of critical elements within MEvo-GAN, namely the multi-scale network, the evolutionary mechanism, and the VGG loss function.

As shown in Table 2, the intact MEvo-GAN configuration demonstrates superior performance across all evaluated metrics, clearly highlighting the significant contributions of the integrated components. Notably, the ablation experiments brought to light instances of gradient explosion in configurations lacking the evolutionary mechanism. This finding underscores the mechanism's pivotal role in bolstering training stability, as depicted in Figure 7.

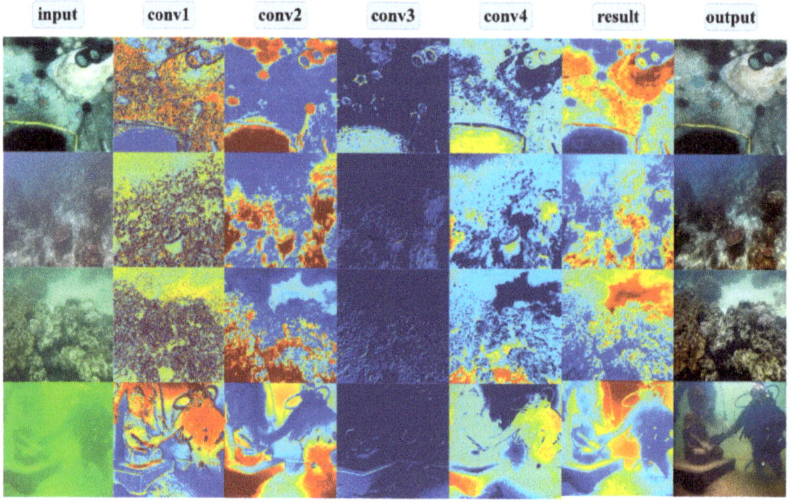

Figure 6. Visualization of feature maps.

Table 2. Comparison of different models on PSNR, SSIM, and UCIQE metrics.

Model	PSNR	SSIM	UCIQE
−w/o multiscale network	19.0107	0.7950	0.6053
−w/o Evo mechanism	20.0486	0.8352	0.5694
−w/o VGG loss	20.8675	0.8251	0.6420
MEvo-GAN	21.2758	0.8662	0.6597

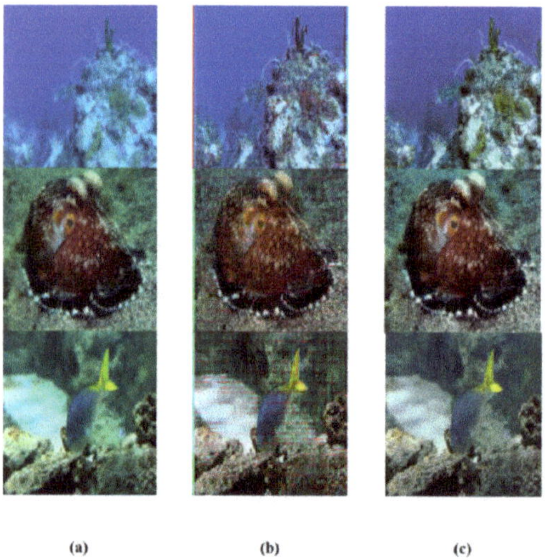

Figure 7. Example of gradient explosion phenomenon during training without integration of evolutionary mechanism. (**a**) Original image; (**b**) training image without incorporation of evolutionary mechanism; (**c**) MEvo-GAN.

Additionally, we employed the SIFT detection algorithm to compare feature point matching before and after image enhancement, as illustrated in Figure 8. When using the SIFT detection algorithm, a higher number of matched feature points indicates that the generation process preserves many key features of the original image, resulting in higher image clarity. The data presented in Figure 8 demonstrate that MEvo-GAN exhibits superior feature-point retention capability.

Moreover, the omission of either the multi-scale network or the VGG perceptual loss function markedly diminished the quality of the resultant images. Specifically, in real underwater environments, images produced without the multi-scale network exhibited noticeable blurriness, particularly in finer details. Similarly, the absence of VGG perceptual loss led to issues like excessive color saturation and texture loss. MSE and MAE losses often result in the generated image being too smooth and lacking in detail and texture. VGG losses retain more detail and texture information, making the generated image visually sharper and more realistic. In contrast, the complete MEvo-GAN architecture synergistically combines these components to yield images that closely resemble real underwater scenes in color accuracy and detail clarity. The Visual comparison of the enhancement effects of different ablation models is shown in Figure 9.

(a) (b) (c)

Figure 8. Enhancement effect feature point matching for different ablation models and the full MEvo-GAN model. (**a**) −w/o-VGG loss; (**b**) −w/o-multiscale network; (**c**) MEvo-GAN.

The results from the ablation study affirm the significant contribution of each component in MEvo-GAN towards its overall efficacy in underwater image enhancement tasks. A notable highlight is MEvo-GAN's exceptional performance in the Underwater Color Image Quality Evaluation (UCIQE) metric, where it substantially outperforms existing methods. This achievement underscores MEvo-GAN's advanced capability in significantly enhancing underwater image quality by effectively reducing color biases and blurriness, improving brightness, and restoring intricate details.

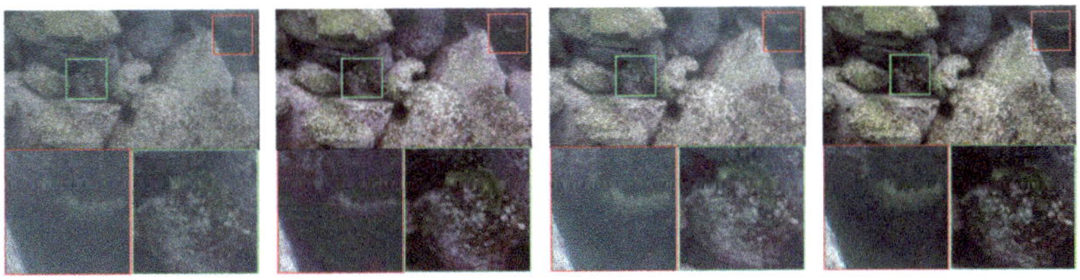

(a) (b) (c) (d)

Figure 9. Visual comparison of the enhancement effects of different ablation models and the full MEvo-GAN model. (**a**) Original image; (**b**) −w/o-VGG loss; (**c**) −w/o-multiscale network; (**d**) MEvo-GAN.

5. Conclusions

The Multi-scale Evolutionary Generative Adversarial Network (MEvo-GAN) has demonstrated remarkable capabilities in enhancing underwater images. Its innovative integration of adversarial learning with evolutionary strategies enables effective multi-scale image processing and optimization. This approach has led to substantial improvements in the visual quality of underwater images. Compared to previously popular methods, MEvo-GAN performed particularly well in the UCIQE metric, which comprehensively evaluates multiple aspects of the image, such as color balance, contrast, and clarity. This further verifies the significant advantages of MEvo-GAN in enhancing the quality of underwater images.

Notably, while MEvo-GAN exhibits exceptional proficiency in color reproduction, it is recognized that the full spectrum of its capabilities in enhancing overall underwater image quality is yet to be fully tapped. Future research endeavors will focus on this aspect, aiming to further elevate the model's efficacy. This will involve delving into optimization of the model structure and integrating attention mechanisms to refine the restoration of image details.

In summary, MEvo-GAN represents a significant stride forward in the realm of underwater image enhancement, thanks to its synergistic use of deep learning and evolutionary strategies. Building upon the robust framework of MEvo-GAN, our future objectives are twofold—to extend the application of this methodology into a broader spectrum of related tasks and to refine the quality of training datasets specific to MEvo-GAN. Specifically, through meticulous selection and fine tuning of hyperparameters such as λ_{vgg}, λ_{cyc}, and λ_{idt}, alongside the incorporation of additional evolutionary algorithms, we anticipate further enhancement of MEvo-GAN's performance and elevation of the level of detail and overall image quality. This strategic approach is anticipated to substantially enhance the overall visual quality and efficiency of MEvo-GAN. Through these advancements, our aim is to not only elevate MEvo-GAN's current capabilities but also to expand its range of practical applications. Such developments are expected to contribute significantly to the field of image processing, highlighting MEvo-GAN's role as a versatile and impactful tool in this domain.

Author Contributions: Conceptualization, F.F.; Funding acquisition, M.F.; Methodology, F.F.; Project administration, F.F. and M.F.; Resources, J.X.; Software, P.L.; Supervision, Z.S.; Writing—original draft, P.L.; Writing—review and editing, P.L. All authors have read and agreed to the published version of the manuscript.

Funding: This paper was supported by the Education Department of Jilin Province (No. JJKH20220770KJ), Jilin Provincial Scientific and Technological Development Program (No. JYDZJ202301ZYTS411), and the Zhongshan Science and Technology Bureau introduced scientific research and innovation team projects (No. CXTD2023005).

Institutional Review Board Statement: Not applicable.

Informed Consent Statement: Not applicable.

Data Availability Statement: The original contributions presented in the study are included in the article; further inquiries can be directed to the corresponding author.

Conflicts of Interest: The authors declare no conflicts of interest.

References

1. Liang, Z.; Zhang, W.; Ruan, R.; Zhuang, P.; Xie, X.; Li, C. Underwater image quality improvement via color, detail, and contrast restoration. *IEEE Trans. Circuits Syst. Video Technol.* **2023**, *34*, 1726–1742. [CrossRef]
2. He, K.; Sun, J.; Tang, X. Single image haze removal using dark channel prior. *IEEE Trans. Pattern Anal. Mach. Intell.* **2010**, *33*, 2341–2353. [PubMed]
3. Chiang, J.Y.; Chen, Y.C. Underwater image enhancement by wavelength compensation and dehazing. *IEEE Trans. Image Process.* **2011**, *21*, 1756–1769. [CrossRef] [PubMed]
4. Galdran, A.; Pardo, D.; Picón, A.; Alvarez-Gila, A. Automatic red-channel underwater image restoration. *J. Vis. Commun. Image Represent.* **2015**, *26*, 132–145. [CrossRef]

5. Drews, P.L.; Nascimento, E.R.; Botelho, S.S.; Campos, M.F.M. Underwater depth estimation and image restoration based on single images. *IEEE Comput. Graph. Appl.* **2016**, *36*, 24–35. [CrossRef]
6. Peng, Y.T.; Cosman, P.C. Underwater image restoration based on image blurriness and light absorption. *IEEE Trans. Image Process.* **2017**, *26*, 1579–1594. [CrossRef] [PubMed]
7. Hou, G.; Li, N.; Zhuang, P.; Li, K.; Sun, H.; Li, C. Non-uniform illumination underwater image restoration via illumination channel sparsity prior. *IEEE Trans. Circuits Syst. Video Technol.* **2023**, 34, 799–814. [CrossRef]
8. Yao, X.; He, F.; Wang, B. Deep learning-based recurrent neural network for underwater image enhancement. In Proceedings of the Sixth Conference on Frontiers in Optical Imaging and Technology: Imaging Detection and Target Recognition, Nanjing, China, 30 April 2024; Volume 13156, pp. 368–378.
9. Zhang, M.; Li, Y.; Yu, W. Underwater Image Enhancement Algorithm Based on Adversarial Training. *Electronics* **2024**, *13*, 2184. [CrossRef]
10. Jiang, X.; Yu, H.; Zhang, Y.; Pan, M.; Li, Z.; Liu, J.; Lv, S. An underwater image enhancement method for a preprocessing framework based on generative adversarial network. *Sensors* **2023**, *23*, 5774. [CrossRef]
11. Guo, Y.; Li, H.; Zhuang, P. Underwater image enhancement using a multiscale dense generative adversarial network. *IEEE J. Ocean. Eng.* **2019**, *45*, 862–870. [CrossRef]
12. Li, J.; Skinner, K.A.; Eustice, R.M.; Johnson-Roberson, M. WaterGAN: Unsupervised generative network to enable real-time color correction of monocular underwater images. *IEEE Robot. Autom. Lett.* **2017**, *3*, 387–394. [CrossRef]
13. Yang, Y.; Lu, H. Single image deraining using a recurrent multi-scale aggregation and enhancement network. In Proceedings of the 2019 IEEE International Conference on Multimedia and Expo (ICME), Shanghai, China, 8–12 July 2019; pp. 1378–1383.
14. Li, Q.Z.; Bai, W.X.; Niu, J. Underwater image color correction and enhancement based on improved cycle-consistent generative adversarial networks. *Acta Autom. Sin.* **2023**, *49*, 820–829.
15. Cong, R.; Yang, W.; Zhang, W.; Li, C.; Guo, C.L.; Huang, Q.; Kwong, S. Pugan: Physical model-guided underwater image enhancement using gan with dual-discriminators. *IEEE Trans. Image Process.* **2023**, *32*, 4472–4485. [CrossRef] [PubMed]
16. Wang, Z.; Shen, L.; Wang, Z.; Lin, Y.; Jin, Y. Generation-based joint luminance-chrominance learning for underwater image quality assessment. *IEEE Trans. Circuits Syst. Video Technol.* **2022**, *33*, 1123–1139. [CrossRef]
17. Li, K.; Fan, H.; Qi, Q.; Yan, C.; Sun, K.; Wu, Q.J. TCTL-Net: Template-free Color Transfer Learning for Self-Attention Driven Underwater Image Enhancement. *IEEE Trans. Circuits Syst. Video Technol.* **2023**. [CrossRef]
18. Wang, C.; Xu, C.; Yao, X.; Tao, D. Evolutionary Generative Adversarial Networks. *IEEE Trans. Evol. Comput.* **2019**, *23*, 921–934. [CrossRef]
19. Chen, S.; Wang, W.; Xia, B.; You, X.; Peng, Q.; Cao, Z.; Ding, W. CDE-GAN: Cooperative dual evolution-based generative adversarial network. *IEEE Trans. Evol. Comput.* **2021**, *25*, 986–1000. [CrossRef]
20. Mu, J.; Zhou, Y.; Cao, S.; Zhang, Y.; Liu, Z. Enhanced evolutionary generative adversarial networks. In Proceedings of the 2020 39th Chinese Control Conference (CCC), Shenyang, China, 27–29 July 2020; pp. 7534–7539.
21. He, C.; Huang, S.; Cheng, R.; Tan, K.C.; Jin, Y. Evolutionary multiobjective optimization driven by generative adversarial networks (GANs). *IEEE Trans. Cybern.* **2020**, *51*, 3129–3142. [CrossRef] [PubMed]
22. Zhang, L.; Zhao, L. High-quality face image generation using particle swarm optimization-based generative adversarial networks. *Future Gener. Comput. Syst.* **2021**, *122*, 98–104. [CrossRef]
23. Liu, F.; Wang, H.; Zhang, J.; Fu, Z.; Zhou, A.; Qi, J.; Li, Z. EvoGAN: An evolutionary computation assisted GAN. *Neurocomputing* **2022**, *469*, 81–90. [CrossRef]
24. Xue, Y.; Zhang, Y.; Neri, F. A method based on evolutionary algorithms and channel attention mechanism to enhance cycle generative adversarial network performance for image translation. *Int. J. Neural Syst.* **2023**, *33*, 2350026. [CrossRef] [PubMed]
25. Zhang, Z.; Chen, L.; Zhang, C.; Shi, H.; Li, H. GMA-DRSNs: A novel fault diagnosis method with global multi-attention deep residual shrinkage networks. *Measurement* **2022**, *196*, 111203. [CrossRef]
26. Mao, X.; Li, Q.; Xie, H.; Lau, R.Y.; Wang, Z.; Paul Smolley, S. Least squares generative adversarial networks. In Proceedings of the IEEE International Conference on Computer Vision, Venice, Italy, 22–29 October 2017; pp. 2794–2802.
27. Mittal, A.; Soundararajan, R.; Bovik, A.C. Making a "completely blind" image quality analyzer. *IEEE Signal Process. Lett.* **2012**, *20*, 209–212. [CrossRef]
28. Nagarajan, V.; Kolter, J.Z. Gradient descent GAN optimization is locally stable. *Adv. Neural Inf. Process. Syst.* **2017**, *30*.
29. Li, C.; Guo, C.; Ren, W.; Cong, R.; Hou, J.; Kwong, S.; Tao, D. An underwater image enhancement benchmark dataset and beyond. *IEEE Trans. Image Process.* **2019**, *29*, 4376–4389. [CrossRef] [PubMed]
30. Yuzhen, L.; Meiyi, L.; Sen, L.; Zhiyong, T. Underwater Image Enhancement Based on Multi-Scale Feature Fusion and Attention Network. *J. Comput.-Aided Des. Comput. Graph.* **2023**, *35*, 685–695.
31. Islam, M.J.; Luo, P.; Sattar, J. Simultaneous enhancement and super-resolution of underwater imagery for improved visual perception. *arXiv* **2020**, arXiv:2002.01155.
32. Saleh, A.; Sheaves, M.; Jerry, D.; Azghadi, M.R. Adaptive uncertainty distribution in deep learning for unsupervised underwater image enhancement. *arXiv* **2022**, arXiv:2212.08983.
33. Han, J.; Shoeiby, M.; Malthus, T.; Botha, E.; Anstee, J.; Anwar, S.; Wei, R.; Petersson, L.; Armin, M.A. Single underwater image restoration by contrastive learning. In Proceedings of the 2021 IEEE iNternational Geoscience and Remote Sensing Symposium IGARSS, Brussels, Belgium, 11–16 July 2021; pp. 2385–2388.

34. Islam, M.J.; Xia, Y.; Sattar, J. Fast underwater image enhancement for improved visual perception. *IEEE Robot. Autom. Lett.* **2020**, 5, 3227–3234. [CrossRef]
35. Naik, A.; Swarnakar, A.; Mittal, K. Shallow-uwnet: Compressed model for underwater image enhancement (student abstract). In Proceedings of the AAAI Conference on Artificial Intelligence, Virtual, 2–9 February 2021; Volume 35, pp. 15853–15854.
36. Ren, T.; Xu, H.; Jiang, G.; Yu, M.; Luo, T. Reinforced swin-convs transformer for underwater image enhancement. *arXiv* **2022**, arXiv:2205.00434.
37. Fabbri, C.; Islam, M.J.; Sattar, J. Enhancing underwater imagery using generative adversarial networks. In Proceedings of the 2018 IEEE International Conference on Robotics and Automation (ICRA), Brisbane, Australia, 21–25 May 2018; pp. 7159–7165.
38. Peng, W.; Zhou, C.; Hu, R.; Cao, J.; Liu, Y. RAUNE-Net: A Residual and Attention-Driven Underwater Image Enhancement Method. *arXiv* **2023**, arXiv:2311.00246.

Journal of
Marine Science and Engineering

MDPI

Article

Deep Learning Based Characterization of Cold-Water Coral Habitat at Central Cantabrian Natura 2000 Sites Using YOLOv8

Alberto Gayá-Vilar [1,*], Alberto Abad-Uribarren [1], Augusto Rodríguez-Basalo [1], Pilar Ríos [2], Javier Cristobo [2] and Elena Prado [1]

[1] Centro Oceanográfico de Santander (COST-IEO), IEO-CSIC, Promontorio San Martín, 39004 Santander, Spain; alberto.abad@ieo.csic.es (A.A.-U.); augusto.rodriguez@ieo.csic.es (A.R.-B.); elena.prado@ieo.csic.es (E.P.)

[2] Centro Oceanográfico de Gijón (COG-IEO), IEO-CSIC, Avda. Principe de Asturias 70bis, 33212 Gijón, Spain; pilar.rios@ieo.csic.es (P.R.); javier.cristobo@ieo.csic.es (J.C.)

* Correspondence: alberto.gaya@ieo.csic.es

Abstract: Cold-water coral (CWC) reefs, such as those formed by *Desmophyllum pertusum* and *Madrepora oculata*, are vital yet vulnerable marine ecosystems (VMEs). The need for accurate and efficient monitoring of these habitats has driven the exploration of innovative approaches. This study presents a novel application of the YOLOv8l-seg deep learning model for the automated detection and segmentation of these key CWC species in underwater imagery. The model was trained and validated on images collected at two Natura 2000 sites in the Cantabrian Sea: the Avilés Canyon System (ACS) and El Cachucho Seamount (CSM). Results demonstrate the model's high accuracy in identifying and delineating individual coral colonies, enabling the assessment of coral cover and spatial distribution. The study revealed significant variability in coral cover between and within the study areas, highlighting the patchy nature of CWC habitats. Three distinct coral community groups were identified based on percentage coverage composition and abundance, with the highest coral cover group being located exclusively in the La Gaviera canyon head within the ACS. This research underscores the potential of deep learning models for efficient and accurate monitoring of VMEs, facilitating the acquisition of high-resolution data essential for understanding CWC distribution, abundance, and community structure, and ultimately contributing to the development of effective conservation strategies.

Keywords: cold-water corals; underwater image processing; deep learning; marine protected areas; Aviles Canyon system; El Cachucho

1. Introduction

Cold-water coral (CWC) reefs, such as those formed by the framework-building scleractinians *Desmophyllum pertusum* (Linneus, 1758) and *Madrepora oculata* (Linneus, 1758), are vital yet vulnerable marine ecosystems (VMEs) renowned for their biodiversity and crucial role in deep-sea environments [1]. The ecological significance of CWC reefs lies in their ability to create complex three-dimensional structures that provide habitat [2], feeding grounds [3], and nursery areas for a diverse array of marine organisms [4,5], thereby enhancing overall biodiversity and biomass in the deep sea. The intricate framework of these reefs also influences critical ecological processes, including larval dispersal, retention, and feeding efficiency, further underscoring their importance in maintaining the health and productivity of deep-sea ecosystems [6–8].

These reefs are found in diverse locations, including continental slopes, seamounts, fjords, and submarine canyons [9,10]. The unique geomorphological characteristics of submarine canyons and seamounts, with their complex topography and steep slopes, offer natural refuges for CWC reefs, shielding them from destructive fishing practices and other anthropogenic disturbances [11,12]. The varied terrain and hydrodynamic conditions associated with these geological formations create a mosaic of habitats that support a rich tapestry of benthic communities, including the iconic CWC reefs.

The scleractinian corals, *D. pertusum* and *M. oculata*, are key ecosystem engineers in the deep sea, constructing the framework of CWC reefs that provide essential habitat for numerous associated species. These corals, often referred to as 'white corals' due to their ahermatypic nature, do not rely on symbiotic algae for nutrition and can thrive in the cold, dark depths of the ocean. The presence of these corals fosters a diverse assemblage of fauna, including bivalves, gastropods, echinoderms, sponges, and worms, many of which utilize the coral skeletons for attachment or as a source of food [13]. The structural complexity of CWC reefs, with their numerous crevices and overhangs, creates microhabitats that support a wide range of ecological niches, further contributing to the high biodiversity associated with these ecosystems [14].

Despite their ecological importance, CWC reefs face numerous threats, including bottom trawling, deep-sea mining, and climate change [15,16]. These anthropogenic pressures pose a significant risk to the integrity and persistence of CWC reefs, potentially leading to habitat degradation, loss of biodiversity, and disruption of ecosystem functions [17]. The slow growth rates and fragile nature of many CWC species make them particularly vulnerable to disturbance, and recovery from damage can take decades or even centuries [18]. The increasing recognition of the threats facing CWC reefs has led to their designation as VMEs by the United Nations General Assembly (Resolution 61/105), highlighting the urgent need for their protection and conservation [1].

Accurate identification and delineation of CWC reefs, particularly from underwater imagery, remains a challenge due to the lack of precise information on coral cover and the associated structural complexity. Traditional methods of monitoring and assessing CWC reefs often rely on labor-intensive manual annotation of images or video footage, which can be time-consuming and prone to subjective interpretation [19]. The vastness and remoteness of deep-sea environments further complicate efforts to obtain comprehensive and representative data on CWC distribution and abundance.

Recent advances in computer vision techniques, particularly deep learning-based object detection and segmentation models, have offered promising solutions to these challenges. These models, trained on large labeled datasets, can automatically identify and delineate coral colonies in underwater images, thereby enabling accurate estimation of coral cover and spatial distribution [20,21]. The application of deep learning in marine ecological research has gained significant traction in recent years, demonstrating its potential to revolutionize the way we study and monitor marine ecosystems. In the realm of underwater object detection, this task is particularly challenging due to the unique characteristics of the marine environment, such as poor visibility, light attenuation, and complex backgrounds. Traditional object detection methods, including the Region-based Convolutional Neural Network (R-CNN) series [22,23], have been explored for underwater applications, but often face limitations in computational efficiency and real-time performance. The emergence of single-stage detectors, such as the You Only Look Once (YOLO) series [24–26], has addressed these limitations, offering a faster and more streamlined approach. YOLO models have demonstrated remarkable success in various domains, including underwater object detection [21]. Their ability to simultaneously predict bounding boxes and class probabilities in a single pass contributes to their efficiency and suitability for real-time applications. Among these models, YOLOv8 has emerged as a powerful tool, demonstrating effectiveness in detecting various marine organisms, including corals in diverse environments [27].

In this study, we leveraged the power of YOLOv8 for the automated detection and segmentation of coral species in remotely operated towed vehicle (ROTV) imagery collected at two Natura 2000 sites in the Cantabrian Sea. Our objectives are threefold: (1) to develop novel methodologies for monitoring CWC VMEs, (2) to assess variability in coral cover across geographically proximate areas and among transects within each area, and (3) to characterize these CWC communities in terms of CWC coverage. By analyzing ROTV imagery, we aim to overcome the challenges associated with manual annotation and obtain accurate quantitative data for improved understanding and management of these ecosystems.

2. Materials and Methods

2.1. Study Area

This study, framed within the INTEMARES project, focuses on two regions of the bathyal rocky outcrops in the Cantabrian Sea, south of the Bay of Biscay (Figure 1): the Avilés Canyon System (ACS) and El Cachucho Seamount (CSM). These areas were selected due to their designation as vulnerable marine ecosystems (VMEs) and their harboring of benthic communities classified as habitat 1170 (Reefs) under the European Union Habitats Directive (92/43/EEC). Of particular interest within these communities are the cold-water coral reefs, which are a focal point of this research.

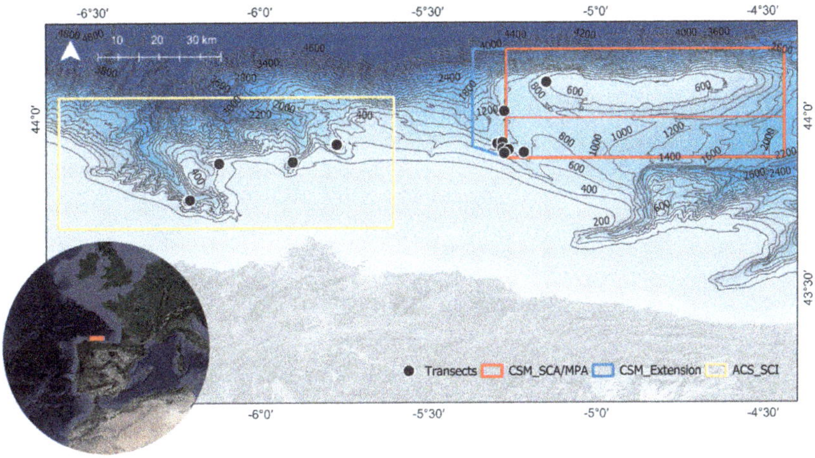

Figure 1. Map of the study area showing the boundaries of ACS and CSM, including its expansion (blue). Points indicate the transects surveyed in the study.

El Cachucho Seamount, designated a Marine Protected Area (MPA) and Special Area of Conservation (SAC) in 2011, is characterized by its complex geomorphology, featuring rocky outcrops and steep slopes [28]. Its summit, known as Le Danois Bank, lies at 425 m depth and predominantly consists of rocky outcrops with sparse sediment cover, contrasting with its inner basin (800–1000 m), where sediment accumulation is higher [12].

The ACS, a Site of Community Importance (SCI) and potential SAC within the Natura 2000 network, extends from the continental shelf to bathyal depths. It is characterized by rocky outcrops with diverse morphologies and relief, some of which exhibit tectonic activity [29]. This complex geomorphology creates a diverse habitat that supports rich benthic biodiversity [14,30].

The presence of key reef-building species, such as *D. pertusum* and *M. oculata*, highlights their ecological importance [31]. These white corals harbor a suite of associated fauna species such as bivalves, gastropods, echinoderms, sponges, and worms and host breeding grounds for fished species. These species do not depend on living corals, but use the skeletal remains as a substrate for fixation or grazing on sessile invertebrates [30]. The primary role of CWC reefs is to function as feeding grounds, refuges, and as substrata for larval settlement, juvenile growth, and as nursery areas. Furthermore, they contribute to the goods and services of the deep sea. Finally, the three-dimensional nature of these reefs increases the structural complexity of these ecosystems, making them particularly vulnerable and, therefore, deserving of special attention in terms of conservation [32]. In the study area, both species co-occur at different depths. *D. pertusum* has been recorded in the Avilés Canyon System (ACS) in a bathymetric distribution range of 342–1473 m; in the NW Atlantic, this is the most abundant and widely distributed construction species. *M. oculata*, in ACS, appears at 342–1660 m and is slow-growing and very vulnerable to trawling; it has construction activity on the continental margins of Europe.

2.2. Data Acquisition

High-resolution underwater imagery was obtained using two ROTV: Politolana [33] and TASIFE [34]. These vehicles are capable of descending to depths of 2000 m and are equipped with a high-resolution camera, bidirectional telemetry, and an acoustic positioning system (Figure 2). The camera was oriented in a zenith position, capturing images of the seabed at five-second intervals, and synchronized with environmental data to ensure the acquisition of comprehensive datasets during each dive.

Figure 2. (**a**) The TASIFE ROTV used in the ECOMARG 2024 survey. (**b**) ROTV Politolana used in the INDEMARES-INTEMARES 2014–2021 surveys. (**c**) Example image obtained by ROTV of the Cantabrian Sea seabed with *D. pertusum* and *M. oculata* colonies.

A total of 19 transects were conducted on different dates in both study areas: 9 in the CSM and 10 in the ACS. To standardize data acquisition, the ROTV Politolana was maintained at a constant distance of 1.5 m from the seabed during all transects, each lasting 20 min and covering an average distance of 460 m. The samplings were conducted at depths ranging from 450 to 1200 m. The Avilés transects were associated with submarine canyon head areas, whereas most of the El Cachucho transects were located in areas adjacent to the seamount. This sampling approach provided a detailed view of the diversity of habitats and communities present in the study area, as well as an assessment of the influence of geomorphology and other environmental factors on species distribution.

2.3. Data Processing and Analysis

In this study, the YOLOv8l-seg model was employed for the detection and segmentation of the coral species, *M. oculata* and *D. pertusum* (Figure 3), due to several key advantages it offers. YOLOv8l-seg is a state-of-the-art model characterized by its unified architecture, capable of performing both object detection and instance segmentation in a single process [26]. This capability is particularly crucial in the analysis of complex underwater imagery, where corals often exhibit irregular shapes and may be partially obscured by other elements in the environment. Moreover, YOLOv8l-seg has consistently demonstrated superior performance compared to previous YOLO versions and other object detection models, such as Faster R-CNN, across a range of computer vision applications [35]. The model's efficiency, accuracy, and ability to handle instance segmentation make it well-suited for the challenges of automating cold-water coral analysis in underwater images.

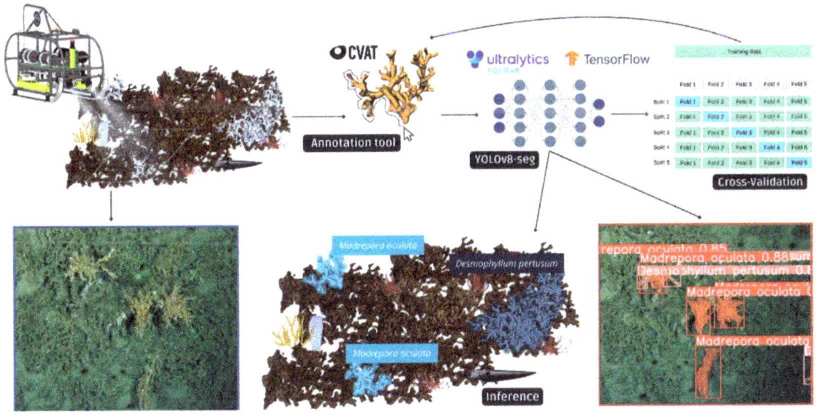

Figure 3. Workflow for cold-water coral analysis: Underwater imagery from a ROTV was annotated in CVAT to train the YOLOv8l-seg model. Five-fold cross-validation ensured model robustness before inferring new imagery, accurately detecting and segmenting coral species.

The YOLOv8l-seg architecture is based on a deep neural network composed of three main components:

- **Backbone (CSPDarknet53):** This component is responsible for extracting relevant features from the input images at different scales. The CSPDarknet53 architecture has proven highly effective in feature extraction for object detection tasks.
- **Neck (Path Aggregation Network, PAN):** This network combines the features extracted by the backbone at different scales, thereby enabling better detection of objects of various sizes. YOLOv8 utilizes a modified PAN structure to optimize this process.
- **Head:** This component performs final detection and segmentation predictions. In the case of YOLOv8l-seg, the head has two branches: one for object detection, predicting bounding boxes and object classes; and another for instance segmentation, generating accurate segmentation masks for each detected object.

The dataset used for model training and validation consisted of 670 coral images collected during various campaigns (see Acknowledgements) at Natura 2000 sites within the central Cantabrian region, representing a batch from each of the 19 transects conducted in the study areas. These images were manually labeled in YOLO format using the CVAT tool. The labeling process was optimized through an iterative approach that combined the training of an initial model with manual correction of the predictions generated by that model, utilizing the "auto_annotate" function of Ultralytics with a YOLOv8 model and SAM "mobile_sam.pt" [36]. Annotations in COCO format were converted to YOLO PyTorch. Model training was performed for 500 epochs with an initial learning rate of 0.01, applying data augmentation techniques to enhance model generalization.

For validation, 20% of the images (128 images) were reserved as an independent validation dataset, and 5-fold cross-validation (K-Fold) was implemented. Model performance was evaluated using metrics such as precision (B, M), recall (B, M), F1 score (B, M), intersection over union (IoU), mAP50, mAP50-95, and fitness. The complete source code, weights, and example data are available at: https://github.com/AlbertoGaya/cold-water-coral-reef/tree/main (27 August 2024).

Using the data extracted from the YOLOv8l-seg model, a comprehensive analysis was conducted to assess coral cover and species distribution in the study areas. The mean percentage of area covered and the number of individuals per species and transect were calculated. Non-parametric statistical tests (Kruskal–Wallis and Dunn's test) were applied to compare coral cover between percentage areas and identify significant differences.

Additionally, non-metric multidimensional scaling (nMDS) was employed to visualize the similarity between transects based on coral reef composition, and hierarchical cluster analysis was used to group the most similar transects, with results presented on a map.

Regarding the experimental setup, all analyses were performed in a Jupyter Lab environment using Python 3.9. The Ultralytics package was employed for model training and inference. The hardware configuration included an Intel Core i7-13700F Processor (16 cores, 2.1 GHz), 16 GB DDR5 RAM (2 × 8 GB, 4800 MHz), and an NVIDIA GeForce RTX 4070 VENTUS 2X E 12G OC GPU.

3. Results

3.1. Model Performance Evaluation

The YOLOv8l-seg model demonstrated robust performance in the detection and segmentation of the target coral species, *M. oculata* and *D. pertusum*. Validation results, both in 5-fold cross-validation and independent validation, are summarized in Table 1.

Table 1. Cross-validation results on bounding boxes (B) and masks (M), and independent validation results.

Validation	Precision (P)	Recall (R)	mAP50	mAP50-95
Cross-validation (B)	0.784	0.703	0.781	0.544
Cross-validation (M)	0.784	0.694	0.769	0.508
Independent validation	0.839	0.749	0.833	0.601

In the independent validation, the model achieved even higher performance, highlighting its ability to generalize to unseen data. Notably, the model exhibited slightly superior performance in detecting and segmenting *D. pertusum* (P = 0.876, R = 0.810) compared to *M. oculata* (P = 0.804, R = 0.693).

Overall, the validation results support the effectiveness of the YOLOv8l-seg model in the automated detection and segmentation of cold-water corals in underwater imagery, suggesting its potential as a valuable tool for monitoring and assessing these vulnerable ecosystems.

While precise runtime measurements were not collected in this study, the utilization of the YOLOv8l-seg model in conjunction with a dedicated GPU (NVIDIA GeForce RTX 4070) enabled efficient processing of the high-resolution imagery, facilitating the timely completion of the analysis.

3.2. Coral Cover Comparison between Study Areas

Descriptive analysis of the data revealed differences in coral cover among transects and study areas. The mean cover of *D. pertusum* was 0.56% ± 0.02% (range: 0–13.29%), while that of *M. oculata* was 0.40% ± 0.01% (range: 0–7.58%). The Kruskal–Wallis test confirmed significant differences in total coral cover both among transects ($p < 0.001$) and between CSM and ACS ($p < 0.001$), with higher cover in the latter.

Non-metric multidimensional scaling (nMDS) and hierarchical cluster analysis (Figure 4) identified three distinct groups based on coral species composition, showing an increasing gradient of cover from group yellow to group red.

Group yellow, with the lowest coral cover (maximum 3% for *M. oculata*), is mainly distributed in CSM. Group blue, with intermediate cover (maximum 6% for *D. pertusum*), is found in both CSM and ACS. Group red, with the highest cover (maximum 13% for *M. oculata*), is exclusively located in the La Gaviera canyon head within the ACS (Figure 5).

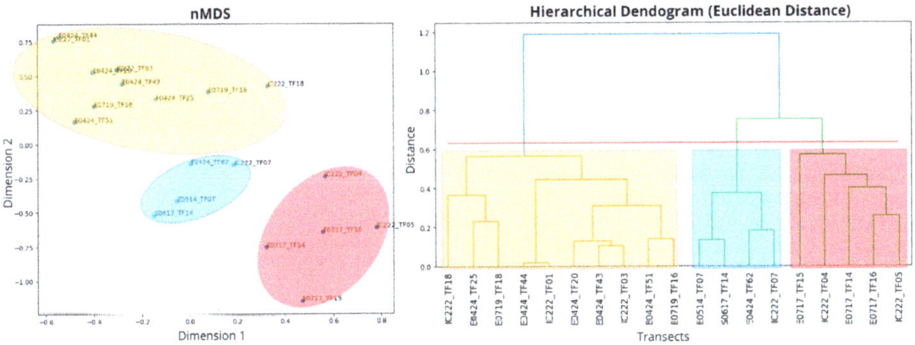

Figure 4. Nonmetric multidimensional scaling (nMDS) of transects based on cold-water coral composition. Points represent transects, and colors indicate groups identified by hierarchical cluster analysis.

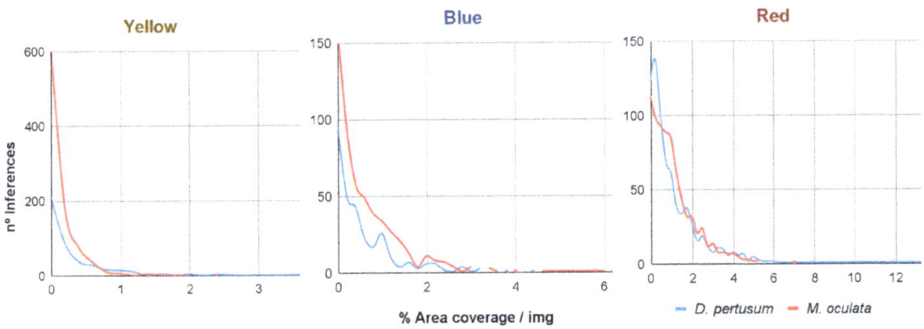

Figure 5. Line graph showing the mean cover (%) of *D. pertusum* and *M. oculata* in each group identified by hierarchical cluster analysis.

4. Discussion

The results of this study demonstrate the effectiveness of deep learning models, such as YOLOv8l-seg, in the automated detection and segmentation of cold-water corals in underwater imagery. The high performance of the model in terms of precision, recall, and mAP (Table 1) highlights its potential as a valuable tool for streamlining the monitoring and assessment of vulnerable ecosystems, overcoming the limitations of laborious manual annotation. These results are consistent with other studies that have successfully applied deep learning techniques for the automated identification of benthic fauna [27,35] and highlight the growing potential of these methods in marine ecological research.

Analysis of the data revealed significant variability in coral cover, not only among different transects, but also between the El Cachucho and Avilés areas. This discrepancy suggests that coral distribution is influenced by local factors, such as substrate density, curvature, and rugosity [37], as evidenced in the La Gaviera canyon head, where particular environmental conditions appear to favor higher coral cover. The identification of three distinct groups of transects based on coral species composition (Figures 4 and 5) supports this hypothesis, showing an increasing gradient of cover from group yellow to group red (La Gaviera).

The variability in coral cover observed even among nearby transects highlights the inherent challenges associated with sampling these VMEs. Cold-water coral habitats often occur in discontinuous patches or are strongly delimited by specific conditions such as substrate type [38], depth, slope, orientation [39], and currents [40]. The fragmented and localized nature of cold-water coral reefs, coupled with the technological and logistical constraints associated with deep-sea research [41], makes it challenging to obtain a comprehensive and representative picture of the distribution and abundance of these species. In this regard, two transects conducted in La Gaviera, located in areas with conditions distinct from those of the reefs, exhibited very different coral cover values, reinforcing the notion of spatial heterogeneity in these ecosystems (Figure 6). Interestingly, all the transects located below 650 m belong to the yellow group. The red group, exclusively located in La Gaviera canyon head, has a mean depth of 773 m, while the blue group ranges between 750 and 1200 m. This suggests a potential depth gradient in coral cover, although the patchy nature of the transects and the limited sample size in different depth ranges require further research to confirm this hypothesis. However, our findings provide valuable preliminary evidence suggesting a potential depth-related pattern in cold-water coral distribution.

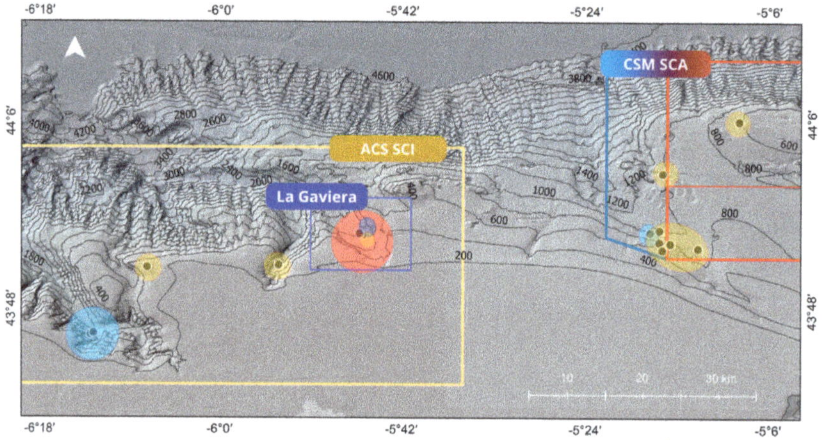

Figure 6. Map of the study area (CSM and ACS) with the locations of the transects color-coded by a group based on hierarchical cluster analysis. La Gaviera canyon head is indicated within the ACS.

The application of deep learning models like YOLOv8l-seg, in conjunction with the collection of detailed environmental and biological data, presents a promising avenue for enhancing our understanding and capacity to protect cold-water coral reefs. The integration of environmental data, such as temperature, salinity, current velocity, and substrate characteristics, into predictive models could help elucidate the complex relationships between physical factors and coral distribution. Such models could also be used to forecast the potential impacts of climate change and other anthropogenic disturbances on these vulnerable ecosystems, informing the development of adaptive management strategies.

The promising results obtained with YOLOv8 in this study highlight the transformative potential of deep learning in facilitating the automated assessment of vulnerable marine ecosystems. The model's efficacy in accurately detecting and segmenting cold-water corals, even within the challenging visual conditions of the deep sea, paves the way for more efficient and comprehensive monitoring efforts. However, we recognize that the inherent complexities of underwater imaging, such as low visibility and color distortion, present ongoing challenges [42]. Future research could explore the integration of advanced image enhancement techniques, leveraging innovations in reinforcement learning [43] or metalens technology [44], to further refine the accuracy and robustness of deep-sea object detection models. The expansion of automated detection and segmentation to encompass a wider range of benthic species, including sponges and gorgonians, would also significantly enhance our understanding of cold-water coral ecosystems' structure and function [32]. Additionally, addressing limitations related to variations in image scale due to fluctuations in ROTV altitude and inconsistencies in data collection arising from disparities in transect design could further improve the precision and comparability of future studies. The continued integration of these advanced techniques with ongoing improvements in data acquisition and processing will undoubtedly enhance our capacity to study, monitor, and ultimately conserve these invaluable ecosystems.

5. Conclusions

The automated analysis of underwater imagery using YOLOv8l-seg has proven to be an effective tool for the detection and segmentation of cold-water coral species, facilitating the assessment and monitoring of these vulnerable ecosystems. Our results reveal significant variability in spatial coral cover, not only among geographically distinct areas but also between nearby transects within the same area, highlighting the inherently patchy and localized nature of these habitats. This heterogeneity underscores the challenges of sampling and monitoring cold-water coral reefs and emphasizes the need for comprehensive, high-resolution surveys to accurately assess their distribution and abundance. The observed depth gradient in coral cover, with a potential optimum range, warrants further investigation to understand the underlying ecological drivers.

The next step in this research involves leveraging the acquired coverage data and associated environmental variables (e.g., temperature, salinity, depth, substrate type, current flow) to develop a predictive model for cold-water coral species distribution. Such a model could help identify key environmental predictors of coral presence and abundance, enabling more targeted and efficient surveys and informing the design of effective conservation and management strategies, particularly in the context of seabed management.

Despite methodological limitations, this study provides valuable insights into the distribution of key species in Natura 2000 sites and lays the groundwork for future research integrating environmental data and expanding the range of species studied, thereby enhancing our ability to understand, manage, and conserve cold-water coral reefs. The continued integration of these advanced techniques with ongoing improvements in data acquisition and processing will undoubtedly enhance our ability to understand, manage, and conserve these invaluable ecosystems.

Author Contributions: A.G.-V.: Conceptualization, Methodology, Software, Validation, Formal analysis, Investigation, Resources, Data curation, Writing—original draft preparation, Writing—review and editing, Visualization, Supervision, Project administration. A.A.-U.: Writing—review and editing, Validation, Investigation, Formal analysis, Data curation. A.R.-B.: Writing—review and editing, Visualization, Validation, Investigation, Formal analysis, Data curation. P.R.: Writing—review and editing, Validation, Investigation. J.C.: Writing—review and editing, Validation, Investigation. E.P.: Conceptualization, Resources, Data curation, Writing—review and editing, Supervision, Project administration, Funding acquisition. All authors have read and agreed to the published version of the manuscript.

Funding: This research received funding from multiple sources, including the European Union's LIFE program through the LIFE IP INTEMARES project (LIFE15 IPE ES 012) and BIODIV project through the Biodiv_A3 Action. BIODIV project: "Scientific and technical advice for the monitoring of marine biodiversity: protected marine areas and species of state competence (2022–2025)", funded by the European Union—NextGenerationEU through the Recovery, Transformation and Resilience Plan and promoted by the Directorate General for Biodiversity, Forests and Desertification of the Ministry for Ecological Transition and the Demographic Challenge and CSIC, through the Spanish Institute of Oceanography (IEO). The APC was funded by the LIFE IP INTEMARES project, specifically through its C2.1 Action focused on developing new methodologies for monitoring MPAs.

Institutional Review Board Statement: Ethical review and approval were waived for this study due to the fact that it involved the analysis of remotely operated underwater vehicle (ROTV) imagery of the seabed in offshore Natura 2000 areas, and did not involve any direct interaction or manipulation of human or animal subjects.

Informed Consent Statement: Not applicable.

Data Availability Statement: The code used in this study is openly available on GitHub at https://github.com/AlbertoGaya/cold-water-coral-reef/tree/main (28 August 2024), which includes validation images and model weights. Additional images or data are available upon reasonable request from the corresponding author.

Acknowledgments: The authors would like to thank the crew and scientific team aboard the R/V Ramón Margalef and Angeles Alvariño from the Spanish Institute of Oceanography, as well as the technicians of the ROTV Politolana and Tasife for their skillful execution of the challenging visual transects in the study area. This research was conducted within the framework of action C.2.1 of the INTEMARES project, which focuses on the development of new monitoring methodologies for marine protected areas. The INTEMARES project was partially funded by the European Commission LIFE + "Nature and Biodiversity" call (LIFE15 IPE ES 012). The authors also acknowledge the support from the iMagine project (funded by the European Union Horizon Europe Programme—Grant Agreement number 101058625).

Conflicts of Interest: The authors declare no conflicts of interest. The funders had no role in the design of the study; in the collection, analyses, or interpretation of data; in the writing of the manuscript; or in the decision to publish the results.

References

1. Food and Agriculture Organization of the United Nations (FAO). *International Guidelines for the Management of Deep-Sea Fisheries in the High Seas*; FAO: Rome, Italy, 2009; p. 73.
2. Sundahl, H.; Buhl-Mortensen, P.; Buhl-Mortensen, L. Distribution and Suitable Habitat of the Cold-Water Corals Lophelia pertusa, Paragorgia arborea, and Primnoa resedaeformis on the Norwegian Continental Shelf. *Front. Mar. Sci.* **2020**, *7*, 213. [CrossRef]
3. D'Onghia, G.; Maiorano, P.; Carlucci, R.; Capezzuto, F.; Carluccio, A.; Tursi, A.; Sion, L. Comparing Deep-Sea Fish Fauna between Coral and Non-Coral "Megahabitats" in the Santa Maria Di Leuca Cold-Water Coral Province (Mediterranean Sea). *PLoS ONE* **2012**, *7*, e44509. [CrossRef] [PubMed]
4. Henry, L.-A.; Navas, J.M.; Hennige, S.J.; Wicks, L.C.; Vad, J.; Murray Roberts, J. Cold-Water Coral Reef Habitats Benefit Recreationally Valuable Sharks. *Biol. Conserv.* **2013**, *161*, 67–70. [CrossRef]
5. Henry, L.-A.; Stehmann, M.F.W.; De Clippele, L.; Findlay, H.S.; Golding, N.; Roberts, J.M. Seamount Egg-Laying Grounds of the Deep-Water Skate *Bathyraja richardsoni*: DEEP-WATER BATHYRAJA RICHARDSONI EGG-LAYING GROUNDS. *J. Fish Biol.* **2016**, *89*, 1473–1481. [CrossRef] [PubMed]
6. Buhl-Mortensen, L.; Vanreusel, A.; Gooday, A.J.; Levin, L.A.; Priede, I.G.; Buhl-Mortensen, P.; Gheerardyn, H.; King, N.J.; Raes, M. Biological Structures as a Source of Habitat Heterogeneity and Biodiversity on the Deep Ocean Margins. *Mar. Ecol.* **2010**, *31*, 21–50. [CrossRef]
7. Orejas, C.; Gori, A.; Rad-Menéndez, C.; Last, K.S.; Davies, A.J.; Beveridge, C.M.; Sadd, D.; Kiriakoulakis, K.; Witte, U.; Roberts, J.M. The Effect of Flow Speed and Food Size on the Capture Efficiency and Feeding Behaviour of the Cold-Water Coral Lophelia Pertusa. *J. Exp. Mar. Biol. Ecol.* **2016**, *481*, 34–40. [CrossRef]
8. Cordes, E.E.; Demopoulos, A.W.J.; Davies, A.J.; Gasbarro, R.; Rhoads, A.C.; Lobecker, E.; Sowers, D.; Chaytor, J.D.; Morrison, C.L.; Weinnig, A.M.; et al. Expanding Our View of the Cold-Water Coral Niche and Accounting of the Ecosystem Services of the Reef Habitat. *Sci. Rep.* **2023**, *13*, 19482. [CrossRef]
9. Henry, L.-A.; Roberts, J.M. Global Biodiversity in Cold-Water Coral Reef Ecosystems. In *Marine Animal Forests: The Ecology of Benthic Biodiversity Hotspots*; Rossi, S., Bramanti, L., Gori, A., Orejas, C., Eds.; Springer International Publishing: Cham, Switzerland, 2017; pp. 235–256. [CrossRef]
10. Rowden, A.A.; Schlacher, T.A.; Williams, A.; Clark, M.R.; Stewart, R.; Althaus, F.; Bowden, D.A.; Consalvey, M.; Robinson, W.; Dowdney, J. A Test of the Seamount Oasis Hypothesis: Seamounts Support Higher Epibenthic Megafaunal Biomass than Adjacent Slopes. *Mar. Ecol.* **2010**, *31*, 95–106. [CrossRef]
11. Fernandez-Arcaya, U.; Ramirez-Llodra, E.; Aguzzi, J.; Allcock, A.L.; Davies, J.S.; Dissanayake, A.; Harris, P.; Howell, K.; Huvenne, V.A.I.; Macmillan-Lawler, M.; et al. Ecological Role of Submarine Canyons and Need for Canyon Conservation: A Review. *Front. Mar. Sci.* **2017**, *4*, 5. [CrossRef]
12. Bourque, J.R.; Demopoulos, A.W.J. The Influence of Different Deep-Sea Coral Habitats on Sediment Macrofaunal Community Structure and Function. *PeerJ* **2018**, *6*, e5276. [CrossRef]
13. Oevelen, D.; Duineveld, G.; Lavaleye, M.; Mienis, F.; Soetaert, K.; Heip, C. The Cold-Water Coral Community as Hotspot of Carbon Cycling on Continental Margins: A Food-Web Analysis from Rockall Bank (Northeast Atlantic). *Limnol. Oceanogr.* **2009**, *54*, 1829–1844. [CrossRef]
14. Ríos, P.; Altuna, Á.; Frutos, I.; Manjón-Cabeza, E.; García-Guillén, L.; Macías-Ramírez, A.; Ibarrola, T.P.; Gofas, S.; Taboada, S.; Souto, J.; et al. Avilés Canyon System: Increasing the Benthic Biodiversity Knowledge. *Estuar. Coast. Shelf Sci.* **2022**, *274*, 107924. [CrossRef]

15. Pinheiro, M.; Martins, I.; Raimundo, J.; Caetano, M.; Neuparth, T.; Santos, M.M. Stressors of Emerging Concern in Deep-Sea Environments: Microplastics, Pharmaceuticals, Personal Care Products and Deep-Sea Mining. *Sci. Total Environ.* **2023**, *876*, 162557. [CrossRef] [PubMed]
16. Morato, T.; González-Irusta, J.-M.; Dominguez-Carrió, C.; Wei, C.-L.; Davies, A.; Sweetman, A.K.; Taranto, G.H.; Beazley, L.; García-Alegre, A.; Grehan, A.; et al. Climate-Induced Changes in the Suitable Habitat of Cold-Water Corals and Commercially Important Deep-Sea Fishes in the North Atlantic. *Glob. Chang. Biol.* **2020**, *26*, 2181–2202. [CrossRef]
17. Weinnig, A.M.; Gómez, C.E.; Hallaj, A.; Cordes, E.E. Cold-Water Coral (Lophelia Pertusa) Response to Multiple Stressors: High Temperature Affects Recovery from Short-Term Pollution Exposure. *Sci. Rep.* **2020**, *10*, 1768. [CrossRef]
18. Maier, S.R.; Bannister, R.J.; van Oevelen, D.; Kutti, T. Seasonal Controls on the Diet, Metabolic Activity, Tissue Reserves and Growth of the Cold-Water Coral Lophelia Pertusa. *Coral Reefs* **2020**, *39*, 173–187. [CrossRef]
19. Beijbom, O.; Edmunds, P.J.; Roelfsema, C.; Smith, J.; Kline, D.I.; Neal, B.P.; Dunlap, M.J.; Moriarty, V.; Fan, T.-Y.; Tan, C.-J.; et al. Towards automated annotation of benthic survey images: Variability of human experts and operational modes of automation. *PLoS ONE* **2015**, *10*, e0130312. [CrossRef]
20. Nawarathne, M.; Kumari, H.M.L.S.; Herath Mudiyanselage, N. Comparative Analysis of Jellyfish Classification: A Study Using YOLOv8 and Pre-Trained Models. In Proceedings of the 2024 International Research Conference on Smart Computing and Systems Engineering (SCSE), Colombo, Sri Lanka, 4 April 2024; p. 6. [CrossRef]
21. Li, J.; Xu, W.; Deng, L.; Xiao, Y.; Han, Z.; Zheng, H. Deep Learning for Visual Recognition and Detection of Aquatic Animals: A Review. *Rev. Aquac.* **2023**, *15*, 409–433. [CrossRef]
22. Ren, S.; He, K.; Girshick, R.; Sun, J. Faster R-CNN: Towards Real-Time Object Detection with Region Proposal Networks. *IEEE Trans. Pattern Anal. Mach. Intell.* **2017**, *39*, 1137–1149. [CrossRef] [PubMed]
23. Song, P.; Li, P.; Dai, L.; Wang, T.; Chen, Z. Boosting R-CNN: Reweighting R-CNN Samples by RPN's Error for Underwater Object Detection. *Neurocomputing* **2023**, *530*, 150–164. [CrossRef]
24. Redmon, J.; Divvala, S.; Girshick, R.; Farhadi, A. You only look once: Unified, real-time object detection. In Proceedings of the IEEE Conference on Computer Vision and Pattern Recognition (CVPR), Las Vegas, NV, USA, 26 June–1 July 2016.
25. Wang, C.-Y.; Bochkovskiy, A.; Liao, H.-Y.M. YOLOv7: Trainable bag-of-freebies sets new state-of-the-art for real-time object detectors. In Proceedings of the IEEE/CVF Conference on Computer Vision and Pattern Recognition (CVPR), Vancouver, BC, Canada, 17–24 June 2023.
26. Terven, J.; Cordova-Esparza, D. A Comprehensive Review of YOLO Architectures in Computer Vision: From YOLOv1 to YOLOv8 and YOLO-NAS. *Mach. Learn. Knowl. Extr.* **2023**, *5*, 1680–1716. [CrossRef]
27. Clark, H.P.; Smith, A.G.; Fletcher, D.M.; Larsson, A.I.; Jaspars, M.; Clippele, L.H.D. New Interactive Machine Learning Tool for Marine Image Analysis. *bioRxiv* **2023**. [CrossRef] [PubMed]
28. Rodríguez-Basalo, A.; Sánchez, F.; Punzón, A.; Gómez-Ballesteros, M. Updating the Master Management Plan for El Cachucho MPA (Cantabrian Sea) Using a Spatial Planning Approach. *Cont. Shelf Res.* **2019**, *184*, 54–65. [CrossRef]
29. Gómez-Ballesteros, M.; Druet Vélez, M.; Muñoz, A.; Arrese-González, B.; Rivera, J.; Sánchez-Delgado, F.; Cristobo, J.; Parra-Descalzo, S.; García-Alegre, A.; González-Pola, C.; et al. Geomorphology of the Avilés Canyon System, Cantabrian Sea (BayofBiscay). *Deep.-Sea Res. Part II Top. Stud. Oceanogr.* **2014**, *106*, 99–117. [CrossRef]
30. Altuna, Á.; Ríos, P. Scleractinia (Cnidaria: Anthozoa) from INDEMARES 2010–2012 Expeditions to the Avilés Canyon System (Bay of Biscay, Spain, Northeast Atlantic). *Helgol. Mar. Res.* **2014**, *68*, 399–430. [CrossRef]
31. García-Alegre, A.; Sánchez, F.; Gómez-Ballesteros, M.; Hinz, H.; Serrano, A.; Parra, S. Modelling and Mapping the Local Distribution of Representative Species on the Le Danois Bank, El Cachucho Marine Protected Area (Cantabrian Sea). *Deep Sea Res. Part II Top. Stud. Oceanogr.* **2014**, *106*, 151–164. [CrossRef]
32. Price, D.M.; Robert, K.; Callaway, A.; Lo Lacono, C.; Hall, R.A.; Huvenne, V.A.I. Using 3D Photogrammetry from ROV Video to Quantify Cold-Water Coral Reef Structural Complexity and Investigate Its Influence on Biodiversity and Community Assemblage. *Coral Reefs* **2019**, *38*, 1007–1021. [CrossRef]
33. Sánchez, F.; Rodríguez, J.M. POLITOLANA, a New Low Cost Towed Vehicle Designed for the Characterization of the Deep-Sea Floor. *Instrum. Viewp.* **2013**, *15*, 69.
34. Martín-García, L.; Prado, E.; Falcón, J.M.; González Porto, M.; Punzón, A.; Martín-Sosa, P. Population Structure of Asconema Setubalense Kent, 1870 at Concepción Seamount, Canary Islands (Spain). Methodological Approach Using Non-Invasive Techniques. *Deep Sea Res. Part 1 Oceanogr. Res. Pap.* **2022**, *185*, 103775. [CrossRef]
35. Gayá-Vilar, A.; Cobo, A.; Abad-Uribarren, A.; Rodríguez, A.; Sierra, S.; Clemente, S.; Prado, E. High-Resolution Density Assessment Assisted by Deep Learning of Dendrophyllia Cornigera (Lamarck, 1816) and Phakellia Ventilabrum (Linnaeus, 1767) in Rocky Circalittoral Shelf of Bay of Biscay. *PeerJ* **2024**, *12*, e17080. [CrossRef]
36. Kirillov, A.; Mintun, E.; Ravi, N.; Mao, H.; Rolland, C.; Gustafson, L.; Xiao, T.; Whitehead, S.; Berg, A.C.; Lo, W.-Y.; et al. Segment Anything. *arXiv* **2023**. [CrossRef]
37. Prado, E.; Rodríguez-Basalo, A.; Cobo, A.; Ríos, P.; Sánchez, F. 3D Fine-Scale Terrain Variables from Underwater Photogrammetry: A New Approach to Benthic Microhabitat Modeling in a Circalittoral Rocky Shelf. *Remote Sens.* **2020**, *12*, 2466. [CrossRef]
38. Somoza, L.; Ercilla, G.; Urgorri, V.; León, R.; Medialdea, T.; Paredes, M.; Gonzalez, F.J.; Nombela, M.A. Detection and Mapping of Cold-Water Coral Mounds and Living *Lophelia* Reefs in the Galicia Bank, Atlantic NW Iberia Margin. *Mar. Geol.* **2014**, *349*, 73–90. [CrossRef]

39. Vinha, B.; Murillo, F.J.; Schumacher, M.; Hansteen, T.H.; Schwarzkopf, F.U.; Biastoch, A.; Kenchington, E.; Piraino, S.; Orejas, C.; Huvenne, V.A.I. Ensemble Modelling to Predict the Distribution of Vulnerable Marine Ecosystems Indicator Taxa on Data-Limited Seamounts of Cabo Verde (NW Africa). *Divers. Distrib.* **2024**, *30*, e13896. [CrossRef]
40. Prado, E.; Abad-Uribarren, A.; Ramo, R.; Sierra, S.; González-Pola, C.; Cristobo, J.; Ríos, P.; Graña, R.; Aierbe, E.; Rodríguez, J.M.; et al. Describing Polyps Behavior of a Deep-Sea Gorgonian, Placogorgia Sp., Using a Deep-Learning Approach. *Remote Sens.* **2023**, *15*, 2777. [CrossRef]
41. Šiaulys, A.; Vaičiukynas, E.; Medelytė, S.; Buškus, K. Coverage Estimation of Benthic Habitat Features by Semantic Segmentation of Underwater Imagery from South-Eastern Baltic Reefs Using Deep Learning Models. *Oceanologia* **2024**, *66*, 286–298. [CrossRef]
42. Li, H.; Zhu, J.; Deng, J.; Guo, F.; Yue, L.; Sun, J.; Zhang, Y.; Hou, X. Visibility Enhancement of Underwater Images Based on Polarization Common-Mode Rejection of a Highly Polarized Target Signal. *Opt. Express* **2022**, *30*, 43973–43986. [CrossRef]
43. Wang, H.; Sun, S.; Chang, L.; Li, H.; Zhang, W.; Frery, A.C.; Ren, P. INSPIRATION: A Reinforcement Learning-Based Human Visual Perception-Driven Image Enhancement Paradigm for Underwater Scenes. *Eng. Appl. Artif. Intell.* **2024**, *133*, 108411. [CrossRef]
44. Liu, X.; Chen, M.K.; Chu, C.H.; Zhang, J.; Leng, B.; Yamaguchi, T.; Tanaka, T.; Tsai, D.P. Underwater Binocular Meta-Lens. *ACS Photonics* **2023**, *10*, 2382–2389. [CrossRef]

Journal of
Marine Science and Engineering

MDPI

Article

Research on the Identification and Classification of Marine Debris Based on Improved YOLOv8

Wenbo Jiang [1,2,*], Lusong Yang [1,2] and Yun Bu [1,2]

1 School of Electrical Engineering and Electronic Information, Xihua University, Chengdu 610039, China; 212022085400046@stu.xhu.edu.cn (L.Y.); buyun@mail.xhu.edu.cn (Y.B.)
2 Sichuan Provincial Key Laboratory of Signal and Information Processing, Xihua University, Chengdu 610039, China
* Correspondence: jiangwenbo@mail.xhu.edu.cn

Abstract: Autonomous underwater vehicles equipped with target recognition algorithms are a primary means of removing marine debris. However, due to poor underwater visibility, light scattering by suspended particles, and the coexistence of organisms and debris, current methods have problems such as poor recognition and classification effects, slow recognition speed, and weak generalization ability. In response to these problems, this article proposes a marine debris identification and classification algorithm based on improved YOLOv8. The algorithm incorporates the CloFormer module, a context-aware local enhancement mechanism, into the backbone network, fully utilizing shared and context-aware weights. Consequently, it enhances high- and low-frequency feature extraction from underwater debris images. The proposed C2f-spatial and channel reconstruction (C2f-SCConv) module combines the SCConv module with the neck C2f module to reduce spatial and channel redundancy in standard convolutions and enhance feature representation. WIoU v3 is employed as the bounding box regression loss function, effectively managing low- and high-quality samples to improve overall model performance. The experimental results on the TrashCan-Instance dataset indicate that compared to the classical YOLOv8, the mAP@0.5 and F1 scores are increased by 5.7% and 6%, respectively. Meanwhile, on the TrashCan-Material dataset, the mAP@0.5 and F1 scores also improve, by 5.5% and 5%, respectively. Additionally, the model size has been reduced by 12.9%. These research results are conducive to maintaining marine life safety and ecosystem stability.

Keywords: marine debris identification and classification; YOLOv8; CloFormer transformer; SCConv; WIoU

1. Introduction

Human domestic waste enters the ocean, where it can be eaten by animals or entangled in marine organisms, affecting ecosystem stability and human health [1,2]. Currently, the state-of-the-art technological approach involves the use of autonomous underwater vehicles equipped with marine debris recognition algorithms to clean up oceanic debris [3,4]. However, due to reasons such as weak underwater light intensity, interference from suspended particles, biological adhesion, and changes in the shape of debris, the quick and accurate identification of marine debris is still an urgent problem to be solved.

Marine debris identification can be roughly divided into traditional and deep learning methods. Traditional methods generally use either sensing technology (e.g., sonar, lidar) or traditional machine learning (e.g., dictionary learning). Initially, sonar or lidar was used to conduct underwater detection directly. Zhang et al. (2010) and Tucker et al. (2011) conducted underwater detection based on different sonar systems. The detection range was improved, but there were problems with low resolution and weak anti-interference [5,6]. Pellen et al. (2012) and Gao et al. (2014) used lidar to detect underwater targets, which improved resolution and anti-interference performance, but the recognition effect was poor, and propagation loss was large [7,8]. With the development of machine learning, dictionary learning has been combined with sonar images in underwater target detection. Azimi-Sadjadi et al. (2017) proposed a subspace-based underwater sonar image detection method that solved the propagation loss but had low generalization ability [9]. Similarly, Lu et al. (2019) improved the recognition rate by using sparse representation to identify underwater targets, but signal loss caused instability in the algorithm's performance [10]. In summary, traditional marine debris identification methods have shortcomings such as poor recognition effects, large propagation loss, weak anti-interference, low generalization ability, and performance instability and are difficult to extend to all scene types.

With the development of deep learning, convolutional neural networks (CNNs) have been widely used in image processing. Valdenegro-Toro (2016) trained a CNN classifier to identify marine debris, but it identified fewer types and overlooked environmental disturbances [11]. Xian et al. (2018) developed an underwater man-made object recognition system. Due to the use of synthetic underwater images, authenticity was lacking, and the model was complex [12]. Hong et al. (2020) used a classifier trained with enhanced data to classify and identify marine debris and applied it to a real environment, but there were few recognition categories [13]. Politikos et al. (2021) utilized region-based CNNs to detect submarine debris in a real environment, expanding marine debris categories, but their approach had a low recognition rate [14]. Wei et al. (2022) proposed an improved U-Net-based architecture, which enriched the semantic segmentation dataset of marine debris. However, the recognition speed was slow [15]. Sinthia et al. (2023) improved the YOLOv8 model to detect marine debris, with good recognition effects but weak generalization ability [16]. In summary, marine debris identification based on deep learning has the advantages of automatic feature learning, adaptability to different types of marine debris, and strong scalability, but it requires a large amount of data support. Current deep learning algorithms have shortcomings such as poor recognition and classification effects, slow recognition speed, complex models, and weak generalization capabilities.

This article proposes a marine debris identification and classification algorithm based on improved YOLOv8. The main contributions are as follows:

(1) Because small targets in marine debris occupy few pixels in an image and have unclear and missing features, the CloFormer module with a dual-branch architecture is introduced in the backbone network to enhance the perception of image information;
(2) The C2f-SCConv module is proposed to enhance feature representation capabilities, addressing the problem that recognition is easily affected by factors such as underwater suspended matter, light intensity, and biological habits, resulting in overlap and damage of debris and organisms in an image, and hence feature confusion and redundancy;
(3) WIoU v3 with weight factors is used as the bounding box loss function to reduce harmful gradients caused by low-quality samples while managing samples of different quality;
(4) Simulation experiments show that the proposed algorithm has strong generalization performance, good recognition and classification effects, fast recognition, and low complexity.

The remaining paper is structured as follows. Section 2 introduces the dataset used in this study and discusses the strengths and limitations of the classic YOLOv8. Section 3 outlines the improvements made to YOLOv8. Section 4 details the experiment and provides an in-depth analysis of the results. Finally, Section 5 summarizes the conclusions of this research.

2. Materials and Methods

2.1. Dataset Preparation

The TrashCan dataset compiled by Hong et al. [17] was used for training. The images were sourced from the Japanese Marine Geosciences and Technology Bureau's deep sea image electronic library. The authors extracted debris data from nearly 1000 videos of various lengths captured by underwater vehicles. The dataset includes 7212 annotated images, labeled into two subsets: TrashCan-Instance and TrashCan-Material.

TrashCan-Instance, with an 84:16 training-validation split, contains 6065 and 1147 images, respectively, and 9540 and 2588 labels across 22 categories. These labels include rov (artificial objects deliberately placed in the scene), plants, eels, unknown instances, nets, cups, bottles, pipes, snack wrappers, and clothing. TrashCan-Material, split into 83:17 training and validation sets, comprises 6008 and 1204 images, respectively, with 9741 and 2595 labels across 16 categories. These labels include plants, fish, eels, metals, plastics, rubbers, wood, fishing gear, paper, and fabric (Table 1).

Table 1. Division and classification statistics of TrashCan dataset.

TrashCan	Training Set	Validation Set	Training Set Label	Validation Set Label	Number of Categories
Instance	6065	1147	9540	2588	22
Material	6008	1204	9741	2595	16

2.2. *Classic YOLOv8 Network*

Released by Ultralytics in 2023, YOLOv8 has four advantages: (1) the anchor-free structure solves problems that anchor boxes may encounter with non-standard-shaped objects; (2) cutting-edge data enhancement technology enhances the robustness and generalization capabilities of the model; (3) adaptive training strategies to optimize the learning rate and balanced loss function can improve model performance; (4) the flexible architecture enables users to easily adjust the structure and parameters to adapt to a variety of target detection tasks.

The YOLOv8 network consists of an Input, Backbone, Neck, and Head, as shown in Figure 1.

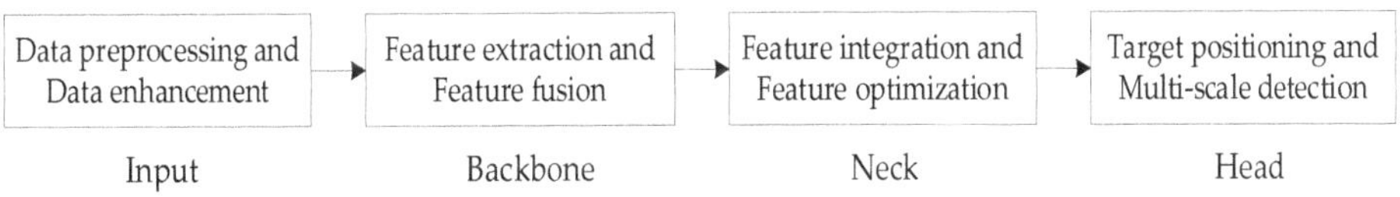

Figure 1. Classic YOLOv8 network structure.

YOLOv8 has shown excellent performance in image recognition in fields such as industrial defects [18], agricultural pests [19], and medical imaging [20], due to its efficient target detection and real-time processing. However, there are three problems in the case of marine debris identification: (1) Because some C2f modules in the backbone network extract features too many times, direct use will lead to degradation of the feature extraction function and cause insufficient fusion of key information during feature fusion. (2) Because the neck C2f modules are located behind the splicing layer, the direct stacking of features from different layers will cause redundancy and interference in the feature integration process and will affect feature optimization. (3) Model performance decreases because the CIoU loss function is unable to handle samples of different quality in the TrashCan dataset.

3. The Proposed Approach

3.1. *Improved YOLOv8 Network*

To address the above-mentioned issues, the classic YOLOv8 network is enhanced, as shown in red in Figure 2. The lightweight CloFormer module [21], featuring context-aware local enhancement, is integrated into backbone layers 4 and 6 to boost the C2f module's feature extraction. This mechanism realizes the deep mining of both high-frequency local information (such as edges and texture of marine debris) and low-frequency global information (such as the overall structure and spatial layout of the image), improving focus on debris features while minimizing background interference. Consequently, feature fusion becomes more accurate and effective. The SCConv [22] module is integrated into the four C2f modules in the neck, optimizing the fine reconstruction of features across spatial and channel dimensions. The former eliminates redundant spatial information to make the feature map more compact and richer in key information, while the latter enables SCConv to further reduce the interference between channels by optimizing the channel correlations, thereby enhancing the overall feature consistency and discrimination. This dual-dimension optimization decreases feature dimension during integration, lightening the computational load on subsequent processing layers and significantly improving feature expression. WIoU v3 is used as the bounding box regression loss function, and its wise gradient gain distribution method is used to reduce the competitiveness of high-quality anchor boxes and the interfering gradients generated by low-quality samples, which improves the recognition effect.

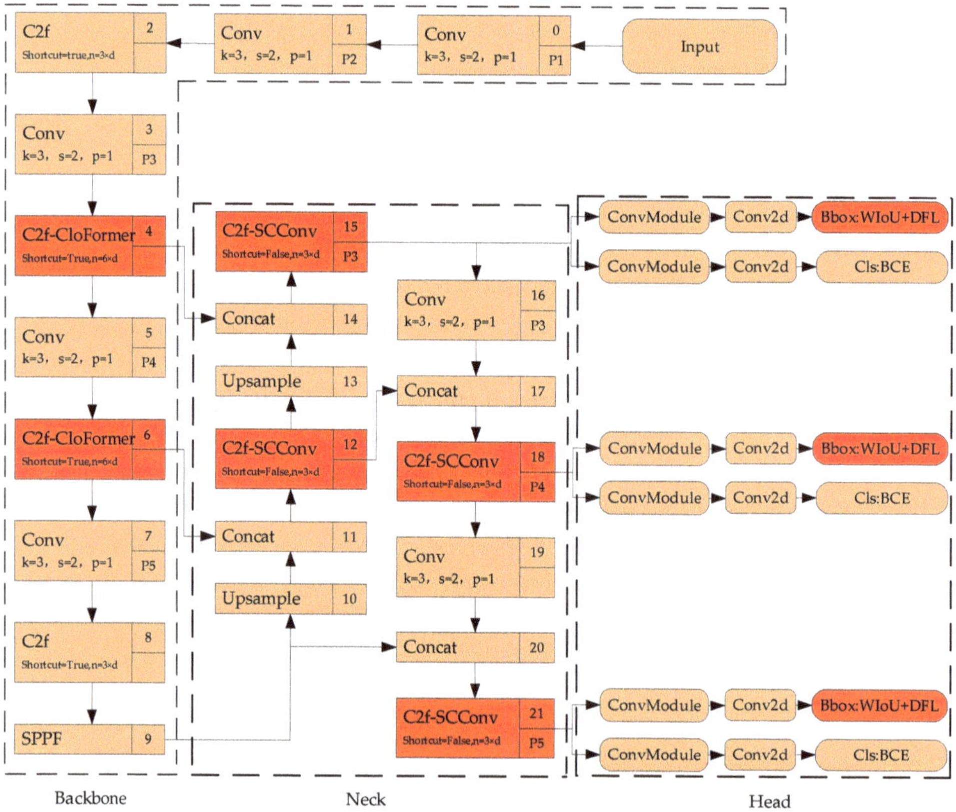

Figure 2. Improved YOLOv8 network structure, where the yellow box represents the classic structure of YOLOv8, and the red box represents the improved structure in our model.

3.2. Improved Backbone C2f Structure

Before the improvement, the input backbone C2f module divided the feature map into two parts. Feature extraction in the bottleneck was limited to small debris targets and struggled to differentiate complex background information (Figure 3a). For the convolution block in its internal bottleneck, the CloFormer module is introduced, realizing the C2f-CloFormer module with a dual-branch architecture, as shown in Figure 3b, which focuses on the small targets themselves and screens useful background information to better understand and represent features in the input image, improving feature extraction capabilities and the feature fusion effect on small targets in marine debris images.

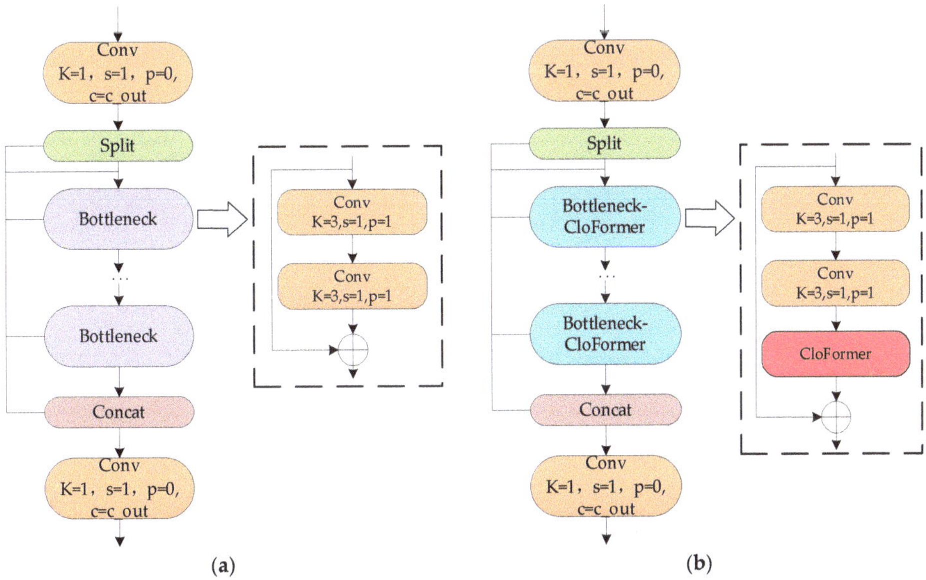

Figure 3. Comparison of the structures of two types of C2f before and after improvement, where the purple box represents the classic Bottleneck, and blue box represents the improved Bottleneck-CloFormer, the yellow box represents the convolutional layer, the green box represents the Split operation, and the red box represents the Concat operation. (**a**) Classic backbone C2f structure. (**b**) Improved backbone C2f-CloFormer structure.

The Clo block structure in CloFormer is shown in Figure 4. The global branch uses downsampling for K (key) and V (value) and ordinary attention for Q (query), K, and V to capture low-frequency global information. The local branch adopts the attention-style convolution operator AttnConv. To aggregate high-frequency local information, depth-wise convolution (DWconv) with shared weights is implemented to extract local representations. The Hadamard product of Q and K is computed, followed by transformations to generate context-aware weights that enhance local features. The outputs of the global branch and local branch are fused, allowing the model to capture both high- and low-frequency information. This process is defined in Formulas (1)–(5).

$$Y_{global} = Attn(Q, Pool(K), Pool(V)) \tag{1}$$

$$Q, K, V = F_c(P_{in}) \tag{2}$$

$$V_o = DWconv(V), Q_o = DWconv(Q), K_o = DWconv(K) \tag{3}$$

$$Y_{local} = Tanh\left(\frac{F_c(Swish(F_c(Q_o \odot w_o)))}{\sqrt{d}}\right) \odot V_o \tag{4}$$

$$Y_o = F_c\Big(Concat\Big(\Big(Y_{global}, Y_{local}\Big)\Big)\Big) \tag{5}$$

where Y_{global} is the output of the global branch; Attn is the attention mechanism; Pool is downsampling; P_{in} is the input of AttnConv; F_c is a fully connected layer; V_o, Q_o, and K_o are the outputs after depth-wise convolution; Y_{local} is the output of the local branch; Tanh and Swish are activation functions; $\odot$ is the Hadamard product; d is the number of token channels; and Y_o is the integrated output of the local and global branches.

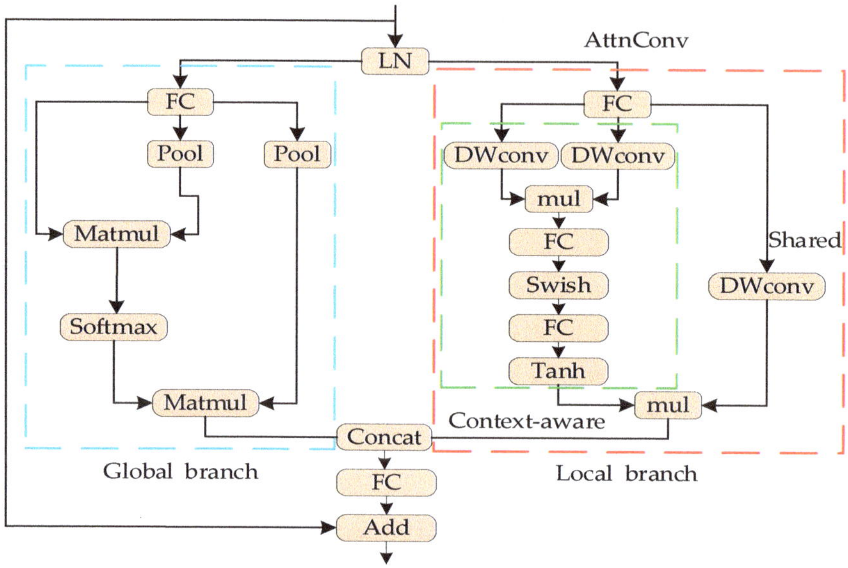

Figure 4. The internal structure of Clo block in CloFormer, where the blue dashed box represents the Global branch module, the red dashed box represents the Local branch module, and the green dashed box represents the Context−aware module.

3.3. Improved Neck C2f Structure

The classic neck C2f module has a significant increase in feature dimension during feature integration, complicating the feature representation problem during feature optimization, as shown in Figure 5a. Multiple SCConv modules were introduced to replace the bottleneck in the C2f module, creating a multi-branch structure called the C2f-SCConv module. This module reduces redundant information and enhances feature representation, as illustrated in Figure 5b. This can increase the representation ability of features in feature integration and make feature optimization more flexible through self-adaptive adjustment of the spatial structure and channel relationship of features, which can improve the recognition of overlapping targets and damaged targets in marine debris images.

Figures 6 and 7 show the spatial reconstruction unit (SRU) and channel reconstruction unit (CRU), respectively, which constitute SCConv. The SRU suppresses the spatial redundancy of feature maps through Separate-Reconstruct, and the CRU reduces channel redundancy through Split-Transform-Fuse.

As shown in Figure 6, the SRU first separates these information-rich feature maps from those with less information corresponding to the spatial content and uses cross-reconstruction to fully combine the two information features after weighting to obtain cross-reconstructed features X^{w1} and X^{w2}, which are connected to obtain a spatially refined feature map, X^{w}.

As shown in Figure 7, the CRU divides the feature map X^{w} into the main part, the upper road, and a supplementary part, the lower road; adds group-wise convolution (GWC) and point-wise convolution (PWC) in the upper road to obtain Y_1; and splices the lower road to obtain Y_2 after point convolution. Through concatenation and operations such as softmax on S1 and S2 after global average pooling (P), we obtain refined channel feature Y.

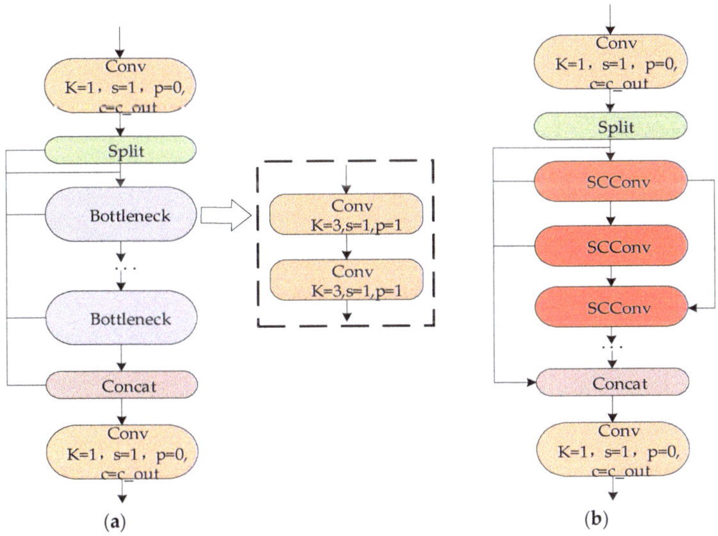

Figure 5. Comparison of two types of neck C2f structures before and after improvement, where the red-box represents a multi-branch structure composed of multiple SCConv modules. (**a**) Classic neck C2f structure. (**b**) Improved neck C2f-SCConv structure.

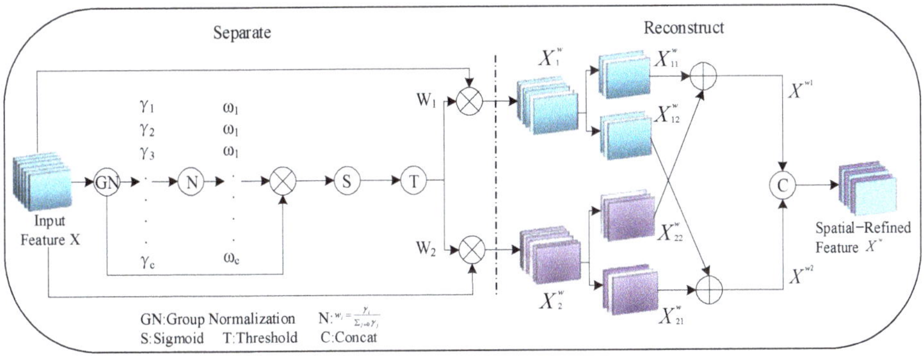

Figure 6. The internal structure of the spatial reconstruction unit.

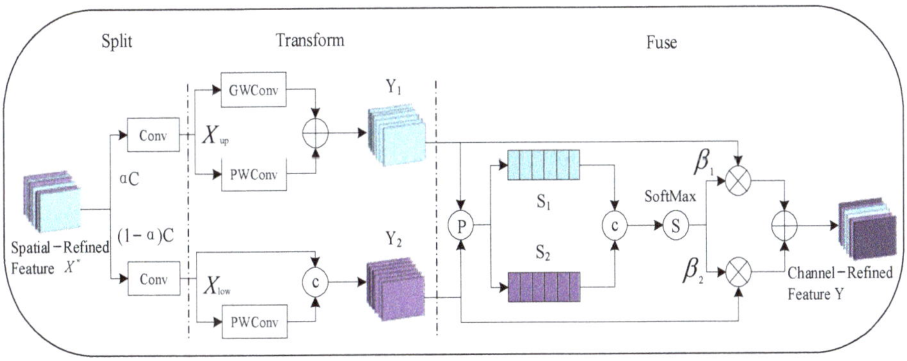

Figure 7. The internal structure of the channel reconstruction unit.

3.4. Loss Function

YOLOv8 uses DFL and CIoU to calculate the regression loss of the bounding box. The CIoU and bounding box regression loss functions are defined as Formulas (6) and (7) [23,24].

$$CIoU = IoU - \frac{\rho^2(b, b^{gt})}{c^2} - \alpha v \tag{6}$$

$$L_{CIoU} = 1 - IoU + \frac{\rho^2(b, b^{gt})}{c^2} + \alpha v \tag{7}$$

where IoU is the overlap rate between the predicted and real boxes, $\rho^2(b,b^{gt})$ is the Euclidean distance between their center points, α is a balance parameter, v indicates the consistency of the aspect ratio, and c is the diagonal distance of the minimum enclosed area covering two bounding boxes.

CIoU has three disadvantages: (1) overemphasizing a factor in formula (6) may cause bias in the model during training; (2) CIoU may not accurately reflect the quality of the prediction frame when the target is occluded by other objects; and (3) it is difficult for the model to learn the correct prediction method for targets with extreme aspect ratios (e.g., eels, pipes, wood).

CIoU does not consider the impact of low-quality samples in the training data. We replace it with WIoU v3 [25] in the YOLOv8 model of marine debris identification. WIoU v3 employs a unique gradient allocation mechanism that assigns smaller gradient gains to small and large outliers, effectively identifying prediction errors caused by poor sample quality and reducing harmful gradients generated by low-quality samples. In addition, WIoU v3 adjusts the gradient size and direction, enabling the model to prioritize high-quality representative samples during optimization. The improved loss function is shown in Formulas (8)–(10).

$$R_{WIoU} = \exp\left(\frac{(x - x_{gt})^2 + (y - y_{gt})^2}{\left(W_g^2 + H_g^2\right)^*}\right) \in [1, e] \tag{8}$$

$$L_{WIoUv1} = R_{WIoU} L_{IoU},\ L_{IoU} \in [0, 1] \tag{9}$$

$$L_{WIoUv3} = rL_{WIoUv1}, r = \frac{\beta}{\delta\alpha^{\beta-\delta}}, \beta = \frac{L^*_{IoU}}{\overline{L_{IoU}}} \in [0, +\infty) \tag{10}$$

where x and y are the center point coordinates of the prediction box, x_{gt} and y_{gt} are the center point coordinates of the real box, W_g and H_g are the respective width and height of the minimum bounding box, * indicates the separation operation, R_{WIoU} is the amplification factor based on the center point distance between the predicted box and the real box, L_{IoU} is the IoU loss, r is the gradient gain, α and δ are hyperparameters, β is the outlier degree, L^*_{IoU} is the monotonic focusing coefficient, and $\overline{L_{IoU}}$ is the average loss value.

4. Experiment and Results

4.1. Experimental Environment and Parameter Settings

The experimental platform was built on Python 3.7, PyTorch 1.10, CUDA 10.2, and CUDNN 7.6, using a 24-GB NVIDIA GeForce RTX 3090 equipped with Windows 10. The model was trained using Stochastic Gradient Descent (SGD) optimization, with respective initial and final learning rates of 0.001 and 0.0001; weight decay and momentum set to 0.0003 and 0.91, respectively; and a batch size of 16. During the training process of up to 200 epochs, if the model did not have better results within 20 consecutive epochs, the training ended early.

4.2. Evaluation Indicator and Training Process

To comprehensively and objectively evaluate the proposed algorithm, training was carried out based on the instance and material of the TrashCan dataset, and the accuracy, recall, F1 score, frames per second (FPS), model size (Size), mAP@0.5, mAP@0.75, and mAP [0.5:0.95] were used to evaluate the performance of the model, where mAP@0.5 and mAP@0.75 are the mean average accuracies on all categories with respective IoU thresholds of 0.5 and 0.75; and mAP [0.5:0.95] is the average degree of the mAP under different IoU thresholds when the IoU threshold is increased from 0.5 to 0.95 in steps of 0.05. The relevant formulas are shown in Formulas (11)–(14).

$$\text{Precision} = \frac{\text{TP}}{\text{TP} + \text{FP}} \tag{11}$$

$$\text{Recall} = \frac{\text{TP}}{\text{TP} + \text{FN}} \tag{12}$$

$$\text{F1} = \frac{2 \times \text{Recall} \times \text{Precision}}{\text{Recall} + \text{Precision}} \tag{13}$$

$$\text{mAP} = \frac{\sum_{i=1}^{n} \text{AP}_i}{\text{N}}, \text{AP} = \int_0^1 \text{P(R)dR} \tag{14}$$

where TP (true positive) is the number of correctly predicted positive cases, FP (false positive) is the number of negative cases predicted as positive, and FN (false negative) is the number of positive cases predicted as negative. P is precision and R is recall. AP stands for the average detection precision for one object category. N is the total number of target categories. mAP is the average accuracy for all categories.

The above indicators are employed to assess the trained model. The proposed model is trained as shown in Algorithm 1.

Algorithm 1 Training steps of the model

Basic parameters: θ = (hyper-parameters = {image size = (640, 640, 3), epoch = 200, batch-size = 16, Do not use pre-trained weights, optimizer = SGD with learning rate = 0.001, validation = per epoch, batch-size = 16}
Evaluation indicator: [P, R, F1, mAP@0.5, mAP@0.75, mAP [0.5:0.95], Size (MB), FPS].
Create data loaders for Trashcan_train, Trashcan _val, and Trashcan _test.
Initialize model training:
    For epoch = 1 to total_epochs:
            For iteration = 1 to number_of_iterations_per_epoch.
    Feed Trashcan_train and Trashcan_val into the model.
for iteration ∈ [1, number_of_iterations_per_epoch] **do**
    Feed Trashcan_train and Trashcan_val into the model.
    Split batches into images M and labels N.
    Forward M through the model to get output tensor T.
    Compute the loss L and apply it to the label tensor N and the output tensor T.
    The gradient of the loss L relative to the model parameters is obtained.
    Update the model weights using the SGD optimizer.
    Obtain model = α
end for
Load α for model testing
Initialize the predicted list X and the ground truth label list Y as empty.
for the batch in the Trashcan test
    Split batches into images M and labels N.
    Pass M through α to get output tensor T.
    Add the output tensor T and label tensor N to the lists X and Y, respectively.
end for
Output [P, R, F1, mAP@0.5, mAP@0.75, mAP [0.5:0.95], Size (MB), FPS].

4.3. TrashCan-Instance Dataset Simulation Analysis

To validate the proposed model, it was trained and evaluated on the TrashCan-Instance dataset and compared with current algorithms, including a similar single-stage algorithm and a two-stage algorithm biased toward accuracy. The results are shown in Tables 2 and 3.

Table 2. Comparison of multi-threshold mAP performance on TrashCan-Instance dataset.

Model	mAP@0.5 (%)	mAP@0.75 (%)	mAP [0.5:0.95] (%)	Reference
Improved Mask R-CNN	65.00	48.10	44.10	Deng et al. (2021) [26]
IEM	63.09	48.38	44.03	Ali et al. (2022) [27]
YOLACT	58.80	42.50	37.70	Corrigan et al. (2023) [28]
MLDet	68.90	55.10	49.20	Ma et al. (2023) [29]
Improved YOLOv8	72.00	57.80	51.60	This study

Table 3. Comprehensive evaluation of model performance on the TrashCan-Instance dataset.

Model	Size (MB)	mAP@0.5 (%)	FPS	Reference
Faster R-CNN	795	55.40	18	Hong et al. (2020) [17]
SSD	205	58.12	78	Liu et al. (2016) [30]
YOLOTrashCan	214	65.01	36	Zhou et al. (2022) [31]
Improved YOLOv5	17.2	67.00	61	Liu et al. (2023) [32]
Improved YOLOv8	43.2	72.00	66	This study

(1) The comparison algorithms in Table 2 improve accuracy in different ways. As can be seen in the table, our model has superior mAP@0.5, mAP@0.75, and mAP [0.5:0.95] values, with overall improvements of 4.5%, 4.9%, and 4.9%, respectively, relative to the best-performing MLDet [29].

(2) Hong [17] uses a two-stage algorithm with high accuracy, and SSD [30], YOLOTrash-Can [31], and Improved YOLOv5 [32] are fast single-stage algorithms. As can be seen from Table 3, the mAP@0.5 of the proposed model is the best, the size is second only to that reported by the Improved YOLOv5 [32], the FPS is second only to that of SSD [30], and the difference is not great.

(3) It can be seen from Tables 2 and 3 that the proposed model achieves a balance in evaluation indicators and realizes better recognition and classification effects with less running time.

To intuitively compare the effectiveness of each improvement, we used a YOLOv8m baseline to verify the effectiveness of the proposed model. The results of these ablation experiments are shown in Table 4, from which it can be seen that, except for the FPS of the proposed model, which is slightly worse than the original, other indicators are improved. The precision, recall, F1 score, and mAP@0.5 are increased by 6%, 5%, 6%, and 5.7%, respectively, and the size is reduced by 6.4 MB. Figure 8 shows that the proposed model achieves high accuracy with fewer training cycles, indicating faster convergence and enhanced recognition performance.

A comparison of the mAP@0.5 for each category between the improved YOLOv8 model and the unimproved YOLOv8m model in the TrashCan-Instance dataset is shown in Figure 9.

(1) As can be seen from Figure 9, the recognition rate of the proposed model exceeds 40% in the categories of clothing and crabs, and it is improved to varying degrees in the other 16 categories.

(2) The recognition rates of can, branch, wreckage, and tarp declined slightly, perhaps due to the higher feature discrimination of the improved model for other small target categories and low feature discrimination caused by being too sensitive to some labels in target cans, which are smaller than the other object types in the images. Moreover,

due to the large number of categories in the dataset and the low number of labels for branch, wreckage, and tarp, the model may be biased toward learning categories with a large number of labels during training, resulting in a slight reduction in the recognition rate.

To evaluate the proposed model's capability to handle low-resolution images and resist interference, this study selected low-quality data with noise from the TrashCan-Instance dataset and conducted heatmap experiments, as depicted in Figure 10.

(1) The heat maps in Figure 10b,c demonstrate that the proposed model displays brighter and more concentrated warm colors for the target, effectively capturing important features from low-resolution images and providing more accurate target position information.
(2) In the heatmap in Figure 10c, the proposed model exhibits reduced brightness in noisy areas, indicating a significant decrease in the impact of suspended particle noise during image processing, showcasing the model's anti-interference ability in noisy environments.

Table 4. Performance of ablation experiments conducted by integrating different modules with the TrashCan-Instance dataset.

Group	Model	Precision	Recall	F1	Size (MB)	mAP@0.5 (%)	FPS
1	YOLOv8m	0.73	0.62	0.67	49.6	66.30	89
2	+ CloFormer	0.78	0.62	0.69	46.9	68.00	61
3	+ SCConv	0.79	0.63	0.70	45.7	68.90	73
4	+ WIoU v3	0.75	0.62	0.68	49.6	67.50	89
5	+ CloFormer + SCConv	0.80	0.65	0.72	43.2	71.10	66
6	+ CloFormer + WIoU v3	0.79	0.63	0.70	46.9	69.00	61
7	+ SCConv + WIoU v3	0.76	0.64	0.70	45.7	69.30	73
8	Our Model	0.79	0.67	0.73	43.2	72.00	66

Note: + represents the modules added by each group.

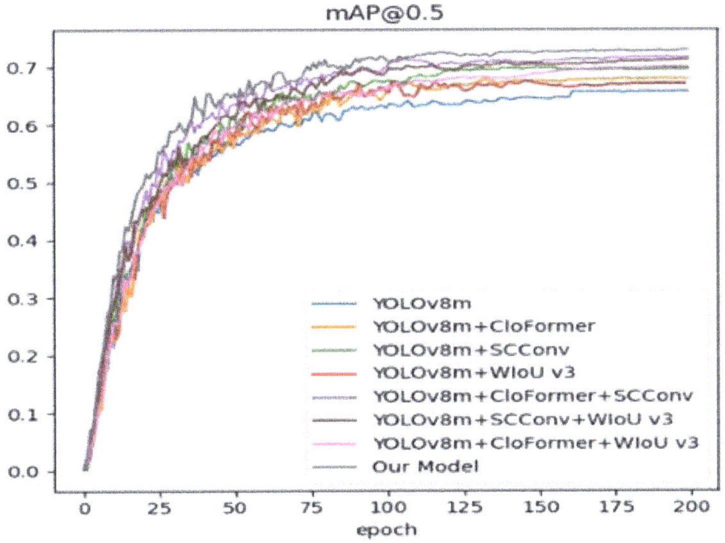

Figure 8. Training curves on the TrashCan-Instance dataset.

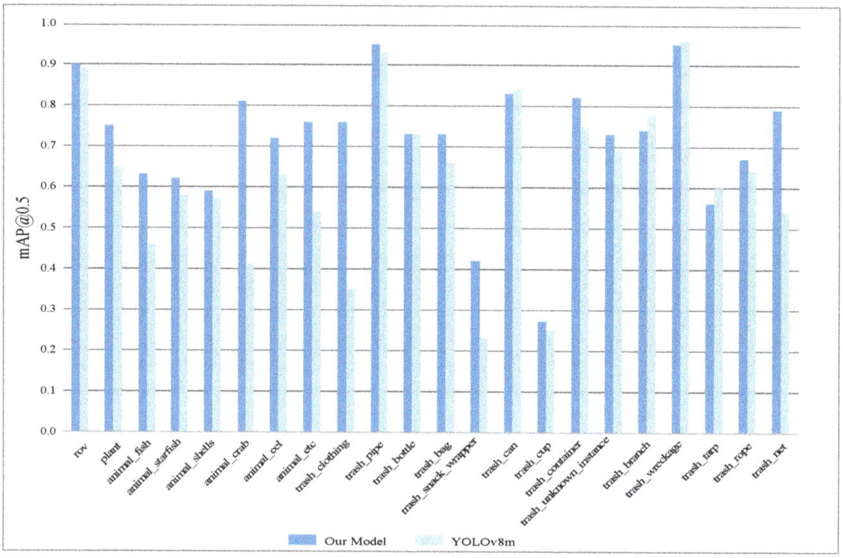

Figure 9. Compare the two models on the TrashCan-Instance dataset for mAP@0.5 on each category.

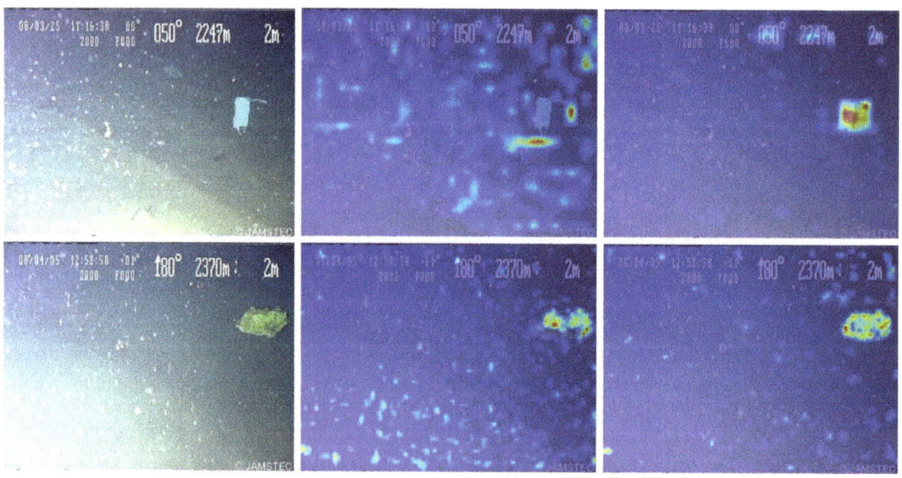

(a) (b) (c)

Figure 10. (**a**) Original images. Comparison of the heatmaps for (**b**) YOLOv8m and (**c**) the proposed model on the TrashCan-Instance dataset, where the brighter color indicates higher attention and the darker color indicates lower attention.

To visualize the marine debris recognition effect of the model, it was compared with the YOLOv8m model on the TrashCan-Instance dataset, with results as shown in Figure 11. The frame line indicates the position of an object in an image, and the object category and recognition rate are indicated above the data.

(a)

(b)

(c)

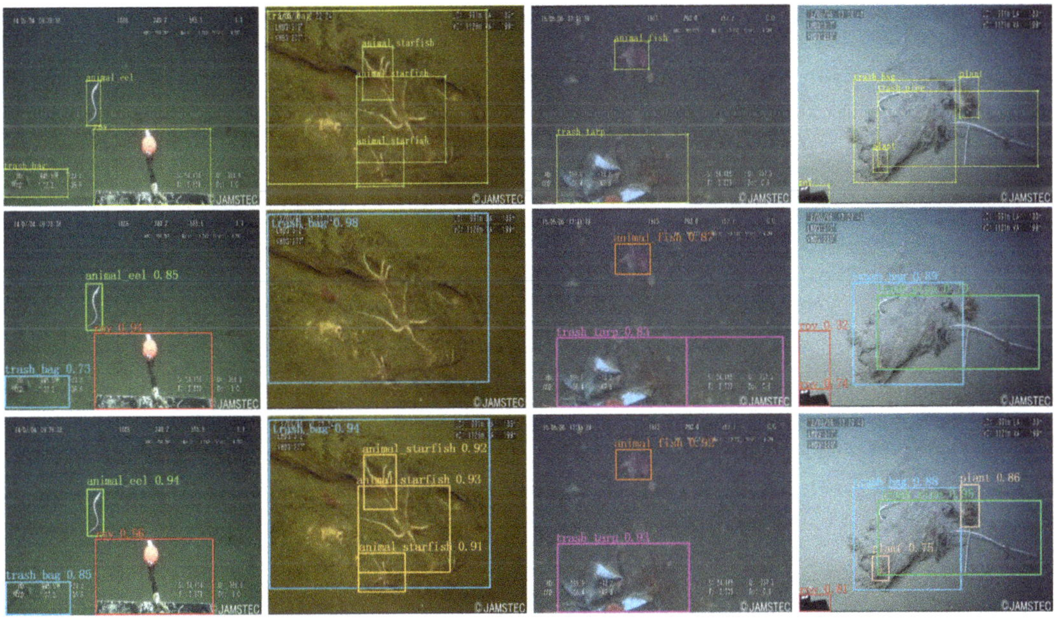

Figure 11. (**a**) Original images. Comparison of the recognition results of (**b**) YOLOv8m and (**c**) the proposed model on the Trashcan-Instance dataset.

(1) From Figure 11b,c, it can be seen that the YOLOv8m model missed the detection of starfish and plants and mistakenly detected tarp and rov. However, the recognition rate of the proposed model was slightly less than that of YOLOv8m for bags. In all other categories, this model improved the recognition rate, with no missed or false detections.
(2) Overall, the proposed model showed good results, better than those of the YOLOv8m model on the TrashCan-Instance dataset, and could accurately identify and classify multiple targets and information-damaged targets in a variety of complex underwater environments.

4.4. Simulation Analysis of TrashCan-Material Dataset

To verify the generalization performance of the proposed model, experiments similar to those in Section 4.3 were performed on the TrashCan-Material dataset, with results as shown in Tables 5 and 6.

(1) The comparison models in Table 5 address various problems in the recognition of marine debris images, using different algorithms. It can be seen that all indicators of the proposed model exceed those of the comparison models, with improvements of 5%, 6.3%, and 3.5% compared to the existing best MLDet [29], respectively.
(2) Table 6 shows the comparison of three algorithms with the proposed model in terms of speed and accuracy. It can be seen that the mAP@0.5 of the proposed model exceeds those of the compared models, and the size is much smaller. The FPS of the proposed model is 13 less than that of SSD [30] but is higher than those of the other two models.
(3) From the analysis of Tables 5 and 6, the proposed model achieved comparatively good results in terms of recognition speed and effect.

To visually assess the improvement effect of the proposed model, it was compared with the unimproved YOLOv8m model in ablation experiments on the TrashCan-Material dataset, with results as shown in Table 7. While the FPS of the proposed model was slightly worse, other indicators improved. Precision increased by 3%, recall by 6%, F1 score by 5%, and mAP@0.5 by 5.5%, and size was reduced by 6.4 MB. In addition, Figure 12 demonstrates that the proposed model also performs well in recognition.

Table 5. Comparison of multi-threshold mAP performance on TrashCan-Material dataset.

Model	mAP@0.5 (%)	mAP@0.75 (%)	mAP [0.5:0.95] (%)	Reference
IEM	56.70	38.68	36.11	Ali et al. (2022) [27]
EfficientDets	27.80	20.90	18.60	Zocco et al. (2023) [33]
ERL-Net	58.90	/	37.00	Dai et al. (2024) [34]
GCC-Net	61.20	/	41.30	Dai et al. (2024) [35]
MLDet	63.50	45.70	42.30	Ma et al. (2023) [29]
Improved YOLOv8	66.70	48.60	43.80	This study

Table 6. Comprehensive evaluation of model performance on the TrashCan-Material dataset.

Model	Size (MB)	mAP@0.5 (%)	FPS	Reference
Mask R-CNN	795	54.00	21	Hong et al. (2020) [17]
SSD	194	55.80	84	Liu et al. (2016) [30]
YOLOTrashCan	214	58.66	37	Zhou et al. (2022) [31]
Improved YOLOv8	43.2	66.70	71	This study

Table 7. Performance of ablation experiments conducted by incorporating different modules into the TrashCan-Material dataset.

Group	Model	Precision	Recall	F1	Size (MB)	mAP@0.5 (%)	FPS
1	YOLOv8m	0.70	0.58	0.63	49.6	61.20	90
2	+ CloFormer	0.69	0.59	0.63	46.9	63.40	69
3	+ SCConv	0.71	0.59	0.64	45.7	63.60	79
4	+ WIoU v3	0.69	0.58	0.63	49.6	63.30	90
5	+ CloFormer + SCConv	0.74	0.60	0.66	43.2	65.30	71
6	+ CloFormer + WIoU v3	0.71	0.60	0.65	46.9	63.90	69
7	+ SCConv + WIoU v3	0.72	0.61	0.66	45.7	64.90	79
8	Our Model	0.73	0.64	0.68	43.2	66.70	71

Note: + represents the modules added by each group.

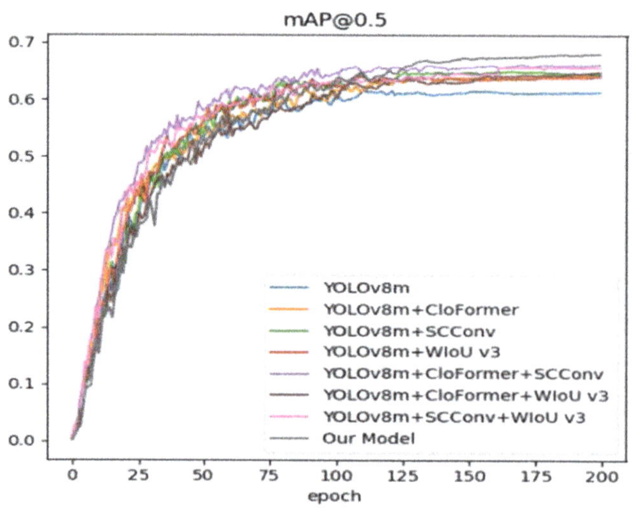

Figure 12. Training curves on the TrashCan-Material dataset.

A comparison between the mAP@0.5 for each category of the improved YOLOv8 model and the unimproved YOLOv8m model in the TrashCan-Material dataset is shown in Figure 13, from which we make the following observations:

(1) Using the proposed model, the recognition rate of paper was the most increased, by 25%, and the rates for the other 12 categories were improved to varying degrees;
(2) The recognition rates of fabric, other trash, and plastic decreased slightly, perhaps due to insufficient learning of the features of these categories; however, the improved module improved its ability to distinguish between other categories, resulting in the model being significantly affected by the similarity of features in the three categories of fabric, other garbage, and plastic when training; this, in turn, led to a slight decrease in the recognition rate.

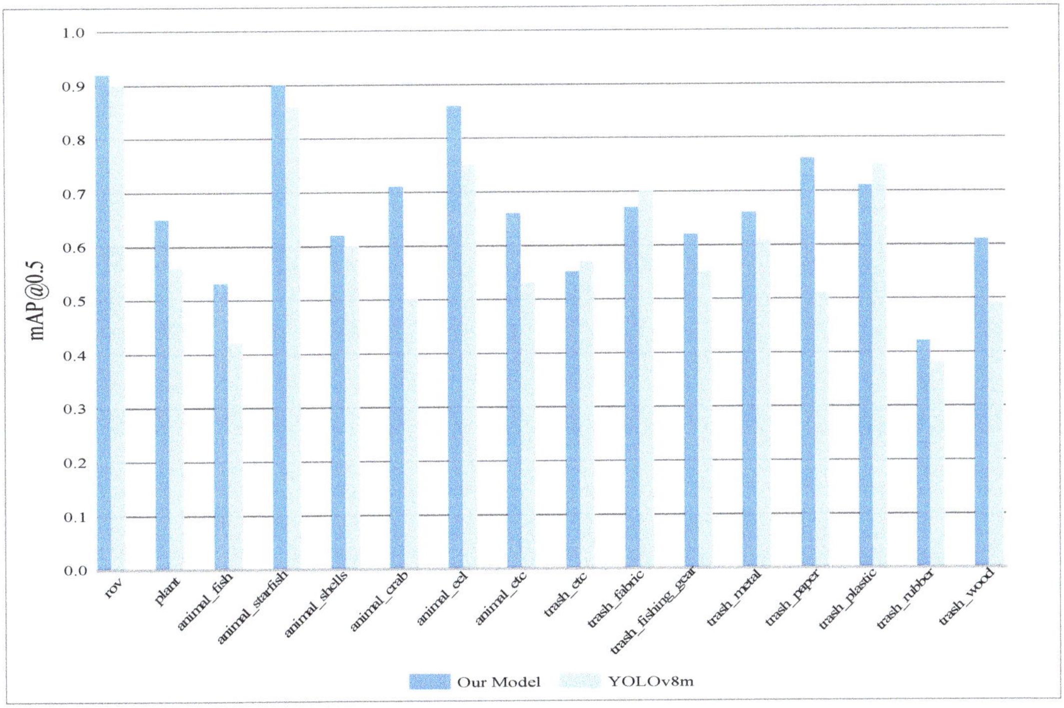

Figure 13. Comparison of the two models on the TrashCan-Material dataset for mAP@0.5 in each category.

Figure 14 depicts the heatmap experiment conducted on the TrashCan-Material dataset. Consistent with the findings in Section 4.3, the proposed model demonstrates a capacity to manage low resolution and exhibit strong anti-interference.

To illustrate the effectiveness of the proposed algorithm, the visualization results on the TrashCan-Material dataset test are shown in Figure 15. We make the following observations:

(1) If there are multiple, small, and overlapping targets, the unmodified YOLOv8m is prone to miss detection of fish, other animals, starfish, and eels, while the proposed model does not have this problem and can identify all with a high recognition rate;
(2) Small target starfish that are not marked in the original image can be identified by the proposed model, indicating good feature learning potential.

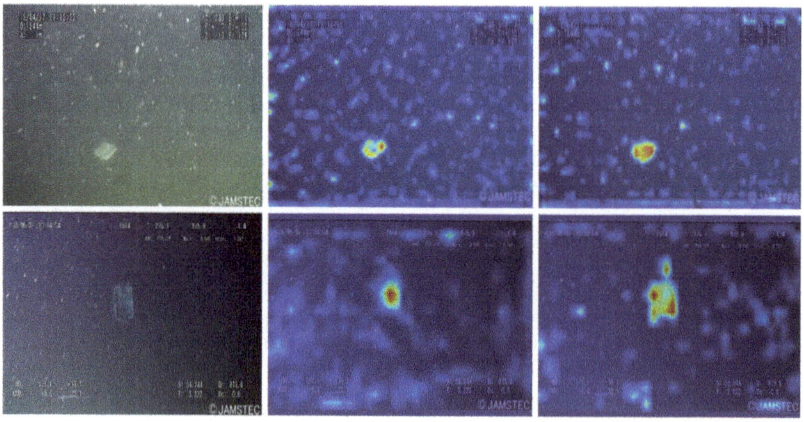

(**a**) (**b**) (**c**)

Figure 14. (**a**) Original images. Comparison of the heatmaps for (**b**) YOLOv8m and (**c**) proposed model on the TrashCan−Material dataset, where the brighter color indicates higher attention and the darker color indicates lower attention.

(**a**) (**b**) (**c**)

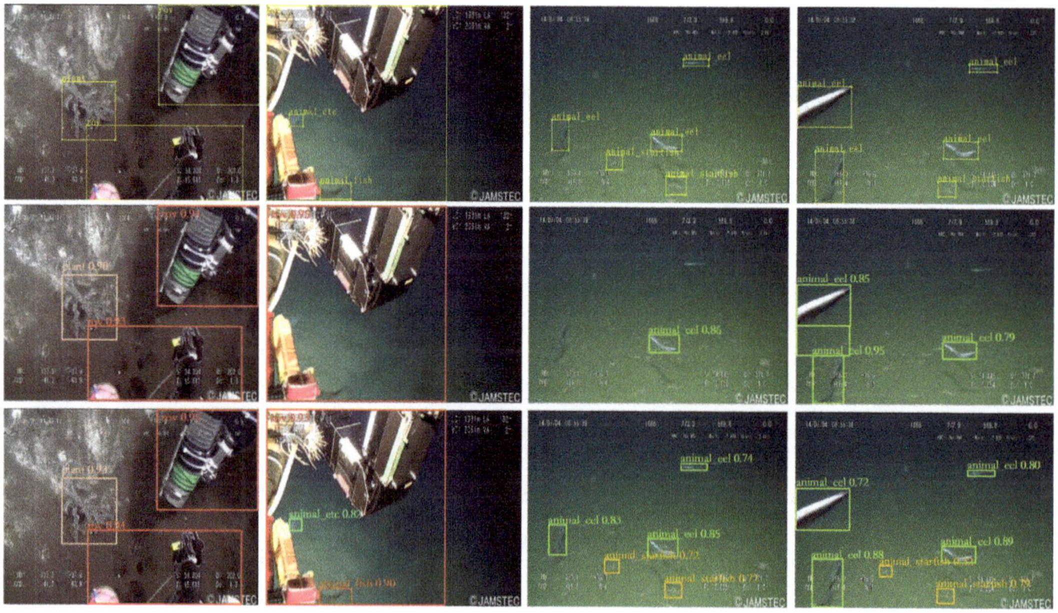

Figure 15. (**a**) Original images. Comparison of the recognition results of (**b**) YOLOv8m and (**c**) proposed model on the Trashcan−Material dataset.

4.5. Analysis and Summary

We observe the following from the test results on the TrashCan-Instance and TrashCan-Material datasets:

(1) From Sections 4.3 and 4.4, it can be seen that the proposed model shows excellent performance on both subsets, indicating good generalization ability.

(2) The constructed network model and loss function have certain effects on the extraction and fusion of marine debris image features, the integration and optimization of

features, and the suppression of harmful gradients, reflecting certain progress in terms of recognition and classification effects, recognition speed, and model complexity. In addition, the model exhibits a certain ability to identify targets in low-resolution images while resisting interference.

(3) The FPS of the proposed model is slightly lower than that of the classic and unimproved models, reflecting the trade-off of reasoning efficiency for better recognition results. Size has not reached the optimal level, possibly because no more lightweight improvements have been made to the YOLOv8 model. Subsequently, we will consider applying model pruning technology to remove weights or connections that contribute little to model performance.

5. Conclusions

This study proposes a marine debris recognition and classification algorithm based on an improved YOLOv8, addressing issues of poor recognition and classification performance, slow recognition speed, model complexity, and weak generalization capabilities of existing marine debris. Experimental results indicate that the proposed model achieves a mAP@0.5 and speed of 72% and 66 FPS, respectively, on the TrashCan-Instance dataset, and an mAP@0.5 and speed of 66.70% and 71 FPS, respectively, on the TrashCan-Material dataset. Our design also reduced the model size by 12.9%. Visual assessment reveals effective recognition and classification in complex and variable underwater environments, significantly minimizing missed and false detections.

Future research will target categories affected by feature similarity and those with limited labels. By employing image enhancement and model pruning techniques, we aim to tackle identification challenges arising from poor original image quality and large model sizes, further enhancing marine debris identification and classification.

Author Contributions: Conceptualization, W.J. and L.Y.; methodology, L.Y. and Y.B.; validation, W.J. and L.Y.; data curation, W.J. and L.Y.; investigation, L.Y. and Y.B.; writing—original draft preparation, L.Y.; writing—review and editing, W.J.; supervision, W.J.; funding acquisition, W.J. All authors have read and agreed to the published version of the manuscript.

Funding: This work was supported by the Xihua University Science and Technology Innovation Competition Project for Postgraduate Students under grant no. YK20240002, Sichuan Science and Technology Program under grant no. 2021JDJQ0027, and the Natural Science Foundation of China under grant no. 61875166.

Institutional Review Board Statement: Not applicable.

Informed Consent Statement: Not applicable.

Data Availability Statement: The data presented in this study are available upon request from the corresponding author.

Acknowledgments: Wenbo Jiang would like to acknowledge the Sichuan Provincial Academic and Technical Leader Training Plan and the Overseas Training Plan of Xihua University (September 2014–September 2015, University of Michigan, Ann Arbor, MI, USA).

Conflicts of Interest: The authors declare that they have no conflicts of interest.

Abbreviations

YOLOv8	You Only Look Once Version 8
CloFormer	Clo block Transformer
WIoU v3	Wise Intersection over Union version 3
C2f	Coarse to Fine
SPPF	Spatial Pyramid Pooling with less FLOPs
CIoU	Complete Intersection of Union
DFL	Distribution Focal Loss

BCE	Binary Cross-Entropy
Bbox	Bounding Box
Cls	Classification

References

1. Yu, J.; Liu, J. Exploring governance policy of marine fishery litter in China: Evolution, challenges and prospects. *Mar. Pollut. Bull.* **2023**, *188*, 114606. [CrossRef] [PubMed]
2. Patra, S.; Khurshid, M.; Basir, A.; Mishra, P.; Ramanamurthy, M.V. Marine litter management: A sustainable action plan and recommendations for the South Asian Seas region. *Mar. Policy* **2023**, *157*, 105854. [CrossRef]
3. Zhang, B.; Ji, D.; Liu, S.; Zhu, X.; Xu, W. Autonomous underwater vehicle navigation: A review. *Ocean Eng.* **2023**, *273*, 113861. [CrossRef]
4. Chen, Z.; Jiao, W.; Ren, K.; Yu, J.; Tian, Y.; Chen, K.; Zhang, X. A survey of research status on the environmental adaptation technologies for marine robots. *Ocean Eng.* **2023**, *286*, 115650. [CrossRef]
5. Zhang, L.; Huanq, J.; Jin, Y.; Hau, Y.; Jianq, M.; Zhang, Q. Waveform diversity based sonar system for target localization. *J. Syst. Eng. Electron.* **2010**, *21*, 186–190. [CrossRef]
6. Tucker, J.D.; Azimi-Sadjadi, M.R. Coherence-based underwater target detection from multiple disparate sonar platforms. *IEEE J. Ocean. Eng.* **2011**, *36*, 37–51. [CrossRef]
7. Pellen, F.; Jezequel, V.; Zion, G.; Le Jeune, B. Detection of an underwater target through modulated lidar experiments at grazing incidence in a deep wave basin. *Appl. Opt.* **2012**, *51*, 7690–7700. [CrossRef] [PubMed]
8. Gao, J.; Sun, J.; Wang, Q. Experiments of ocean surface waves and underwater target detection imaging using a slit Streak Tube Imaging Lidar. *Optik* **2014**, *125*, 5199–5201. [CrossRef]
9. Azimi-Sadjadi, M.R.; Klausner, N.; Kopacz, J. Detection of underwater targets using a subspace-based method with learning. *IEEE J. Ocean. Eng.* **2017**, *42*, 869–879. [CrossRef]
10. Yao, L.; Du, X. Identification of underwater targets based on sparse representation. *IEEE Access* **2019**, *8*, 215–228. [CrossRef]
11. Valdenegro-Toro, M. Submerged marine debris detection with autonomous underwater vehicles. In Proceedings of the 2016 International Conference on Robotics and Automation for Humanitarian Applications (RAHA), Amritapuri, India, 18–20 December 2016; pp. 1–7.
12. Yu, X.; Xing, X.; Zheng, H.; Fu, X.; Huang, Y.; Ding, X. Man-made object recognition from underwater optical images using deep learning and transfer learning. In Proceedings of the 2018 IEEE International Conference on Acoustics, Speech and Signal Processing (ICASSP), Calgary, AB, Canada, 15–20 April 2018; pp. 1852–1856.
13. Hong, J.; Fulton, M.; Sattar, J. A generative approach towards improved robotic detection of marine litter. In Proceedings of the 2020 IEEE International Conference on Robotics and Automation (ICRA), Paris, France, 31 May–4 June 2020; pp. 10525–10531.
14. Politikos, D.V.; Fakiris, E.; Davvetas, A.; Klampanos, I.A.; Papatheodorou, G. Automatic detection of seafloor marine litter using towed camera images and deep learning. *Mar. Pollut. Bull.* **2021**, *164*, 111974. [CrossRef]
15. Wei, L.; Kong, S.; Wu, Y.; Yu, J. Image semantic segmentation of underwater garbage with modified U-Net architecture model. *Sensors* **2022**, *22*, 6546. [CrossRef] [PubMed]
16. Sinthia, A.K.; Rasel, A.A.; Haque, M. Real-time Detection of Submerged Debris in Aquatic Ecosystems using YOLOv8. In Proceedings of the 2023 26th International Conference on Computer and Information Technology (ICCIT), Cox's Bazar, Bangladesh, 13–15 December 2023; pp. 1–6.
17. Hong, J.; Fulton, M.; Sattar, J. Trashcan: A semantically-segmented dataset towards visual detection of marine debris. *arXiv* **2020**, arXiv:2007.08097.
18. Song, X.; Cao, S.; Zhang, J.; Hou, Z. Steel Surface Defect Detection Algorithm Based on YOLOv8. *Electronics* **2024**, *13*, 988. [CrossRef]
19. Uddin, M.S.; Mazumder, M.K.A.; Prity, A.J.; Mridha, M.F.; Alfarhood, S.; Safran, M.; Che, D. Cauli-Det: Enhancing cauliflower disease detection with modified YOLOv8. *Front. Plant Sci.* **2024**, *15*, 1373590. [CrossRef]
20. Lalinia, M.; Sahafi, A. Colorectal polyp detection in colonoscopy images using yolo-v8 network. *Signal Image Video Process.* **2024**, *18*, 2047–2058. [CrossRef]
21. Fan, Q.; Huang, H.; Guan, J.; He, R. Rethinking local perception in lightweight vision transformer. *arXiv* **2023**, arXiv:2303.17803.
22. Li, J.; Wen, Y.; He, L. SCConv: Spatial and channel reconstruction convolution for feature redundancy. In Proceedings of the IEEE/CVF Conference on Computer Vision and Pattern Recognition (CVPR), Vancouver, BC, Canada, 18–22 June 2023; pp. 6153–6162.
23. Zheng, Z.; Wang, P.; Liu, W.; Li, J.; Ye, R.; Ren, D. Distance-IoU loss: Fasterand better learning for bounding box regression. In Proceedings of the AAAI Conference on Artificial Intelligence, New York, NY, USA, 7–12 February 2020; pp. 12993–13000.
24. Li, X.; Wang, W.; Wu, L.; Chen, S.; Hu, X.; Li, J.; Tang, J.; Yang, J. Generalized focal loss: Learning qualified and distributed bounding boxes for dense object detection. In Proceedings of the IEEE Conference on Computer Vision and Pattern Recognition (CVPR), Seattle, WA, USA, 14–19 June 2020.
25. Tong, Z.; Chen, Y.; Xu, Z.; Yu, R. Wise-IoU: Bounding box regression loss with dynamic focusing mechanism. *arXiv* **2023**, arXiv:2301.10051.

26. Deng, H.; Ergu, D.; Liu, F.; Ma, B.; Cai, Y. An embeddable algorithm for automatic garbage detection based on complex marine environment. *Sensors* **2021**, *21*, 6391. [CrossRef]
27. Ali, M.; Khan, S. Underwater object detection enhancement via channel stabilization. In Proceedings of the 2022 International Conference on Digital Image Computing: Techniques and Applications (DICTA), Sydney, Australia, 30 November–2 December 2022; pp. 1–8.
28. Corrigan, B.C.; Tay, Z.Y.; Konovessis, D. Real-time instance segmentation for detection of underwater litter as a plastic source. *J. Mar. Sci. Eng.* **2023**, *11*, 1532. [CrossRef]
29. Ma, D.; Wei, J.; Li, Y.; Zhao, F.; Chen, X.; Hu, Y.; Yu, S.; He, T.; Jin, R.; Li, Z.; et al. MLDet: Towards efficient and accurate deep learning method for Marine Litter Detection. *Ocean. Coast. Manag.* **2023**, *243*, 106765. [CrossRef]
30. Liu, W.; Anguelov, D.; Erhan, D.; Szegedy, C.; Reed, S.; Fu, C.Y.; Berg, A.C. Ssd: Single shot multibox detector. In Proceedings of the Computer Vision–ECCV 2016: 14th European Conference, Amsterdam, The Netherlands, 11–14 October 2016; Proceedings, Part I 14. pp. 21–37.
31. Zhou, W.; Zheng, F.; Yin, G.; Pang, Y.; Yi, J. Yolotrashcan: A deep learning marine debris detection network. *IEEE Trans. Instrum. Meas.* **2022**, *72*, 1–12. [CrossRef]
32. Liu, J.; Zhou, Y. Marine debris detection model based on the improved YOLOv5. In Proceedings of the 2023 3rd International Conference on Neural Networks, Information and Communication Engineering (NNICE), Guangzhou, China, 24–26 February 2023; pp. 725–728.
33. Zocco, F.; Lin, T.C.; Huang, C.I.; Wang, H.C.; Khyam, M.O.; Van, M. Towards more efficient efficientdets and real-time marine debris detection. *IEEE Rob. Autom. Lett.* **2023**, *8*, 2134–2141. [CrossRef]
34. Dai, L.; Liu, H.; Song, P.; Tang, H.; Ding, R.; Li, S. Edge-guided representation learning for underwater object detection. *CAAI Trans. Intell. Technol.* 2024. [CrossRef]
35. Dai, L.; Liu, H.; Song, P.; Liu, M. A gated cross-domain collaborative network for underwater object detection. *Pattern Recognit.* **2024**, *149*, 110222. [CrossRef]

Journal of
*Marine Science
and Engineering*

MDPI

Article

Application and Analysis of the MFF-YOLOv7 Model in Underwater Sonar Image Target Detection

Kun Zheng [1], Haoshan Liang [2], Hongwei Zhao [1,*], Zhe Chen [3,4,*], Guohao Xie [3,4,5], Liguo Li [3], Jinghua Lu [3] and Zhangda Long [3]

1 Graduate School, Guilin University of Electronic Technology, Guilin 541004, China; zhengkunaiguet@gmail.com
2 School of Life and Environmental Sciences, Guilin University of Electronic Technology, Guilin 541004, China; nonooz@163.com
3 School of Information and Communication, Guilin University of Electronic Technology, Guilin 541004, China; 22172301007@mails.guet.edu.cn (G.X.); 22022303068@mails.guet.edu.cn (L.L.); 19994402197@163.com (J.L.); 18276053521@163.com (Z.L.)
4 Cognitive Radio and Information Processing Key Laboratory Authorized by China's Ministry of Education Foundation, Guilin University of Electronic Technology, Guilin 541004, China
5 School of Ocean Engineering, Guilin University of Electronic Technology, Beihai 536065, China
* Correspondence: zhaohw@guet.edu.cn (H.Z.); chenzhe@mail.nwpu.edu.cn (Z.C.)

Abstract: The need for precise identification of underwater sonar image targets is growing in areas such as marine resource exploitation, subsea construction, and ocean ecosystem surveillance. Nevertheless, conventional image recognition algorithms encounter several obstacles, including intricate underwater settings, poor-quality sonar image data, and limited sample quantities, which hinder accurate identification. This study seeks to improve underwater sonar image target recognition capabilities by employing deep learning techniques and developing the Multi-Gradient Feature Fusion YOLOv7 model (MFF-YOLOv7) to address these challenges. This model incorporates the Multi-Scale Information Fusion Module (MIFM) as a replacement for YOLOv7's SPPCSPC, substitutes the Conv of CBS following ELAN with RFAConv, and integrates the SCSA mechanism at three junctions where the backbone links to the head, enhancing target recognition accuracy. Trials were conducted using datasets like URPC, SCTD, and UATD, encompassing comparative studies of attention mechanisms, ablation tests, and evaluations against other leading algorithms. The findings indicate that the MFF-YOLOv7 model substantially surpasses other models across various metrics, demonstrates superior underwater target detection capabilities, exhibits enhanced generalization potential, and offers a more dependable and precise solution for underwater target identification.

Keywords: deep learning; underwater images; underwater target recognition; sonar images; improved YOLO

1. Introduction

The need for precise recognition in underwater target detection is growing rapidly, encompassing fields such as marine resource exploitation, subsea construction, and oceanic ecosystem monitoring. Sonar image target recognition plays a vital role in these applications [1–5]. Nevertheless, the intricate underwater environment, low-quality sonar image data, and limited sample sizes pose significant challenges to conventional image recognition algorithms, hindering accurate identification. Consequently, there is a pressing need to develop more effective methods to enhance sonar image target recognition performance.

This research aims to address these issues by leveraging deep learning techniques. The main goal is to create a robust underwater sonar image target recognition system that can handle environmental complexities, improve target identification precision, and resolve the problem of limited data samples. The innovative aspect of this research lies in the creation of the Multi-Gradient Feature Fusion YOLOv7 model (MFF-YOLOv7) using deep learning techniques. This model introduces several key improvements: a novel Multi-Scale Information Fusion Module replaces YOLOv7's SPPCSPC, enabling better feature capture of varying target sizes in sonar images; RFAConv substitutes the Conv in the two CBSs following ELAN; and the SCSA mechanism is incorporated at three junctions between the backbone and head to enhance the model's ability to handle underwater environmental complexities, focus on relevant target recognition features, and improve recognition accuracy. These enhancements are expected to significantly improve YOLOv7's performance in sonar image recognition, contributing substantially to underwater target detection and offering more reliable and efficient solutions for various applications.

Sonar images originate from imaging sonar. When operating as an active sonar system, the process is as follows:

1. The sonar system emits sound waves.
2. The sound waves pass through the water, reflect off underwater targets, and return.
3. The reflected echoes return to the sonar system.
4. Images are formed through the complex processing of these echoes.

The underwater environment's complexity and inherent unpredictability make the imaging process susceptible to medium-related influences. Echo signals often encounter issues such as attenuation and distortion, resulting in sonar images with reduced contrast and resolution, indistinct target boundaries, and barely discernible features [6–9].

Conventional methods for recognizing targets in sonar images primarily rely on features based on pixels, grayscale values, or preconceived notions about the targets [10,11], often resulting in limited accuracy. In recent times, the field of computer vision has been revolutionized by deep learning, which has subsequently advanced underwater target detection. Deep learning-based target detection approaches are typically categorized into two-stage and one-stage algorithms. Two-stage algorithms first identify potential regions containing targets, followed by classification and localization within these areas. For instance, Villon et al. [12] employed convolutional neural networks to swiftly identify fish in marine images, achieving a 94.9% accuracy rate. Guo et al. [13] utilized deep residual networks to recognize sea cucumbers with an 89.5% accuracy rate. Dai et al. [14] created a dual-branch backbone network called GCC-Net, which uses both enhanced and original images as input to train underwater target detectors. However, these methods are computationally intensive, slow, and require numerous candidate regions. Moreover, when dealing with complex sonar images, there is still potential for improving accuracy.

Unlike two-stage methods, single-stage algorithms like the YOLO [15–21] family eliminate the need for a candidate region network, instead generating prediction boxes directly on the input image for target detection. Muksit [22] and colleagues introduced the YOLO-Fish algorithm in 2022, achieving 76.56% accuracy in identifying 20 distinct fish species in their habitats. In 2024, Liu et al. [23] developed a modular underwater enhancement component that could be integrated into YOLOv5, resulting in a 2.6% increase in mAP on the DUO dataset. Lei et al. [24] enhanced YOLOv5's backbone network by incorporating the Swin Transformer. Although it achieved a small increase in mAP, it significantly increased the model volume. Although the YOLO algorithm has a speed advantage, it occasionally has missed or false detections in complex environments with high noise points and dense small targets, raising concerns about its stability.

Traditional image recognition algorithms struggle with accurately processing low-resolution sonar image data, which involves a complex procedure. Even sophisticated deep learning models have room for enhancement when dealing with intricate sonar images. For example, the YOLO framework encounters several challenges, including poor image quality in complex underwater settings, challenges in detecting small and closely grouped targets, and issues with both missed detections and false positives. Although YOLOv9 and YOLOv10 excel in optical image processing, for achieving high accuracy with minimal resources, YOLOv7 proves more advantageous for sonar images. This is because underwater sonar targets are typically small-scale and accompanied by increased noise, and

YOLOv7 demonstrates superior recognition capabilities for small targets. Consequently, this study aims to enhance YOLOv7 to address these challenges. It introduces MFF-YOLOv7, which improves noise processing in low-resolution scenarios and enhances detection of dense, small targets.

In the field of underwater target image recognition, the traditional YOLOv7 model has significant limitations when dealing with high-noise and unclear sonar images, such as the complex underwater environment, various target sizes, numerous interfering information in sonar images, low imaging resolution, and small and dense targets. These factors can cause missed and false detections in the model. To solve the above problems, we propose the MFF-YOLOv7 model. First, we introduce the original Multi-Scale Information Fusion Module (MIFM) to replace SPPCSPC. The MIFM can better fuse multi-scale information and enhance the model's processing ability of features at different scales. Even in complex underwater scenes, it can accurately identify targets of various sizes, thereby effectively solving the problem of processing features at different scales due to the large size differences in underwater targets. Secondly, we replace the Conv in the CBS following ELAN with RFAConv. Given the high noise and unclearness of sonar images, the existing feature extraction methods have deficiencies, while the RFAConv has better feature extraction capabilities and is more adaptable to specific types of sonar image data. It can significantly improve the model's learning and representation of sonar image features, enabling it to extract useful target features from noise better.

The SCSA mechanism should be implemented at the three junctions where the backbone connects to the head. In underwater target recognition, sonar images often contain numerous interfering elements, which can lead to the model being influenced by irrelevant information. By employing the SCSA mechanism, the model can prioritize crucial feature information and minimize the impact of unrelated data. This allows the model to concentrate more effectively on target recognition-related features when transferring information from the backbone to the head, thus enhancing the model's recognition accuracy. To assess the efficacy of the proposed approach, we conducted rigorous comparative evaluations using various real-world sonar image datasets, including URPC [25], SCTD [26], and UATD [27].

To conclude, the core innovations of this paper are embodied in the following key contributions:

1. The traditional YOLOv7 model has many limitations when dealing with high-noise and unclear sonar images, such as the complex underwater environment, various target sizes, numerous interfering information, low imaging resolution, and small and dense targets. These factors can easily lead to missed detections and false detections. To address these issues, we have designed the MFF-YOLOv7 model.
2. The Multi-Scale Information Fusion Module (MIFM) has been introduced and implemented to enhance the YOLOv7 model. This module excels at integrating information from various scales, thereby improving the model's ability to process features at different levels, particularly in complex underwater environments where target dimensions fluctuate. The MIFM demonstrates robust fusion capabilities, overcoming the constraints of conventional modules and effectively capturing characteristics of targets with diverse sizes. Additionally, the MIFM can dynamically adjust its focus on targets of varying scales based on the actual underwater conditions, enabling intelligent resource allocation. This mechanism substantially enhances the precision of sonar image target identification while minimizing instances of missed and false detections.
3. Rigorous comparative evaluations were conducted on the real-world sonar image datasets URPC, SCTD, and UATD. The results indicate that the MFF-YOLOv7 model performs exceptionally well in these two datasets. It demonstrates good performance on specific datasets, exhibits strong generalization ability, and can adapt to sonar image recognition tasks in different scenarios.

The article's subsequent sections are organized as follows: Section 2 focuses on introducing YOLOv7 and several enhanced modules, including MIFM, RFAConv, and the SCSA mechanism, along with the architecture and enhancements of the MFF-YOLOv7 network. These module improvements aim to enhance the model's effectiveness in recognizing targets in underwater sonar images. Section 3 offers a comparative evaluation of MFF-YOLOv7 against leading sonar image recognition technologies and presents findings from multiple real-world datasets. Section 4 concludes the article by summarizing the research outcomes and providing closing remarks.

2. Background

In underwater target image recognition, the continuous development of related technologies provides strong support for achieving more accurate and efficient target detection. This part will introduce the related work. Firstly, the traditional target detection algorithm YOLOv7 will be expounded, and its improvement direction will be derived. In sequence, the following will introduce YOLOv7, the Multi-Scale Information Fusion Module (MIFM), RFAConv, the SCSA mechanism, and the ultimately proposed MFF-YOLOv7. Through an in-depth analysis of these technologies, the innovation and breakthroughs of this study in underwater target image recognition will be demonstrated.

2.1. YOLOv7

YOLOv7, as an advanced target detection algorithm, holds an important position in computer vision, especially excelling in target detection tasks. As shown in Figure 1, the structure of YOLOv7 mainly consists of key components such as Backbone, Head, and Prediction. The input image with a size of 640 × 640 × 3 first enters the Backbone part. Here, through a series of elaborately designed network layers, such as the efficient ELAN module, as well as operations like convolution (Conv), batch normalization (BN), activation function (SiLU), etc., a multi-level feature extraction is performed on the image. During the feature extraction, specific structures such as the SPPCSPC module play an important role in feature fusion, capable of integrating feature information at different levels. The image undergoes multiple processing and downsampling operations, such as reducing the feature dimension through operations like Maxpool, thereby gradually forming feature maps of different scales. Subsequently, these features enter the head part for further analysis and processing, and finally, the prediction results are output in the form of three tensors of specific sizes.

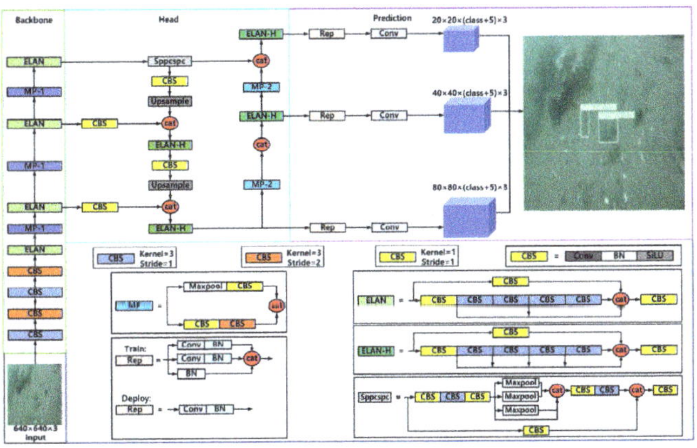

Figure 1. Structure of the YOLOv7.

2.2. Multi-Scale Information Fusion Module (MIFM)

There are many complex challenges in target detection, especially underwater target image recognition. The particularity of the underwater environment makes it difficult for traditional target detection methods to meet the demand for accurate recognition. We introduce the Multi-Scale Information Fusion Module (MIFM) to address these challenges effectively.

The details of MIFM are shown in Figure 2. This module first expands the feature channels through two 1×1 convolution operations, with an expansion ratio of $\gamma = 2$, to increase the feature dimension and provide richer information for subsequent processing. Then, the input features are divided into two parallel paths for processing. One of the paths involves a gating mechanism, and the element-wise product of the features of the two paths enhances the nonlinear transformation, enabling the module to capture the complex relationships between features better. In the lower path, depth wise convolution is used for feature extraction, which can effectively extract features and reduce the computational amount. The module uses two 3×3 dilated convolutions with dilation rates of 2 and 3, respectively. Using these convolutions achieves multi-scale feature extraction, extracting features at different scales and improving its adaptability to targets of different sizes.

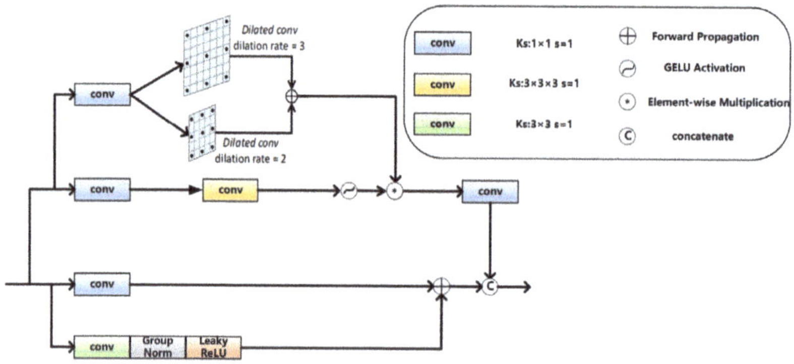

Figure 2. The structure diagram of MIFM.

Given the input tensor $X \in R^{H \times W \times C}$, where H (Height), W (Width), C (Channel), MIFM is formulated as:

$$Up(X) = W_{1\times1}(\varphi(W_{3\times3\times3}W_{1\times1}(X))\Theta(W^2_{3\times3}W_{1\times1}(X) + W^3_{3\times3}W_{1\times1}(X))) \quad (1)$$

$$Down(X) = W_{1\times1}(X) + R(GN(W_{3\times3}(X))) \quad (2)$$

$$X_{out} = Concat(Up(X), Down(X)) \quad (3)$$

Θ represents element-wise multiplication, φ represents GELU nonlinearity, $W^2_{3\times3}$ represents 3×3 dilated convolution with a dilation rate of 2, and $W^3_{3\times3}$ represents 3×3 dilated convolution with a dilation rate of 3. W represents convolution, and the subscript indicates the kernel size of the convolution. R represents the Leaky ReLU activation function, and GN represents the Group Norm.

The Multi-Scale Information Fusion Module (MIFM) has significant underwater target detection advantages. Compared with the original SPPCSPC module of YOLOv7, MIFM can more effectively fuse information of different scales and has stronger adaptability for various underwater target sizes and complex environments. It can automatically adjust the attention to targets of different scales according to the actual situation to intelligently allocate resources, while SPPCSPC could be more flexible in this aspect. In addition, MIFM introduces operations such as gating mechanisms, depthwise convolution, and multi-scale dilated convolution, which can better capture the complex relationships between features, achieve more powerful nonlinear transformations, and enhance the extraction ability of features for targets of different sizes. In contrast, the feature processing method of SPPCSPC is relatively single. In underwater target detection tasks, MIFM significantly improves the accuracy of sonar image target recognition, reduces missed detections and false detections, and performs better in dealing with the challenges of the complex underwater environment.

2.3. RFAConv

In target detection, especially in underwater target image recognition tasks, continuously exploring more effective feature extraction and fusion methods is crucial. To further enhance the model performance and better adapt to the complex underwater environment, we introduce a new type of convolution structure, RFAConv (Recurrent Feature Aggregation Convolution). Traditional convolution methods may have certain limitations when dealing with underwater target detection problems, such as insufficient extraction of multi-scale features and difficulty capturing the complex features of targets. Therefore, the replacement with the RFAConv aims to overcome these problems and improve underwater target detection performance.

The structural diagram of RFAConv (Recurrent Feature Aggregation Convolution) is depicted in Figure 3. RFAConv is crucial for target detection, enhancing network performance by learning an attention map through interaction with receptive field feature information. To mitigate the additional computational burden caused by this interaction, RFAConv employs AvgPool for pooling global information from each receptive field feature. It then facilitates information exchange 1×1 group convolution operations and utilizes softmax to highlight the significance of individual features within the receptive field. Additionally, receptive field attention (RFA) is implemented on the spatial features of the receptive field. This approach not only emphasizes the importance of various features within the receptive field but also considers its spatial characteristics, effectively addressing the issue of convolution kernel parameter sharing. The receptive field spatial features are dynamically generated, and RFA forms a fixed combination with convolution, with both elements working interdependently to boost performance. In essence, RFAConv's unique design substantially improves target detection network performance while efficiently managing computational overhead and parameter count.

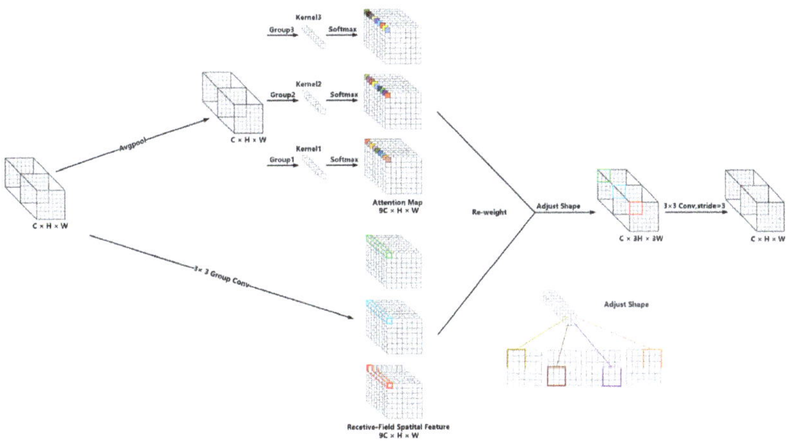

Figure 3. RFAConv Structure diagram.

The calculation of RFAConv can be expressed as:

$$F = Softmax(g^{1\times1}(AvgPool(X))) \times ReLU(Norm(g^{k\times k}(X))), \quad (4)$$

where g represents group convolution, and the superscript indicates the size of the convolution kernel. X represents the input feature map, and the output F is obtained by multiplying the attention map with the transformed receptive field spatial feature. Softmax and ReLU represent activation functions, AvgPool represents the max pooling operation, and Norm is the normalization operation.

Selection and position arrangement of convolution structures are crucial in exploring improvements in target detection models. Different results are produced when RFAConv replaces the CBS (Convolution-BatchNorm-SiLU) structure at different positions. When replacing other CBSs, the effect worsens, while the best effect is achieved when it is placed at the CBS position immediately following the ELAN module in YOLOv7, replacing the Conv with RFAConv.

The reasons for this phenomenon mainly lie in two aspects. On the one hand, the ELAN module plays a key role in feature extraction, and its output features have specific patterns and characteristics. The CBS position immediately following it is crucial for further processing and optimizing these features. The unique recurrent feature aggregation method of RFAConv can better adapt to the characteristics of the output features of the ELAN module, achieving more effective feature fusion and extraction. The CBS at other positions may receive input features that do not match the characteristics of the RFAConv, be unable to exert its advantages fully, and even introduce inappropriate processing, resulting in a worse effect. On the other hand, at this specific position after the ELAN module, RFAConv can collaborate better with the preceding and subsequent modules, better undertake and integrate the features extracted by the upstream modules, and provide a higher-quality feature representation for subsequent processing. Due to the different interaction methods between the modules, a good collaboration effect cannot be achieved at other positions, thereby affecting the model's overall performance.

In summary, placing RFAConv at the CBS position immediately following the ELAN module in YOLOv7 is well-considered. This position can give full play to the advantages of RFAConv, form good coordination with the surrounding modules, significantly enhance the model's ability to extract features of underwater targets, reduce the occurrence of missed detections and false detections, enhance the robustness and adaptability of the model, improve the accuracy and stability of target detection, and thereby achieve better underwater target detection performance.

2.4. Spatial and Channel Synergistic Attention (SCSA) Module

Attention mechanisms are increasingly important in target detection and computer vision. They can help the model focus more precisely on key information, thereby enhancing the features' expression ability and the model's performance. However, current attention mechanisms still have certain limitations in dealing with multi-semantic information and spatial–channel collaboration.

We introduce a brand-new attention mechanism, SCSA (Spatial and Channel Synergistic Attention), to overcome these drawbacks and fully explore the collaboration between spatial and channel attention. SCSA aims to achieve more accurate feature extraction by efficiently integrating multi-semantic spatial information and channel information to obtain better model performance. Next, we will elaborate on the specific structure and working principle of SCSA in detail.

SCSA (Spatial and Channel Synergistic Attention) is a new type of attention mechanism. Its structure is shown in Figure 4. This attention mechanism aims to explore the collaboration between spatial and channel attention. It mainly comprises Shared Multi-Semantic Spatial Attention (SMSA) and Progressive Channel Self-Attention (PCSA).

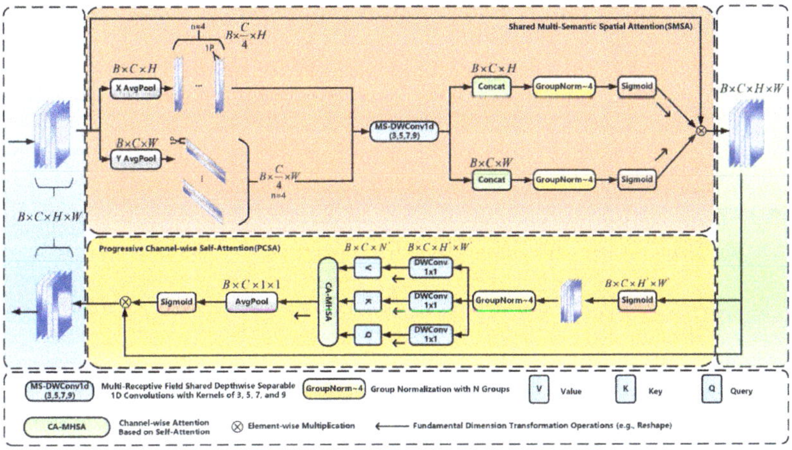

Figure 4. The structure diagram of the SCSA mechanism. The variable n represents the number of groups the sub-features are divided into, and 1P represents a single pixel.

Shared Multi-Semantic Spatial Attention (SMSA) module:

1. Spatial and Channel Decomposition: We break down the given input $X \in R^{B\times C\times H\times W}$ in terms of the height and width dimensions. Global average pooling is then applied to each of these dimensions, which gives rise to two one-way 1D sequence structures, namely $X_H \in R^{B\times C\times W}$ and $X_W \in R^{B\times C\times H}$. In order to capture different spatial distributions as well as contextual relationships, we divide the feature set into K sub-features that are of the same size and are independent of each other. These sub-features are named X_H^i and X_W^i, and each sub-feature has a channel count of $\frac{C}{K}$. In this paper, we have set the default value as $K = 4$. The procedure for decomposing into sub-features is detailed as follows:

$$X_H^i = X_H[:, (i-1)\times\frac{C}{K} : i\times\frac{C}{K}, :] \tag{5}$$

$$X_W^i = X_W[:, (i-1)\times\frac{C}{K} : i\times\frac{C}{K}, :] \tag{6}$$

where X^i represents the i-th sub-feature.

2. Efficient Convolutional Approach: Implement separable one-dimensional convolutions with filter sizes of 3, 5, 7, and 9 across the four sub-features to detect various semantic spatial patterns. Concurrently, employ efficient shared convolutions for alignment to tackle the restricted receptive field issue resulting from splitting features into H and W dimensions and utilizing 1D convolutions. The process of obtaining diverse semantic spatial data are defined as where a denotes the b-th sub-feature.

$$\widetilde{X}_H^i = DWConv1d_{k_i}^{\frac{C}{K}\rightarrow\frac{C}{K}}(X_H^i) \tag{7}$$

$$\widetilde{X}_W^i = DWConv1d_{k_i}^{\frac{C}{K}\rightarrow\frac{C}{K}}(X_W^i) \tag{8}$$

X_i represents the spatial structure information obtained by the i-th sub-feature after the lightweight convolution operation, and k_i represents the convolution kernel applied to the i-th sub-feature.

3. Computing the Spatial Attention Map: Aggregate different semantic sub-features, normalize them using group normalization (GN) of K groups, and then generate

spatial attention through the Sigmoid activation function. The formula for calculating the output feature is:

$$Attn_H = \sigma(GN_H^K(Concat(\tilde{X}_H^1, \tilde{X}_H^2, \ldots, \tilde{X}_H^K)), \quad (9)$$

$$Attn_W = \sigma(GN_W^K(Concat(\tilde{X}_W^1, \tilde{X}_W^2, \ldots, \tilde{X}_W^K)), \quad (10)$$

where σ represents Sigmoid normalization, and GN_K^H and GN_K^W represent Group Norm along the H and W dimensions, respectively.

Progressive Channel-wise Self-Attention (PCSA) module:

Explore the dependencies between channels through convolution operations. We were inspired by using MHSA in ViT to model the similarities between different tokens in calculating spatial attention, combined with the spatial priors modulated by SMSA to calculate the similarities between channels. A progressive compression method is adopted to preserve and utilize the multi-semantic spatial information extracted by SMSA and reduce the computational cost of MHSA. The specific implementation is as follows:

$$X_p = Pool_{(7,7)}^{(H,W)\to(H',W')}(X_s), \quad (11)$$

$$F_{proj} = DWConv1d_{(1,1)}^{C\to C}, \quad (12)$$

$$Q = F_{\mathrm{Proj}}^{Q}(X_p), \quad (13)$$

$$K = F_{\mathrm{Proj}}^{K}(X_p), \quad (14)$$

$$V = F_{\mathrm{Proj}}^{V}(X_p). \quad (15)$$

$Pool_{(7,7)}^{(H,W)\to(H',W')}$ represents the pooling operation with a kernel size of 7×7, adjusting the resolution from (H, W) to (H', W'), and F_{proj} represents the mapping function for generating queries, keys, and values.

Collaboration effect: SCSA guides the learning of channel attention through spatial attention. SMSA extracts multi-semantic spatial information from each feature, providing precise spatial priors for channel attention calculation; PCSA refines the semantic understanding of local sub-features by using the overall feature map X, reducing the semantic differences caused by multi-scale convolutions in SMSA. The final constructed SCSA is:

$$SCSA(X) = PCSA(SMSA(X)). \quad (16)$$

In the underwater sonar image target recognition task, introducing the SCSA (Spatial and Channel Synergistic Attention) attention mechanism is of great significance. The existing attention mechanisms have limitations in dealing with multi-semantic information and spatial-channel collaboration, while SCSA aims to overcome these limitations. Integrating multi-semantic spatial and channel information effectively can achieve more accurate feature extraction, thereby improving the model performance. For underwater sonar image target recognition, the SCSA mechanism offers several benefits. It enhances the model's capacity to concentrate on crucial information, boosts feature representation capabilities, and improves the detection and identification of underwater targets. At the same time, it can effectively integrate multi-semantic spatial information, helping the model to learn higher-quality feature representations and better cope with problems such as various sizes of underwater targets and complex environments. In addition, through the collaboration of space and channel, SCSA can extract target features more accurately and reduce the occurrence of missed and false detections.

Adding the SCSA mechanism at the three connection positions—where the backbone connects to the head—enhances the model's ability to pay attention to and extract features and improve the model's performance in tasks such as target detection and recognition. The

SCSA can effectively integrate multi-semantic spatial information and channel information. Through the collaboration of space and channel, it can better focus on key information and improve the ability to express features. Introducing SCSA at these three connection positions enables the model to better learn and utilize features at different stages and enhance the attention to different scales and semantic information, thereby improving the recognition accuracy of underwater sonar image targets. Incorporating SCSA at these locations can yield several benefits. It can enhance the critical distribution of features, prompting the model to focus more on essential characteristics and improve feature utilization. Additionally, it can enhance feature fusion, allowing for better integration of features from various levels and scales, thereby boosting feature expressiveness and resilience. Furthermore, it can enhance the model's generalization capabilities by learning more representative features, mitigating overfitting risks, and improving the model's adaptability across diverse datasets and tasks.

2.5. MFF-YOLOv7

In the field of underwater target image recognition, the conventional YOLOv7 model exhibits significant limitations when confronted with challenges such as complex underwater environments, diverse target sizes, abundant interfering information in sonar images, low imaging resolution, and small, densely clustered targets. Consequently, it is susceptible to instances of missed detections and false detections. To address these issues, we propose the MFF-YOLOVv7 model, which aims to enhance the performance and accuracy of the model in underwater target image recognition. Figure 5 illustrates the structural diagram of MFF-YOLOv7, demonstrating its distinctive design.

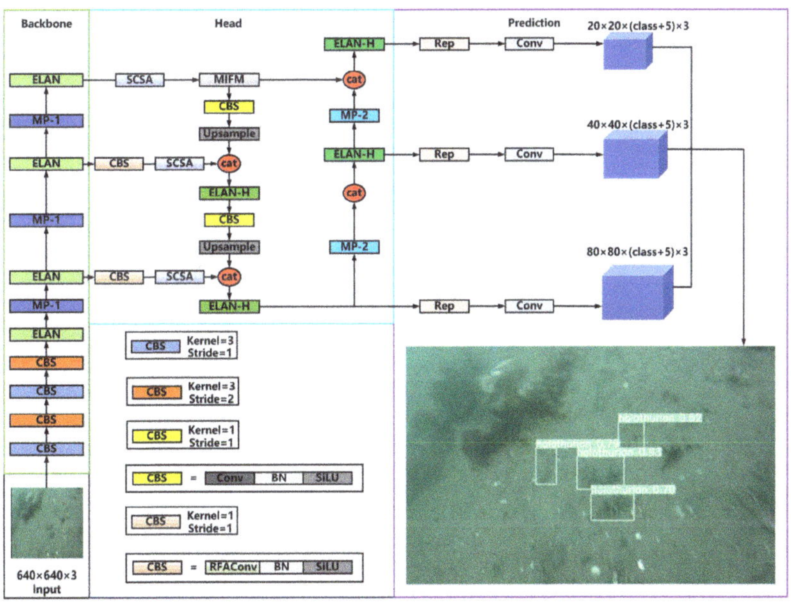

Figure 5. The MFF-YOLOv7 network.

The traditional YOLOv7 model needs to improve when dealing with the problems of complex underwater environments and various target sizes. To solve this problem, the MFF-YOLOv7 model introduces the original Multi-Scale Information Fusion Module (MIFM) to replace SPPCSPC MIFM can fuse multi-scale information more effectively and enhance the model's ability to process features of different scales. Applying the Multi-Scale Information Fusion Module (MIFM) enables the model to identify targets of various sizes in complex underwater scenes accurately. It effectively solves the problem that the traditional module makes it difficult to effectively process features of different scales due to the large size differences in underwater targets. By introducing MIFM, our model can better adapt to the diversity of underwater environments and improve the ability to detect and recognize targets.

Considering the characteristics of sonar images, such as high noise and unclearness, the existing feature extraction methods need to be improved when dealing with such images. Therefore, we replace the Conv in the CBS immediately following ELAN with RFAConv. RFAConv has a better feature extraction ability and is more adaptable to specific types of sonar image data. It can significantly improve the model's learning and representation of sonar image features, enabling it to extract useful target features from the noise better. This improvement helps to enhance the model's performance when processing low-quality sonar images and reduces missed and false detections caused by image quality issues.

In underwater target recognition, sonar images often contain numerous interfering information, which makes the model prone to be disturbed by irrelevant information, thereby affecting recognition accuracy. To solve this problem, we introduce the SCSA mechanism at the three connection positions where the backbone connects to the head. The SCSA mechanism can help the model pay more attention to important feature information and reduce the interference of irrelevant information. By enabling the model to better focus on the features related to target recognition when transferring features from the backbone to the head, the SCSA mechanism can significantly improve the model's recognition accuracy and enhance the model's robustness in complex underwater environments.

The MFF-YOLOv7 model has successfully solved the problems faced by the traditional YOLOv7 model in underwater target image recognition through innovative improvements such as introducing MIFM, replacing Conv with RFAConv, and introducing the SCSA mechanism. These improvements have significantly enhanced the model's ability to fuse multi-scale information, extract features of sonar images, and pay attention to important information, thereby greatly improving the recognition accuracy and robustness of the model. Our improvements provide a more effective solution for underwater target image recognition and have the potential to achieve better results in practical applications.

3. Experimental Design and Experimental Analysis

This section demonstrates the effectiveness and generalization of the proposed method through underwater sonar image target detection experiments. In this section we briefly introduce the datasets and evaluation metrics used in the experiments and the experimental settings. Subsequently, a comparison experiment of attention mechanisms will be conducted to prove the advantages of the SCSA mechanism. Next, ablation experiments will be carried out to prove the effectiveness of each improvement. Then, improvements should be made compared with other mainstream algorithms to prove the superiority of the improved algorithm. Finally, other datasets will be verified to prove the generalization of the improved algorithm.

3.1. Experimental Environment

To verify the model's effectiveness, we conducted comparative experiments on the URPC, SCTD and UATD datasets to verify the detection effect of the model. The operating system is Windows 11, the deep learning framework is PyTorch 1.4.0, the CPU is Intel Core i7 12700H, the memory is 32 GB, and the GPU is NVIDIA GeForce GTX 3070TI.

3.2. Experimental Indicators

The assessment criteria were Precision, Recall, MAP using the following formulas:

$$Precision = \frac{TP}{TP + FP}, \tag{17}$$

$$Recall = \frac{TP}{TP + FN}. \tag{18}$$

Here, TP refers to the number of samples that are correctly judged as positive examples. It represents the situation of successfully identifying the target category during the prediction or classification process. FP refers to the number of samples that are wrongly judged as positive examples. It indicates the situation where the model wrongly classifies negative examples as positive ones. FN refers to the number of samples that are wrongly judged as negative examples. It represents the situation where the model wrongly classifies positive examples as negative ones and misses them.

mAP is an abbreviation of average accuracy and an indicator of recognition accuracy in target recognition. When there are multiple classes to be detected or classified, each class has its own average precision (AP). mAP is the average of the average precisions of all these classes. mAP provides a comprehensive metric to measure the performance of a multi-class classification model and can comprehensively reflect the accuracy of the model on different classes.

$$AP = \int_0^1 P(r)dr, \tag{19}$$

$$mAP = \frac{\sum_{n=1}^{N} AP_n}{N}, \tag{20}$$

where p represents precision, r represents recall, p is regarded as a function with r as a parameter, n is the number of object categories, and AP_n represents the average precision of the neural network when recognizing specific types of targets. MAP has two evaluation metrics, namely MAP@0.5 and MAP@0.5:0.95. MAP@0.5 represents the average precision when the Intersection over the Union (IoU) threshold is 0.5 and is an evaluation metric with relatively low detection requirements. MAP@0.5:0.95 is the average precision calculated at multiple IoU thresholds (starting from 0.5, increasing by 0.05 as the step size up to 0.95), representing an evaluation metric with higher detection requirements. It can more comprehensively evaluate the performance of the model under different precision requirements. The above indicators are used to measure the accuracy of the neural network.

Intersection over Union (IoU) is an important metric widely used in fields such as computer vision to measure the degree of overlap between two sets (usually two regions represented by bounding boxes in images). It is calculated by dividing the intersection (i.e., the overlapping part) of the two sets by the union (all included parts), and the calculation formula is:

$$IOU = \frac{A \cap B}{A \cup B}. \tag{21}$$

Its value range is between 0 and 1. A value of 0 indicates no overlap between the two sets, such as two detection boxes corresponding to different objects in object detection; a value of 1 indicates complete coincidence, that is, the positions of the detection boxes are exactly the same and correspond to the same object. In the object detection task, it is mainly used to evaluate the matching degree between the predicted object bounding box and the real bounding box. By setting a threshold (such as 0.5) we determine that if it is greater than this threshold, the predicted box is considered to accurately detect the target; if it is lower, it may be a wrong or inaccurate detection.

3.3. Experimental Results and Analysis of the URPC Dataset

The URPC dataset encompasses 10,875 images, which are categorized into four subsets: sea urchin, sea cucumber, scallop, and starfish. The distribution of samples among these subsets is highly unbalanced. As shown in Figure 6a, the category statistics chart reveals that the sea urchin subset has the largest number of samples, followed by the starfish and sea cucumber subsets, with the scallop subset having the fewest.

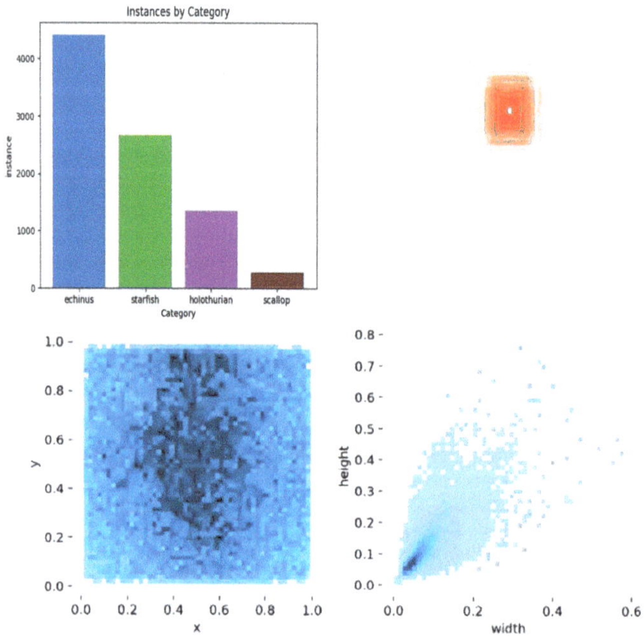

(a)

(b)

Figure 6. The sample information of the URPC dataset. (**a**) presents statistical information such as category distribution, box size, centroid position, and aspect ratio of the dataset for analyzing data characteristics. (**b**) shows example images from the dataset, presenting the actual situation of underwater scenes.

This dataset is divided into a training set and a test set in an 8:2 ratio to support the training and testing of the proposed algorithm. After this division, 8700 images are allocated for training and 2175 for testing. However, to further improve the model's performance and generalization ability, a validation set is also required. The division into training, validation, and test subsets is based on the following considerations. Firstly, to maintain the representativeness of each subset, the samples are divided proportionally according to the original distribution of each category. Secondly, in order to ensure that the model can be effectively evaluated and fine-tuned, a sufficient number of samples are reserved for the validation set.

The dataset showcases a range of intricate scenarios, including visual obstruction caused by clustered underwater organisms, variations in illumination, and image distortion resulting from motion capture. These challenging conditions accurately represent the underwater environment and improve the model's ability to generalize. The box plot indicates that the dimensions of the target boxes are relatively uniform. The normalized target location map suggests that targets are primarily concentrated horizontally but more dispersed vertically. The standardized target size map reveals that target dimensions are fairly consistent, with most being small in size. Sample images from the URPC dataset are displayed in Figure 6b.

3.3.1. Attention Mechanism-Contrast Trials

Conducting the attention comparison experiment on the UPRC dataset is crucial for in-depth research and evaluation of the performance of different attention mechanisms. This experiment focuses on analyzing the attention comparison experiment of YOLOv7, aiming to prove the advantages of the SCSA mechanism through detailed data and result demonstrations.

It can be seen from Table 1 that APechinus, APstarfish, APholothurian, and APscallop represent the average precision of different underwater targets, respectively. YOLOv7-SCSA performs well in these four indicators. APechinus has a value of 86.0%, which is higher than the 85.0% of other models like YOLOv7. For instance, we can use APechinus to illustrate this comparison. The result shows that the SCSA mechanism can extract features more precisely when handling sea urchin targets.

Table 1. Experimental contrasts of attention mechanisms based on YOLOv7.

Method	$AP_{echinus}$	$AP_{starfish}$	$AP_{holothurian}$	$AP_{scallop}$	*mAP*@0.5	*mAP*@0.5:0.95	Precision (%)	Recall (%)
YOLOv7 [20]	85.0%	89.6%	78.5%	35.4%	72.1%	42.6%	82.1	66.9
YOLOv7-CBAM [28]	85.3%	89.8%	79.3%	34.5%	72.2%	42.6%	82.0	67.1
YOLOv7-ECA [29]	85.0%	89.5%	79.5%	36.2%	72.6%	42.8%	82.1	68.1
YOLOv7-SE [30]	85.5%	89.4%	78.6%	35.7%	72.3%	42.7%	83.0	66.4
YOLOv7-SimAM [31]	85.2%	89.9%	79.2%	36.2%	72.7%	42.8%	82.0	67.1
YOLOv7-Biformer [32]	85.3%	90.1%	79.8%	36.3%	72.9%	42.9%	81.4	66.5
YOLOv7-SCSA [33]	**86.0%**	**90.8%**	**84.7%**	**37.7%**	**74.8%**	**43.5%**	**84.9**	**68.9**

As a consequence, our strategy can be improved. For APstarfish, the value of YOLOv7-SCSA is also relatively high, indicating that this mechanism is also effective in starfish target detection. Regarding APholothurian and APscallop, SCSA also shows advantages, indicating that it is adaptable to different types of underwater targets and can perform effective feature extraction for the characteristics of different targets.

Both mAP@0.5 and mAP@0.5:0.95 are two metrics that reflect the comprehensive detection capabilities of a model under different Intersections over Union (IoU) thresholds. YOLOv7-SCSA achieves 74.8% in mAP@0.5, which is higher than other models. The fact that YOLOv7-SCSA achieves 74.8% in mAP@0.5, which is higher than that of other models, implies that under relatively lower detection requirements, the SCSA can enhance the overall detection effect of the model for various targets. Moreover, in terms of mAP@0.5:0.95, its value reaches 43.5%, which is significantly better than other models. Since this metric considers multiple IoU thresholds, it demonstrates that the SCSA can play a good role under different precision requirements. It has a strong generalization ability and can adapt to various complex detection scenarios, thus providing powerful support for the comprehensive evaluation of the model's performance.

The Precision metric represents the precision rate, which is the proportion of samples that are truly positive among those predicted as positive examples. The Precision rate of YOLOv7-SCSA is 84.9%, which is higher than that of other models. The fact that the Precision rate of YOLOv7-SCSA is 84.9%, which is higher than that of other models, indicates that during the detection process, the SCSA mechanism can identify targets more accurately, thereby reducing the occurrence of false positives. It can effectively focus on key information and make more precise judgments on targets, enhancing the accuracy and reliability of the model. Such a high Precision rate is of crucial importance for underwater target image recognition tasks. Especially in situations where it is necessary to distinguish between different target types accurately, the SCSA can provide a more reliable basis for decision-making for the model.

The Recall indicator represents the recall rate, the proportion of samples that are positive examples and are predicted as positive examples. The recall rate of YOLOv7-SCSA is 68.9%, which is higher than that of other models. The higher performance metrics (such as, the higher precision rate or other relevant evaluation indicators mentioned before) indicate that the SCSA mechanism can better detect targets and reduce the number of missed reports. In underwater target image recognition, due to factors such as complex environments and diverse targets, missed detections are a common problem. By better focusing on crucial information and improving the feature expression ability, SCSA can effectively enhance the coverage ability of the model for targets, ensuring that more targets are correctly detected, thereby improving the practicability and effectiveness of the model.

The SCSA mechanism can better focus on essential information, improve the feature expression ability, and extract target features more accurately by effectively integrating multi-semantic spatial information and channel information, thereby significantly improving the performance of the YOLOv7 model in the underwater target image recognition task. This conclusion is consistent with the theoretical expectations of the attention mechanism and provides a more effective solution for underwater target image recognition.

3.3.2. Ablation Experiment

In the underwater target image recognition research, to deeply explore the influence of different modules on the model performance, we use the UPRC dataset to conduct module ablation experiments. This experiment aims to evaluate the importance and contribution of each module in the underwater target recognition task by purposefully removing or replacing specific modules in the model. Through conducting rigorous module ablation experiments on the UPRC dataset, we expect to more accurately understand the mechanism of action of different modules, thereby providing a solid basis for further optimizing the model performance.

Table 2 of the ablation experiment clearly shows the importance of different modules for the YOLOv7 model in underwater target image recognition. The base model (YOLOv7), when no new module is added, has an accuracy of 82.1%, a recall rate of 66.9%, a mAP@0.5 of 72.1%, and a mAP@0.5:0.95 of 42.6%, providing a baseline performance for subsequent comparisons. When only the RFAConv module is added, the accuracy increases to 85.1%, and the mAP@0.5 changes to 73.1%, etc., indicating that this module, with its better feature extraction ability, plays a positive role in dealing with the problems of high noise and unclearness of sonar images, and improves the model's ability to learn and represent target features. Then, when both the RFAConv and SCSA modules are added simultaneously, the accuracy further improves to 88.2%, and all indicators have significantly improved, highlighting the importance of the SCSA mechanism in reducing the interference of irrelevant information and enabling the model to focus on target features.

Table 2. Ablation comparison of model performance improvements on the URPC dataset.

Model	RFAConv	SCSA	MIFM	Precision	Recall	mAP@0.5	mAP@0.5:0.95
YOLOv7				82.1%	66.9%	72.1%	42.6%
	√			85.1%	67.2%	73.1%	42.7%
	√	√		88.2%	69.3%	75.9%	44.3%
	√	√	√	89.9%	73.0%	79.1%	45.0%

The √ represents that this module is used in the model.

Finally, when the three RFAConv, SCSA, and MIFMs are added, the model performance is at its best. The accuracy is 89.9%, the recall rate is 73.0%, the mAP@0.5 is 79.1%, and the mAP@0.5:0.95 is 45.0%. The MIFM solves the problem of significant differences in underwater target sizes by better fusing multi-scale information. The three modules work together to make the model perform outstandingly in the underwater target image recognition task. The excellent performance shown in aspects like higher precision rates, better mAP values at different thresholds, and other remarkable achievements fully prove the respective advantages of the RFAConv, SCSA, and MIFMs and their collaboration effect. This provides a clear direction and a solid basis for further optimizing the underwater target image recognition model.

As shown in Figure 7, by comparing the two confusion matrices, it can be seen that MFF-YOLOv7 shows advantages in multiple aspects. First, for each category of targets, the accuracy of MFF-YOLOv7 has generally improved. For example, the accuracy of the "echinus" category has increased from 0.84 to 0.9. Secondly, the added modules help to reduce the misclassification rate. The RFAConv module improves the feature extraction ability, enabling the model to handle the high noise and unclearness of sonar images better; the SCSA mechanism focuses on important feature information and reduces the interference of irrelevant information. Especially when facing numerous interfering pieces of information in underwater target recognition, it can improve the recognition accuracy of the model; the MIFM effectively solves the problem that the traditional module is challenging to process features of different scales due to the significant size differences in underwater targets by better fusing multi-scale information. These advantages make MFF-YOLOv7 more accurate and reliable in the underwater target image recognition task, providing a more effective model choice for research in this field.

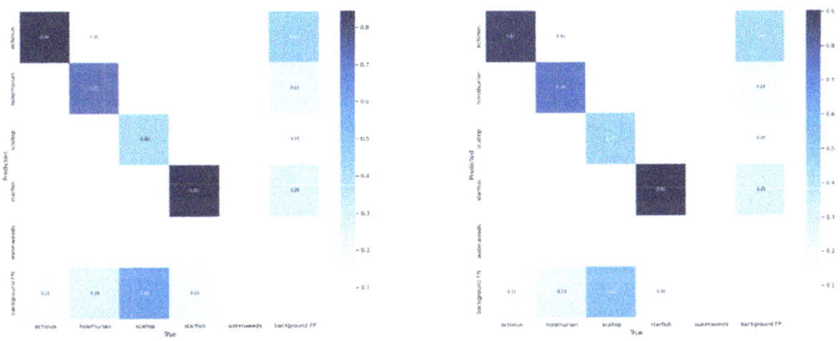

(a) (b)

Figure 7. The confusion matrices of the models under the URPC dataset are as follows: (**a**) is the confusion matrix of YOLOv7, and (**b**) is the confusion matrix of MFF-YOLOv7.

Figure 8a The Precision-Recall (PR) curve of YOLOv7 shows different characteristics in different categories. The precision value of the "echinus" sea urchin target is 0.850, indicating a relatively high detection precision at different recall rate levels. The "starfish" target's precision is 0.896, showing good detection performance. However, the precision of the "scallop" target is only 0.354, indicating that the performance in detecting this type of target needs to be improved. The precision of the "holothurian" sea cucumber target is 0.785, which is at a medium level. Considering all categories, the average precision is 0.721. From the overall shape, the precision shows a downward trend as the recall rate increases. The situation requires further analysis of the specific numerical changes in different recall rate intervals better to understand the performance of YOLOv7 in various situations.

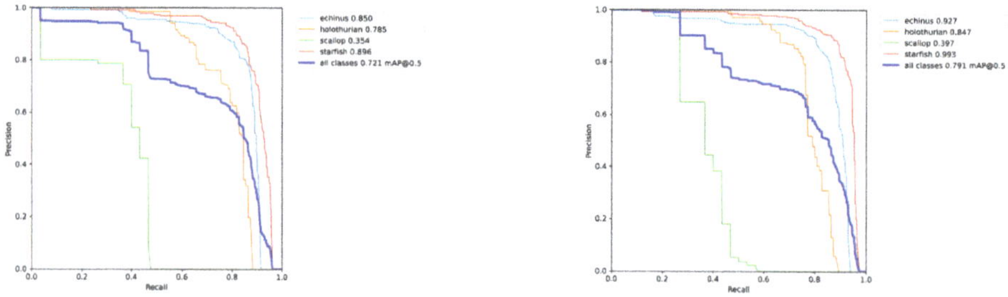

(**a**) (**b**)

Figure 8. The PR curve of the model under the URPC dataset (**a**) is the PR curve of YOLOv7 and (**b**) is the PR curve of MFF-YOLOv7.

Figure 8b The PR curve of MFF-YOLOv7 has improved in all categories. The precision of the "echinus" sea urchin target has increased to 0.927, which is significantly higher than that of YOLOv7. The "starfish" target's precision has reached 0.993, almost approaching perfect detection precision, with excellent performance. The precision of the "scallop" scallop target is 0.397. Although it is still relatively low, it has made particular progress compared to YOLOv7. The precision of the "holothurian" sea cucumber target is 0.847, a noticeable improvement. Overall, the average precision has increased to 0.791. From the overall PR curve, the curve of MFF-YOLOv7 is closer to the upper right corner, meaning that it has higher precision at the same recall rate or higher recall rate at the same precision, showing its advantage in the underwater target image recognition task.

Overall, MFF-YOLOv7 is superior to YOLOv7 in all categories and the comprehensive average precision, fully demonstrating the improvement effect of the added three modules (RFAConv, SCSA, and MIFM) on the model performance. The improvement amplitudes of precision in different categories vary. For example, the "starfish" category significantly improves amplitude, while the "scallop" category has improved but is still relatively low, suggesting that subsequent research needs to further optimize the model for scallop targets. It can be seen from the shape and position of the PR curve that MFF-YOLOv7 can maintain a high precision into a broader range of recall rates, which is very valuable for balancing detection precision and recall rate in practical applications. To sum up, by analyzing the PR curves of these two models, the performance advantages of MFF-YOLOv7 after adding the three modules in the underwater target image recognition task are clarified, providing an essential basis for further improving and optimizing the model.

Figure 9 presents a visual comparison of the performance of YOLOv7 and MFF-YOLOv7 in multiple underwater scenes, which provides valuable insights into the capabilities of both models.

In the context of multi-target detection, YOLOv7 exhibits several limitations. As shown in the figure, in complex underwater environments, it often struggles to accurately detect and distinguish between multiple targets. Small-sized and occluded targets pose particular challenges for YOLOv7. The model is prone to missed detections, where it fails to identify some of the targets present in the scene. Additionally, false detections occur, where it incorrectly identifies objects as targets. This is mainly due to the complex nature of underwater sonar images, which have low resolution and contain various interfering factors. The indistinct target discrimination ability of YOLOv7 further exacerbates these issues, making it difficult to accurately classify different types of targets. Achieving a balanced detection accuracy and recall rate for different target types is also a challenge for YOLOv7 in such scenarios.

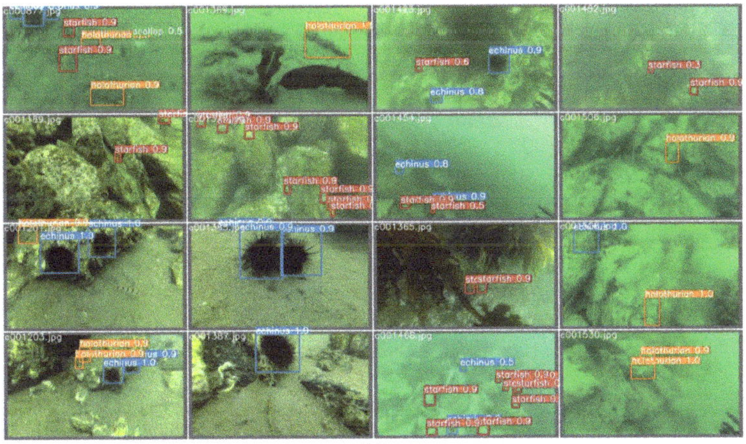

(**a**)

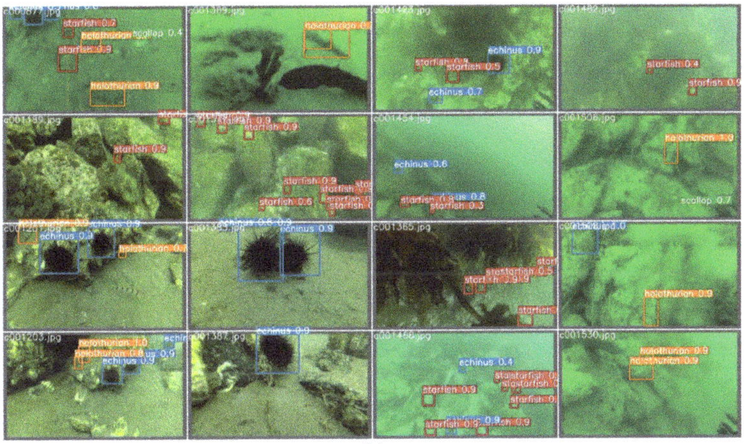

(**b**)

Figure 9. The recognition results of the model in the underwater scenes of the URPC dataset. (**a**) shows the recognition results of the YOLOv7 model, indicating the problems of missed detections, false detections, and interference from the environment in multi-target and single-target detections. (**b**) is the recognition result of the MFF-YOLOv7 model, reflecting the advantages of this model in accurately detecting various targets and balancing accuracy and recall rate.

In contrast, MFF-YOLOv7 shows significant improvements in multi-target detection. The added RFAConv module enhances the model's feature extraction capabilities, especially in the presence of high noise and unclear sonar images, which are typical in underwater environments. This allows the model to better capture the characteristics of targets. The SCSA mechanism focuses the model's attention on the relevant features of each target, reducing the interference from irrelevant information and improving the discrimination between different targets. The MIFM effectively fuses multi-scale information, enabling the model to handle targets of various sizes, including small-sized and occluded ones. As a result, MFF-YOLOv7 can accurately detect a greater number of targets and achieve a more balanced detection accuracy and recall rate across different target types, thereby enhancing the overall multi-target detection performance.

For single-target detection, YOLOv7 is highly susceptible to the complex factors in the underwater environment. Changes in lighting and water turbidity can significantly affect its ability to extract meaningful features from a single target. As a consequence, it has difficulties in accurately identifying single targets with indistinct features. The low annotation probability of such targets further compounds the problem, as the model has less training data to learn from. This leads to unstable detection results, with the performance varying depending on the environmental conditions.

MFF-YOLOv7, on the other hand, demonstrates higher accuracy and stability in single-target detection. The SCSA mechanism enables the model to focus precisely on the features of a single target, effectively reducing the influence of environmental interference. This allows the model to accurately identify single targets even in challenging underwater conditions. For single targets with indistinct features, the combined effect of the RFAConv and MIFMs leads to more effective feature extraction. This, in turn, improves the detection success rate, as the model can better capture the subtle characteristics of the target.

The improved MFF-YOLOv7 offers several notable advantages. Firstly, it provides higher detection accuracy in both multi-target and single-target detection scenarios, effectively reducing false and missed detections. Secondly, it exhibits better adaptability to the complex environment of multiple underwater scenes, including variations in lighting, water turbidity, significant differences in target sizes, and occlusions. This adaptability makes it a more reliable solution for underwater target detection in real-world applications. Furthermore, it achieves a more balanced performance across different types of targets, which is crucial for comprehensive underwater target recognition. The overall improvement in detection performance not only provides more accurate results but also serves as a foundation for more advanced underwater imaging and sensing applications. Additionally, the stable detection results of MFF-YOLOv7 ensure the reliability of the data generated, which is essential for subsequent analysis and decision-making processes.

In summary, the visual evidence in Figure 9, along with the detailed analysis, clearly demonstrates that MFF-YOLOv7 is significantly superior to YOLOv7 in multi-target and single-target detection and recognition in multiple underwater scenes. This superiority validates the effectiveness of the proposed modifications and highlights the potential of MFF-YOLOv7 for various underwater target detection applications.

3.3.3. Contrasting Experiments with the Other Algorithms

In the research of underwater target detection, it is essential to introduce the MFF-YOLOV7 model and compare it with other models. On the one hand, the underwater environment is complex and changeable. The forms and sizes of targets vary, and they are often affected by various factors such as lighting, water flow, and turbidity. By comparing it with other models, the performance of MFF-YOLOV7 in dealing with these complex challenges can be clarified better to evaluate its feasibility and reliability in practical applications. On the other hand, different target detection models adopt different algorithms and techniques. The advantages and disadvantages of various methods can be deeply understood through comparison, providing directions for further improvement and optimization of MFF-YOLOV7.

As shown in Table 3, the performance of different target detection models is compared on the URPC dataset. Among them, YOLOv5s has average performance in various indicators, with an accuracy of 80.4%, a recall rate of 65.5%, a mAp@0.5 of 68.9%, and a mAp@0.5:0.95 of 38.1%, indicating that there is room for improvement in its detection performance. YOLOv5m has made particular progress compared to YOLOv5s, with an accuracy of 82.6%, but the overall performance still needs to improve. YOLOv7 has a relatively balanced performance in multiple indicators, with an accuracy of 82.1%, a recall rate of 66.9%, a mAp@0.5 of 72.1%, and a mAp@0.5:0.95 of 42.6%, but there is still potential for improvement. YOLOv7-Tiny has a particular advantage with an accuracy of 82.7%, but the recall rate of 63.6% and mAp@0.5:0.95 of 36.9% are relatively low. YOLOv7-SDBB performs well in some indicators, with an accuracy of 82.0%, a recall rate of 66.8%, a mAp@0.5 of

72.4%, and a mAp@0.5:0.95 of 43.4%, but there is a gap compared to the excellent models. YOLOv8n has weak overall performance, with an accuracy of 80.1%, a recall rate of 64.8%, a mAp@0.5 of 68.6%, and a mAp@0.5:0.95 of 38.6%. YOLOv9 is comparable to YOLOv7 in some indicators, with an accuracy of 82.1%, a recall rate of 64.8%, a mAp@0.5 of 71.0%, and a mAp@0.5:0.95 of 42.0%, and there is also room for further improvement.

Table 3. Performance comparison of target detection models on the URPC dataset.

Model	Precision (%)	Recall (%)	mAp@0.5 (%)	mAp@0.5:0.95 (%)
YOLOv5s [34]	80.4	65.5	68.9	38.1
YOLOv5m [35]	82.6	65.2	69.5	41.2
YOLOv7 [20]	82.1	66.9	72.1	42.6
YOLOv7-Tiny [36]	82.7	63.6	69.6	36.9
YOLOv7-SDBB [20]	82.0	66.8	72.4	43.4
YOLOv8n [37]	80.1	64.8	68.6	38.6
YOLOv9 [21]	82.1	64.8	71.0	42.0
MFF-YOLOv7	89.9	73.0	79.1	45.0

Our MFF-YOLOv7 model performs outstandingly in various indicators on the URPC dataset. The accuracy is as high as 89.9%, which is significantly higher than other models, and means it can identify targets more accurately and thus reduce false alarms. The recall rate is 73.0%, higher than most models, which can detect targets better and thus reduce missed alarms. The mAp@0.5 is 79.1%, and mAp@0.5:0.95 is 45.0%, leading among all models, which fully demonstrates that this model has excellent performance under different Intersection over Union (IoU) thresholds and can evaluate the detection ability of the model more comprehensively. Compared with other target detection models on the URPC dataset, the MFF-YOLOv7 model shows obvious advantages and achieves better accuracy, recall rate, and average precision, providing a more effective solution for tasks such as underwater target detection.

3.4. Experimental Results and Analysis of the SCTD

The experimental data used in this study comes from the Sonar Common Target Detection Dataset (SCTD) collected and organized by Zhou Yan, containing 596 images. The sonar targets in these images were labeled using the open-source annotation tool LabelImg. The annotated dataset mainly includes three main types of targets: sunken ships (461 images), crashed aircraft (90 images), and human bodies (45 images). Random rotation and Gaussian blurring were applied to balance the original dataset and alleviate the problem of sample imbalance. In the final photos, these three main targets contain 512, 454, and 397 instances, respectively. The dataset is divided into a training set, validation set, and test set in the ratio of 7:1:2. Figure 10 shows the schematic diagrams of these three typical sonar image targets.

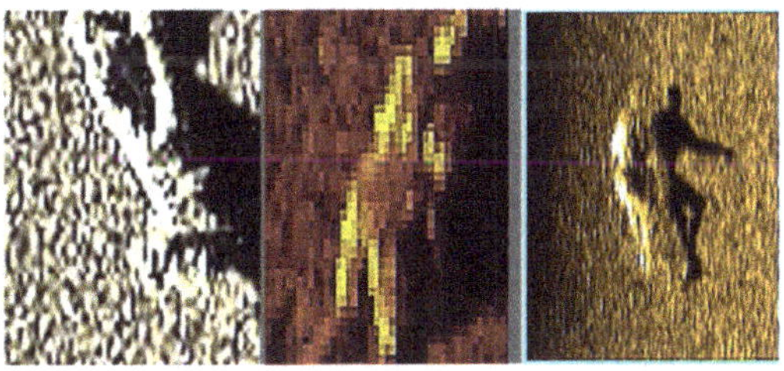

(a)human

Figure 10. *Cont.*

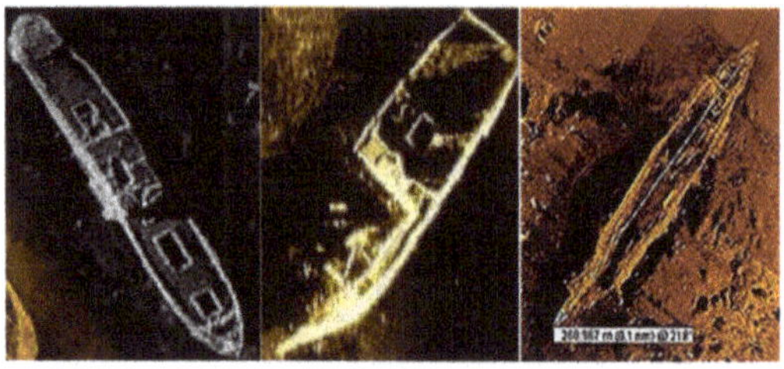

(b)ship

(c)aircraft

Figure 10. The SCTD sonar images dataset.

Model Generalizability Experiment

Accurately identifying targets in sonar images in underwater target detection is crucial for various applications. We need to conduct extensive experimental verification to ensure that the proposed model can perform well in different scenarios. Among them, experiments on the SCTD were conducted to verify the model's generalization. The SCTD contains various underwater sonar images covering various environments and target types. Through experiments on this dataset, we can evaluate whether the model can accurately identify and classify targets when facing complex and changeable actual situations, thereby verifying the model's generalization ability and providing strong support for its reliability in practical applications.

Table 4 shows that our MFF-YOLOv7 model has obvious advantages and excellent generalization among various underwater target detection models.

Table 4. Accuracy of various underwater target detection models built in the SCTD.

Method	AP_{ship}	AP_{plane}	AP_{human}	mAp@0.5	mAp@0.5:0.95
SSD [38]	86.2%	86.8%	86.1%	86.4%	43.0%
Faster R-CNN [17]	88.2%	86.8%	87.3%	87.5%	43.8%
YOLOv3 [17]	87.3%	89.0%	86.1%	87.5%	47.8%
YOLOv4 [18]	89.2%	87.9%	87.3%	88.2%	49.6%
YOLOv5 [39]	90.2%	90.1%	89.9%	90.1%	56.6%
YOLOv7 [20]	89.2%	89.0%	89.9%	89.3%	54.0%
YOLOv8 [37]	89.2%	90.1%	89.9%	89.7%	54.7%
MFF-YOLO v7	**96.3%**	**99.9%**	**99.9%**	**98.7%**	**63.2%**

Compared with other models, the MFF-YOLOv7 model has significantly improved all indicators. In terms of APship (average precision of ship detection), APplane (average precision of aircraft detection), and APhuman (average precision of human detection), the MFF-YOLOv7 model has reached 96.3%, 99.9%, and 99.9%, respectively, far higher than other models. In terms of mAp@0.5 (average precision when the IoU threshold is 0.5) and mAp@0.5:0.95 (average precision when the IoU threshold is between 0.5 and 0.95), the MFF-YOLOv7 model has reached 98.7% and 63.2%, respectively, also significantly ahead of other models.

This indicates that the MFF-YOLOv7 model can more accurately detect various underwater targets, including ships, aircraft, and humans. It has a higher vital generalization ability and can adapt to different underwater target detection tasks. In contrast, other models such as SSD, Faster R-CNN, YOLOv3, YOLOv4, YOLOv5, YOLOv7, and YOLOv8 perform relatively poorly in these indicators. Our MFF-YOLOv7 model has significant advantages and excellent generalization in underwater target detection and can provide more reliable and accurate detection results for related applications.

MFF-YOLOv7 shows an outstanding performance on the SCTD. It can be seen from the PR curve in Figure 11 that the average precision of all categories is as high as 0.987 when the Intersection over the Union threshold is 0.5, which shows that the model as a whole has extremely high detection accuracy. Specifically for each category, the average detection precision of ships is 0.963, and the detection effect is good. However, with the human and aircraft categories at 0.999, it is almost close to perfect detection.

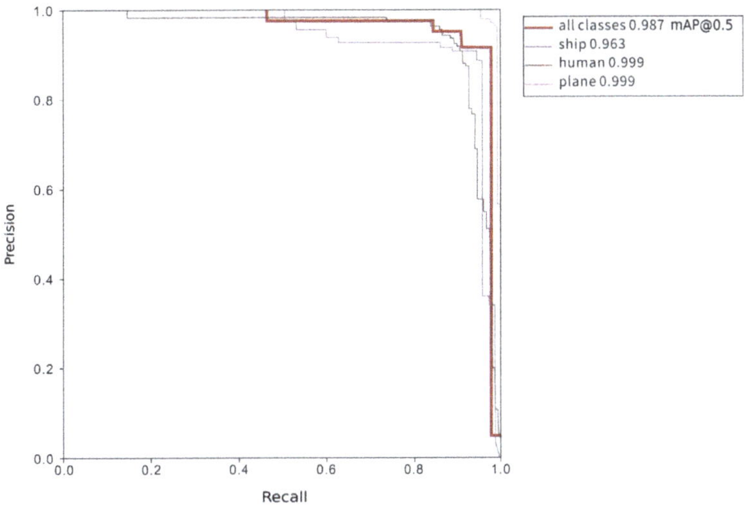

Figure 11. MFF-YOLOv7 PR plots of the SCTD.

Overall, this model performs well in detecting various types of targets, has a high overall average precision, has strong generalization ability and accuracy, and can accurately identify and detect different types of underwater targets. Generally speaking, MFF-YOLOv7 performs outstandingly in underwater sonar image target detection and provides an efficient and reliable solution for this field.

3.5. Generalization Experiment

3.5.1. UATD Datasets

In today's field of scientific research, the performance verification of models is a key link in promoting technological development. To explore the performance of the model in complex situations comprehensively and deeply, we carefully selected the UATD dataset to conduct experiments.

The UATD dataset is quite considerable in size, consisting of 9000 sonar images. Its data collection was accomplished with the aid of a Multi-Beam Forward-Looking Sonar (MFLS), and the adopted Tritech Gemini 1200ik sonar has high-resolution characteristics and can flexibly switch the operating frequency between 720 kHz (long distance) and 1200 kHz (short distance) for targets at different distances.

One of the highlights of this dataset lies in the authenticity of its collection environment. Some of the data are from Jinshitan, Dalian, where the water depth is in the shallow water environment ranges from 4 to 10 m; another part is from Haoxin Lake, Maoming, with a maximum water depth of 4 m. The data collected in such real marine and shallow water environments undoubtedly add more practical significance and challenges to the experiments.

More importantly, the UATD dataset covers a rich and diverse range of object categories, including ten types such as Cube, Ball, Cylinder, Human Body, Plane, Circle Cage, Square Cage, Metal Bucket, Tire, and ROV. These objects are imaged in the underwater environment with low visibility, and the complexity of their sonar images can be imagined. Precisely because the UATD dataset has the notable characteristics of low visibility and multiple objects, it becomes an ideal choice for us to verify the superiority of the model in complex situations. Experiments based on this dataset will also provide an extremely valuable basis and profound insights for the performance evaluation of the model.

The dataset is already split in training, testing and validation sets comprising 7600, 800, and 800 sonar images, respectively.

3.5.2. Experimental Results and Analysis

Table 5 presents a performance comparison between various deep learning models in the literature with MFF-YOLOv7, all trained on the UATD dataset [27]. Among all these models, MFF-YOLOv7 demonstrated superior performance, achieving the best precision, recall, and mAP50.

Table 5. Comparison of Performance of different object detection models and MFF-YOLOv7 on UATD Dataset.

Model	Precision (%)	Recall (%)	mAp@0.5 (%)
RetinaNet [40]	63.2	62.4	62.5
Faster R-CNN [17]	74.3	75.3	75.1
YOLOv3 [17]	85.4	82.1	79.1
SDD Net [41]	81.3	79.7	80.2
YOLO-DCN [42]	86.2	83.4	80.5
YOLOv3SPP [43]	91.1	93.0	92.2
YOLOv8 [37]	85.4	81.0	83.3
MFF-YOLOv7	91.2	88.9	87.2

In the comparison of different object detection models, the precision of MFF-YOLOv7 reached 91.2%. Compared with 63.2% of RetinaNet and 74.3% of FasterRCNN, it has a relatively significant advantage, indicating that in the detection results, the proportion of correct predictions as positive examples is relatively higher, that is, the performance of detection accuracy is excellent, and the false alarm situation is relatively less. At the same time, compared with YOLOv3SPP (91.1%), which also has a high precision, MFF-YOLOv7 is slightly better, which reflects its excellent ability to accurately identify targets.

The recall rate of MFF-YOLOv7 is 88.9%. Compared with models such as 62.4% of RetinaNet and 81.0% of YOLOv8, it can better find out the actual positive examples, meaning that under the UATD dataset, the possibility of missing targets is relatively lower, and it has a good performance in comprehensively detecting all relevant targets, and has a wide coverage of targets.

The mAp@0.5 of MFF-YOLOv7 reached 87.2%. Compared with models such as 62.5% of RetinaNet and 75.1% of FasterRCNN, it has obvious advantages. Furthermore, when compared with other common and better-performing models such as YOLOv3 (79.1%) and YOLO-DCN (80.5%), it is also at a relatively high level. This reflects that its average precision at different recall rates is good. When considering the detection accuracy and recall situation comprehensively, the overall performance is relatively prominent, and it can complete the object detection task more stably and with high quality.

Overall, by comparing with multiple different object detection models on the UATD dataset, MFF-YOLOv7 has shown its superiority in several key indicators such as precision, recall rate, and mean average precision, proving its excellent performance in the complex underwater object detection scene (corresponding to the characteristics of the UATD dataset).

Figure 12 shows the confusion matrix of the UATD dataset obtained by comparing the predicted labels with the actual labels.

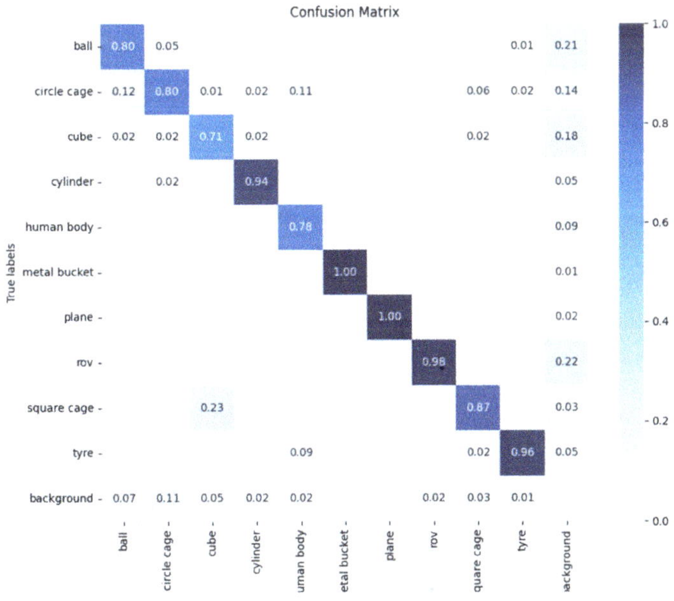

Figure 12. Confusion Matrix Results for UATD Dataset.

Figure 13 shows the label map (Figure 13a) and the predicted label map (Figure 13b). In the label map, different targets such as "cube", "ball", "circle cage", etc., are clearly marked. The predicted label map shows the prediction results of the model for these targets and the corresponding probability values.

In the single-target scenario, such as in sub-images where only "cube" or "ball" exists, the model has significant advantages. When the model predicts a single target, the prediction probability is often high. Taking "cube" as an example, when the sub-image in the label map is marked as "cube", the corresponding predicted label in the predicted label map is "cube", and the probability value is mostly above 0.8, which indicates that the model has high accuracy in identifying single targets and has a strong confidence in recognizing individual targets.

Furthermore, the model performs well in distinguishing different types of single targets. The prediction probability for "ball" is generally high, which means that when dealing with single-target situations, the model can distinguish "cube" and "ball" well and rarely misjudge.

In the multi-target scenario, for example, when there are both "cube" and "circle cage", or "ball" and "circle cage" in the sub-image, the model's performance is satisfactory. It can be seen from the predicted label map that the model can accurately identify multiple targets one by one. For instance, in the sub-image with both "cube" and "circle cage", the model can correctly label these two targets, and the probability value corresponding to each target is within a reasonable and acceptable range.

More importantly, when dealing with multiple targets, the model does not miss any targets. In all sub-images containing multiple targets, the model can completely identify all the targets, reflecting its good performance in the face of a complex multi-target environment.

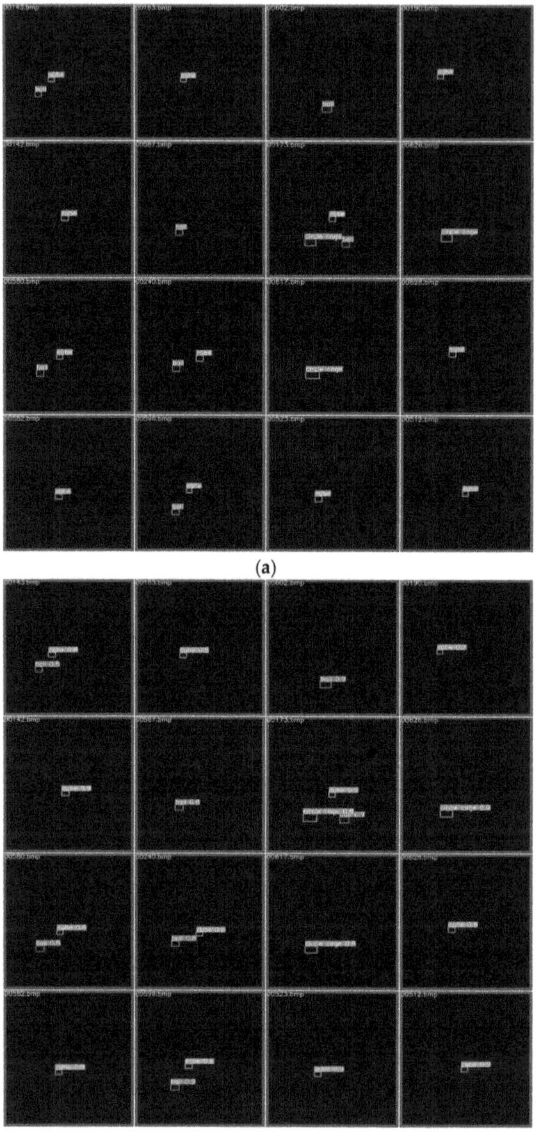

(a)

(b)

Figure 13. (**a**) Test set labels. (**b**) Predicted bounding boxes of the test set.

Overall, whether in single-target or multi-target cases, the labels predicted by the model have a high consistency with the actual labels, and the predicted probability values are mostly at a relatively high level. This characteristic demonstrates the model's outstanding advantages when dealing with single and multi-targets in the UATD dataset. In the complex underwater environment, this superiority is of great significance for target detection and recognition work, which can significantly improve the accuracy and reliability of detection and provide a strong guarantee for related work.

4. Limitations of the Proposed Method

4.1. Data Requirements and Adaptability Challenges

Although the MFF-YOLOv7 model incorporates certain strategies to handle data challenges, it still has notable limitations. The model requires a substantial amount of labeled data for effective training. Despite the efforts to address the small sample size issue, in scenarios where the underwater environment or target characteristics deviate significantly from the training data, the model's performance may degrade. For example, when encountering new types of underwater terrains or previously unseen target species, the model might struggle to accurately detect and classify targets. This indicates that the model's adaptability to novel underwater conditions and target variations needs improvement. Future research should focus on developing techniques to enhance the model's generalization ability across diverse underwater scenarios and target types, potentially through advanced data augmentation methods or unsupervised learning strategies.

4.2. Noise Processing and Computational Efficiency

The MFF-YOLOv7 model has made efforts to address the noise problem in sonar images, but challenges remain. While the introduced modules such as RFAConv and the SCSA mechanism contribute to feature extraction and noise reduction, extremely high noise levels or complex noise patterns in underwater sonar images can still impact the model's accuracy. Additionally, although the model aims to balance performance and computational complexity, its computational requirements are relatively high compared to some lightweight models. This restricts its application in resource-constrained underwater devices with limited processing power and memory. Future work should explore more efficient noise reduction algorithms and optimize the model's architecture to reduce computational overhead, enabling its deployment in a wider range of underwater sensing systems.

4.3. Model Complexity and Generalization Ability

The complexity of the MFF-YOLOv7 model, with its multiple innovative modules like the Multi-Scale Information Fusion Module (MIFM) and the combination of different attention mechanisms, enhances its performance but also brings drawbacks. The intricate architecture increases the difficulty of training and debugging, slows down the model's iteration speed, and reduces its maintainability. Moreover, in cases of severe data imbalance, the model's generalization ability may be compromised. For instance, if certain target classes are severely underrepresented in the training data, the model might not perform optimally for those classes in real-world applications. Future research should strive to simplify the model's structure without sacrificing performance, improve its generalization ability through more effective data handling techniques, and conduct comprehensive evaluations on a broader range of underwater sonar image datasets to ensure its stability and reliability in various practical scenarios.

By understanding these limitations, future research can be directed towards improving the model. This could involve developing more advanced noise suppression techniques, optimizing the model structure to reduce computational costs, exploring data augmentation and balancing strategies to enhance generalization ability, and improving the model's interpretability. These efforts will enable the MFF-YOLOv7 model to play a more significant role in a wider variety of underwater target detection applications and contribute more effectively to the field of underwater imaging and sensing.

4.4. Dataset-Specific Performance and Improvement Directions

The MFF-YOLOv7 model has limitations despite its general superiority. For the URPC dataset, results are subpar, with low recall for scallops. This may stem from the dataset's complex underwater environments and diverse target species, causing the model to struggle in generalization. To boost scallop recall, gathering more diverse, high-quality sonar images of scallops can enrich the dataset and help the model learn scallop features better. Optimizing the model's architecture or training process, like modifying network layers or adjusting algorithms, is also essential to enhance sensitivity to scallop features.

The SCTD shows excellent results, likely due to a better fit between its features and the model's design. However, a deeper analysis is needed to clarify the performance difference.

To improve overall underwater biota detection, other strategies include fine-tuning hyperparameters for target datasets to enhance accuracy and generalization. Advanced data preprocessing such as adaptive noise filtering and normalization can assist in handling input data. Incorporating domain knowledge by customizing feature extraction or attention mechanisms to match underwater organisms' visual cues is beneficial. Understanding these limitations guides future research in developing better noise suppression, optimizing the model for lower computational costs, exploring data augmentation and balancing, and improving model interpretability, enhancing the model's role in underwater target detection and its contribution to underwater imaging and sensing.

5. Conclusions

Our contributions can be summarized as follows:

1. We have introduced a series of new modules to enhance the model's performance. The Multi-Scale Information Fusion Module (MIFM) has replaced the SPPCSPC in YOLOv7. It can integrate multi-scale information better, thereby strengthening the model's ability to handle features of different scales. This improvement has effectively addressed the issue that due to the significant differences in the sizes of underwater targets, traditional modules have difficulties processing features of different scales. It has significantly improved the accuracy of sonar image target recognition accuracy and reduced the occurrences of missed and false detections.
2. The RFAConv has been introduced to replace the Conv in the CBS of ELAN. It boasts better feature extraction capabilities and is more adaptable to sonar images' high-noise and unclear characteristics. As a result, it has significantly enhanced the model's ability to learn and represent the features of sonar images, enabling it to extract helpful target features from noise more effectively.
3. Moreover, the SCSA mechanism has been introduced at three connection positions between the backbone network and the head. It helps the model focus more on important feature information and reduces the interference of irrelevant information. The introduction of the SCSA mechanism at three connection positions between the backbone network and the head further improves the recognition accuracy and robustness of the model, allowing it to focus on target features more accurately in complex underwater environments.

We have carried out detailed experiments on datasets like URPC, SCTD, and UATD, which cover attention mechanism comparison experiments, ablation experiments, and comparison experiments with other mainstream algorithms. Through these efforts, we have fully validated the effectiveness and superiority of the MFF-YOLOv7 model. The MFF-YOLOv7 model has significantly improved all metrics in these experiments. It has shown a more vital generalization ability and the capacity to precisely detect various underwater targets, providing a more reliable and accurate solution for underwater target detection.

Author Contributions: Conceptualization, Z.C. and K.Z.; methodology, H.Z. and G.X.; software, H.L.; validation, Z.C., H.Z. and H.L.; formal analysis, J.L., L.L. and Z.L.; data curation, J.L., L.L. and Z.L.; investigation, K.Z. and G.X.; writing—original draft preparation, K.Z.; writing—review and editing, H.Z. and Z.C.; funding acquisition, Z.C. and H.Z. All authors have read and agreed to the published version of the manuscript.

Funding: This work is supported by Guangxi Science and Technology Base and Talent Project (No. GuikeAD21220098) and the 2021 Open Fund project of the Key Laboratory of Cognitive Radio and Information Processing of the Ministry of Education (No. CRKL210102). We also thank to the support of Beihai City Science and Technology Bureau Project (No. Bei ke He 2023158004) and the Innovation

Project of Guangxi Graduate Education (No. YCSW2024344) and the Innovation Project of GUET Graduate Education (No. 2024YCXS022, 2024YCXS033, 2023YCXS038).

Institutional Review Board Statement: Not applicable.

Data Availability Statement: The data used to support the findings of this study are available from the corresponding author upon request.

Conflicts of Interest: The authors declare no conflicts of interest.

Abbreviations

The following abbreviations are used in this manuscript:

MFF-YOLOv7	Multi-Gradient Feature Fusion YOLOv7 model
MIFM	Multi-Scale Information Fusion Module
SPPCSPC	Spatial Pyramid Pooling Channel Shuffling and Pixel-level Convolution
CBS	Convolution-Batch Normalization-SiLU activation function
ELAN	Efficient Layer Aggregation Network
RFAConv	Recurrent Feature Aggregation Convolution
SCSA	Spatial and Channel Synergistic Attention
UATD	Underwater Acoustic Target Detection Dataset
SCTD	Smaller Common Sonar Target Detection Dataset
URPC	The Underwater Optical Target Detection Intelligent Algorithm Competition 2021 Dataset
GCC-Net	Grouped Channel Composition Network.
YOLO	You Only Look Once
Conv	Convolution
BN	Batch Normalization
SiLU	Sigmoid Linear Unit
RFA	Receptive Field Attention
SMSA	Shared Multi-Semantic Spatial Attention
CBS	Convolution-Batch Normalization-SiLU activation function
PCSA	Progressive Channel Self-Attention
IOU	Intersection Over the Union
CBAM	Convolutional Block Attention Module
ECA	Efficient Channel Attention
SE	Squeeze-and-Excitation
SimAM	A Simple, Parameter-Free Attention Module for Convolutional Neural Networks
Biformer	Bidirectional Interactive Attention Transformer
Faster R-CNN	Faster Region-based Convolutional Neural Networks

References

1. Ahmad-Kamil, E.; Zakaria, S.Z.S.; Othman, M.; Chen, F.L.; Deraman, M.Y. Enabling marine conservation through education: Insights from the Malaysian Nature Society. *J. Clean. Prod.* **2024**, *435*, 140554. [CrossRef]
2. Khoo, L.S.; Hasmi, A.H.; Mahmood, M.S.; Vanezis, P. Underwater DVI: Simple fingerprint technique for positive identification. *Forensic Sci. Int.* **2016**, *266*, e4–e9. [CrossRef] [PubMed]
3. Fan, X.; Lu, L.; Shi, P.; Zhang, X. A novel sonar target detection and classification algorithm. *Multimed. Tools Appl.* **2022**, *81*, 10091–10106. [CrossRef]
4. Yin, Z.; Zhang, S.; Sun, R.; Ding, Y.; Guo, Y. Sonar image target detection based on deep learning. In Proceedings of the 2023 International Conference on Distributed Computing and Electrical Circuits and Electronics (ICDCECE), Ballar, India, 29–30 April 2023.
5. Wang, X.; Yuen, K.F.; Wong, Y.D.; Li, K.X. How can the maritime industry meet Sustainable Development Goals? An analysis of sustainability reports from the social entrepreneurship perspective. *Transp. Res. Part D Transp. Environ.* **2020**, *78*, 102173. [CrossRef]
6. Vijaya Kumar, D.T.T.; Mahammad Shafi, R. A fast feature selection technique for real-time face detection using hybrid optimized region based convolutional neural network. *Multimed. Tools Appl.* **2022**, *82*, 1–14. [CrossRef]
7. Girshick, R.; Donahue, J.; Darrell, T.; Malik, J. Rich feature hierarchies for accurate object detection and semantic segmentation. In Proceedings of the IEEE Conference on Computer Vision and Pattern Recognition, Columbus, OH, USA, 23–28 June 2014; pp. 580–587.

8. Girshick, R. Fast R-CNN. In Proceedings of the IEEE International Conference on Computer Vision, Santiago, Chile, 7–13 December 2015; pp. 1440–1448.
9. Huang, C.; Zhao, J.; Zhang, H.; Yu, Y. Seg2Sonar: A Full-Class Sample Synthesis Method Applied to Underwater Sonar Image Target Detection, Recognition, and Segmentation Tasks. *IEEE Trans. Geosci. Remote Sens.* **2024**, *62*, 5909319. [CrossRef]
10. Zhou, T.; Si, J.; Wang, L.; Xu, C.; Yu, X. Automatic detection of underwater small targets using forward-looking sonar images. *IEEE Trans. Geosci. Remote Sens.* **2022**, *60*, 4207912. [CrossRef]
11. Xi, J.; Ye, X.; Li, C. Sonar image target detection based on style transfer learning and random shape of noise under zero shot target. *Remote Sens.* **2022**, *14*, 6260. [CrossRef]
12. Villon, S.; Mouillot, D.; Chaumont, M.; Darling, E.S.; Subsol, G.; Claverie, T.; Villéger, S. A deep learning method for accurate and fast identification of coral reef fishes in underwater images. *Ecol. Inform.* **2018**, *48*, 238–244. [CrossRef]
13. Guo, X.; Zhao, X.; Liu, Y.; Li, D. Underwater sea cucumber identification via deep residual networks. *Inf. Process. Agric.* **2019**, *6*, 307–315. [CrossRef]
14. Dai, L.; Liu, H.; Song, P.; Liu, M. A gated cross-domain collaborative network for underwater object detection. *Pattern Recognit.* **2024**, *149*, 110222. [CrossRef]
15. Redmon, J.; Divvala, S.; Girshick, R.; Farhadi, A. You only look once: Unified, real-time object detection. In Proceedings of the IEEE Conference on Computer Vision and Pattern Recognition, Las Vegas, NV, USA, 27–30 June 2016; pp. 779–788.
16. Redmon, J.; Farhadi, A. YOLO9000: Better, faster, stronger. In Proceedings of the IEEE Conference on Computer Vision and Pattern Recognition, Honolulu, HI, USA, 21–26 July 2017; pp. 7263–7271.
17. Redmon, J.; Farhadi, A. Yolov3: An incremental improvement. *arXiv* **2018**, arXiv:1804.02767.
18. Bochkovskiy, A.; Wang, C.Y.; Liao, H.Y.M. Yolov4: Optimal speed and accuracy of object detection. *arXiv* **2020**, arXiv:2004.10934.
19. Li, C.; Li, L.; Jiang, H.; Weng, K.; Geng, Y.; Li, L.; Ke, Z.; Li, Q.; Cheng, M.; Nie, W.; et al. YOLOv6: A single-stage object detection framework for industrial applications. *arXiv* **2022**, arXiv:2209.02976.
20. Wang, C.Y.; Bochkovskiy, A.; Liao, H.Y.M. YOLOv7: Trainable bag-of-freebies sets new state-of-the-art for realtime object detectors. In Proceedings of the IEEE/CVF Conference on Computer Vision and Pattern Recognition, Vancouver, BC, Canada, 17–24 June 2023; pp. 7464–7475.
21. Wang, C.-Y.; Yeh, I.-H.; Liao, H.-Y.M. YOLOv9: Learning What You Want to Learn Using Programmable Gradient Information. *arXiv* **2024**, arXiv:2402.13616.
22. Al Muksit, A.; Hasan, F.; Emon, M.F.H.B.; Haque, M.R.; Anwary, A.R.; Shatabda, S. YOLO-Fish: A robust fish detection model to detect fish in realistic underwater environment. *Ecol. Inform.* **2022**, *72*, 101847. [CrossRef]
23. Liu, Z.; Wang, B.; Li, Y.; He, J.; Li, Y. UnitModule: A light-weight joint image enhancement module for underwater object detection. *Pattern Recognit.* **2024**, *151*, 110435. [CrossRef]
24. Lei, F.; Tang, F.; Li, S. Underwater target detection algorithm based on improved YOLOv5. *J. Mar. Sci. Eng.* **2022**, *10*, 310. [CrossRef]
25. Zhou, Y.; Chen, S.; Wu, K.; Ning, M.; Chen, H.; Zhang, P. SCTD1.0: Common Sonar Target Detection Dataset. *Ship Sci. Technol.* **2021**, *43*, 54–58.
26. Dong, J.; Yang, M.; Xie, Z.; Cai, L. Overview of Underwater Image Object Detection Dataset and Detection Algorithms. *J. Ocean. Technol.* **2022**, *41*, 60–72.
27. Xie, K.; Yang, J.; Qiu, K. A dataset with multibeam forward-looking sonar for underwater object detection. *Sci. Data* **2022**, *9*, 739. [CrossRef] [PubMed]
28. Woo, S.; Park, J.; Lee, J.; Kweon, I. CBAM: Convolutional Block Attention Module. *arXiv* **2018**, arXiv:1807.06521.
29. Wang, Q.; Wu, B.; Zhu, P.; Li, P.; Zuo, W.; Hu, Q. ECA-Net: Efficient Channel Attention for Deep Convolutional Neural Networks. *arXiv* **2020**, arXiv:1910.03151v4.
30. Hu, J.; Shen, L.; Albanie, S.; Sun, G.; Wu, E. Squeeze-and-Excitation Networks. *arXiv* **2019**, arXiv:1709.01507v4.
31. Yang, L.; Zhang, R.; Li, L.; Xie, X. SimAM: A Simple, Parameter-Free Attention Module for Convolutional Neural Networks. In Proceedings of the 38th International Conference on Machine Learning, Online, 18–24 July 2021; pp. 11863–11874.
32. Zhu, L.; Wang, X.; Ke, Z.; Zhang, W.; Lau, R. BiFormer: Vision Transformer with Bi-Level Routing Attention. *arXiv* **2023**, arXiv:2303.08810v1.
33. Si, Y.; Xu, H.; Zhu, X.; Zhang, W.; Dong, Y.; Chen, Y.; Li, H. SCSA: Exploring the Synergistic Effects Between Spatial and Channel Attention. *arXiv* **2024**, arXiv:2407.05128v1.
34. Wu, W.; Luo, X. Sonar Object Detection Based on Global Context Feature Fusion and Extraction. In Proceedings of the 2024 12th International Conference on Intelligent Control and Information Processing (ICICIP), Nanjing, China, 8–10 March 2024; pp. 195–202.
35. Mehmood, S.; Irfan Muhammad, H.U.H.; Ali, S. Underwater Object Detection from Sonar Images Using Transfer Learning. In Proceedings of the 2024 21st International Bhurban Conference on Applied Sciences Technology (IBCAST), Murree, Pakistan, 20–23 August 2024; pp. 1–2.
36. Xue, G.; Zhang, J.; Wang, K.; Ma, D.; Weichen, P.; Hu, S.; Yang, Z.; Liu, T. Application of YOLOv7-tiny in the detection of steel surface defects. In Proceedings of the 3rd International Conference on Computer, Artificial Intelligence and Control Engineering, Xi'an, China, 26–28 January 2024; pp. 718–723. [CrossRef]
37. Glenn, J. Yolov8. Available online: https://github.com/ultralytics/ultralytics/tree/main (accessed on 16 November 2024).

38. Liu, W.; Anguelov, D.; Erhan, D.; Szegedy, C.; Reed, S.; Fu, C.Y.; Berg, A.C. Ssd: Single shot multibox detector. In Proceedings of the Computer Vision–ECCV 2016: 14th European Conference, Amsterdam, The Netherlands, 11–14 October 2016; Proceedings, Part I 14; Springer: Berlin/Heidelberg, Germany, 2016; pp. 21–37.
39. Glenn, J. Yolov5 Release v7.0. Available online: https://github.com/ultralytics/yolov5/tree/v7.0 (accessed on 16 November 2024).
40. Wang, Z.; Guo, J.; Zeng, L.; Zhang, C.; Wang, B. MLFFNet: Multilevel feature fusion network for object detection in sonar images. *IEEE Trans. Geosci. Remote Sens.* **2022**, *60*, 5119119.
41. Lin, T.Y.; Goyal, P.; Girshick, R.; He, K.; Dollár, P. Focal loss for dense object detection. In Proceedings of the IEEE International Conference on Computer Vision, Venice, Italy, 22–29 October 2017; pp. 2980–2988.
42. Hou, J. Underwater Detection using Forward-Looking Sonar Images based on Deformable Convolution YOLOv3. In Proceedings of the 2024 4th International Conference on Neural Networks, Information and Communication (NNICE), Gaungzhou, China, 19–21 January 2024; pp. 490–493.
43. Pebrianto, W.; Mudjirahardjo, P.; Pramono, S.H.; Rahmadwati; Setyawan, R.A. YOLOv3 with Spatial Pyramid Pooling for Object Detection with Unmanned Aerial Vehicles. *arXiv* **2023**, arXiv:2305.12344.

Journal of
Marine Science and Engineering

MDPI

Article

SS-YOLO: A Lightweight Deep Learning Model Focused on Side-Scan Sonar Target Detection

Na Yang [1], Guoyu Li [2], Shengli Wang [1], Zhengrong Wei [1,*], Hu Ren [1], Xiaobo Zhang [1] and Yanliang Pei [3]

1 College of Ocean Science and Engineering, Shandong University of Science and Technology, Qingdao 266590, China; yangna@sdust.edu.cn (N.Y.); shlwang@sdust.edu.cn (S.W.); 202383190015@sdust.edu.cn (H.R.); zxb@sdust.edu.cn (X.Z.)
2 Qingdao Xiushan Mobile Mapping Co., Ltd., Qingdao 266590, China; liguoyu@supersurs.com
3 First Institute of Oceanography of Ministry of Natural Resources, Qingdao 266061, China; peiyanliang@fio.org.cn
* Correspondence: wzr@sdust.edu.cn

Abstract: As seabed exploration activities increase, side-scan sonar (SSS) is being used more widely. However, distortion and noise during the acoustic pulse's travel through water can blur target details and cause feature loss in images, making target recognition more challenging. In this paper, we improve the YOLO model in two aspects: lightweight design and accuracy enhancement. The lightweight design is essential for reducing computational complexity and resource consumption, allowing the model to be more efficient on edge devices with limited processing power and storage. Thus, meeting our need to deploy SSS target detection algorithms on unmanned surface vessel (USV) for real-time target detection. Firstly, we replace the original complex convolutional method in the C2f module with a combination of partial convolution (PConv) and pointwise convolution (PWConv), reducing redundant computations and memory access while maintaining high accuracy. In addition, we add an adaptive scale spatial fusion (ASSF) module using 3D convolution to combine feature maps of different sizes, maximizing the extraction of invariant features across various scales. Finally, we use an improved multi-head self-attention (MHSA) mechanism in the detection head, replacing the original complex convolution structure, to enhance the model's ability to focus on important features with low computational load. To validate the detection performance of the model, we conducted experiments on the combined side-scan sonar dataset (SSSD). The results show that our proposed SS-YOLO model achieves average accuracies of 92.4% (mAP 0.5) and 64.7% (mAP 0.5:0.95), outperforming the original YOLOv8 model by 4.4% and 3%, respectively. In terms of model complexity, the improved SS-YOLO model has 2.55 M of parameters and 6.4 G of FLOPs, significantly lower than those of the original YOLOv8 model and similar detection models.

Keywords: side-scan sonar (SSS); YOLOv8; lightweight design; partial convolution; multi-head self-attention; feature fusion

1. Introduction

In modern marine surveying, side-scan sonar (SSS) technology is widely used [1–3]. With the increase in marine development activities, the application of SSS technology has been continuously deepening. It is now widely used in various fields such as underwater target search, seabed geomorphological surveys, pipeline inspections, and marine environmental studies. The principle of side-scan sonar equipment is to transmit acoustic pulses to the seabed or underwater targets, through sensors mounted on an unmanned surface vessel (USV) or an autonomous underwater vehicle (AUV). The acoustic pulses transmitted outward will be reflected, and will generate scattered echo when they encounter underwater objects or the seabed [4]. The intensity of the scattered echo is influenced by factors such as the shape and material of the target, while the return time difference is affected by the position of the target. SSS records the scattered echos, which are used to generate a two-dimensional sonar image of the underwater surface. Thus, indicating the position and status of the target.

In recent decades, manual extraction has been the primary method for target identification in SSS. These manual approaches include support vector machine (SVM) [5,6], singular value decomposition (SVD) [7,8], and independent component analyses (ICA) [9,10], among others. They rely on image pixel features, image variations, grayscale thresholds, or known prior information, along with manually designed filters, to accomplish underwater target detection tasks. However, in addition to being laborious and resource-intensive, many approaches have drawbacks such inadequate robustness, excessive complexity, and low detection accuracy [11].

Because of the remarkable performance and broad application of convolutional neu-ral networks (CNN) in standard optical images, scholars have recently begun to pay more attention to the use of deep learning techniques for target detection in SSS images. Kong et al. [12] proposed a model via the dual-path network and the fusion transition module to conduct feature extraction. In addition, this model uses a dense connection technique to enhance multi-scale prediction, enabling low signal-to-noise ratio and small effective sample sizes for object location and classification. Wang et al. [11] proposed the use of multi-scale convolution and attention mechanisms with global reception fields to obtain sonar image multi-scale semantic features and enhance the correlation between features. Zhang et al. [13] improved the YOLOv7 network by adding the Contextual Transformer (CoT) module in the backbone and the Coordinate Attention (CA) module in the neck to enhance the model's feature extraction capabilities. Additionally, they reconstructed the combined network features based on the BiFPN structure. Wen et al. [14] highlights the Swin-Transformer for dynamic attention and global modeling, which enhanced model attention to the desired regions inside SSS images containing large amounts of irrelevant information. They also incorporated a feature scaling factor into the model to address the uncertainty of geometric features in SSS target characteristics. In their subsequent research, Wen et al. [15] proposed a new YOLOv7 model that combines attention mechanisms and scaling factors for underwater SSS target detection. This model builds on their previous findings by incorporating scaling factors into the improved YOLOv7 architecture, addressing the issue of low detection accuracy caused by the uncertainty of geometric features in SSS targets. The new YOLOv7 model achieved a 2.58% increase in mAP accuracy compared to their earlier YOLOv7 model, and a 9.28% increase compared to the original YOLOv7 model.

Mittal et al. [16] evaluated the performance of several traditional lightweight object detection models on well-known datasets such as MS-COCO and PASCAL-VOC, demonstrating the effectiveness of these models on edge devices. They listed the applications of lightweight object detection models in areas such as autonomous driving, robotic vision, intelligent transportation, and industrial quality inspection. They also suggested that future research should continue to optimize lightweight object detection models to improve their practicality and performance on edge devices. Liu et al. [17] proposed a lightweight object detection algorithm for robots based on an improved YOLOv5. They introduced C3Ghost and GhostConv modules into the YOLOv5 backbone network to reduce model parameters while maintaining detection speed and accuracy. Additionally, DWConv modules were incorporated into the YOLOv5 neck network to further decrease model parameters and enhance feature fusion speed. Lang et al. [18] proposed a lightweight object detection framework called MSF-SNET for the real-time detection of airborne remotely sensed images in resource-limited situations. The framework uses SNET as a backbone network to reduce parameter and computational complexity while enhancing small object detection through multi-scale feature fusion. Zhang et al. [19] proposed MSFF-Net, a Multi-scale Spatiotemporal Feature Fusion Network, for video saliency prediction. The network uses 3D convolutions and a Bi-directional Temporal-Spatial Feature Pyramid (BiTSFP) to fully exploit spatiotemporal features. An Attention-Guided Fusion (AGF) mechanism and Frame-wise Attention (FA) module are introduced to adaptively learn fusion weights and emphasize useful frames. MSFF-Net outperforms existing methods in accuracy on DHF1K, Hollywood-2, and UCF-sports datasets.

In summary, the current challenges in target detection tasks for SSS images include a limited number of dataset samples, high model complexity, and poor detection accuracy. SSS images typically have low resolution, complex seabed backgrounds, and a variety of underwater targets, making object detection particularly difficult. Traditional object detection methods are often inefficient in these special environments, especially on edge devices where computational resources and storage space are limited, exacerbating these issues. To address these challenges, lightweight object detection models and feature fusion techniques have become key research directions. The lightweight model greatly reduces computational complexity and resource consumption, making object detection on edge devices more efficient. Feature fusion technology can effectively integrate multi-scale information to compensate for the recognition difficulties caused by low resolution and a complex background. Therefore, combining lightweight object detection with feature fusion technology can ensure high detection accuracy and improve efficiency and real-time performance, making the model more suitable for real-time object detection tasks in resource-limited environments. Single-stage object detection models are more suited for embedding in hardware devices to enable real-time target recognition in SSS images because they typically offer higher accuracy, quicker speed, and better real-time performance [20]. The YOLOv8 [21,22] network is used as the foundational model in this paper. It may be used for tasks like instance segmentation, object identification, and image categorization. The streamlined design of YOLOv8 makes it easily adjustable to a variety of hardware platforms, from cloud to edge devices, and suited to a wide range of applications [23].

The following are this study's primary contributions:

(1) We combined self-collected SSS data with public datasets to create a new side-scan sonar dataset (SSSD) dataset. Through data augmentation, we expanded the original dataset, addressing the issue of a limited sample size and few target types in SSS data. This dataset can now be used for performance evaluation of various sonar target detection methods.

(2) To reduce the complexity of the convolution process, we introduced a combination of the partial convolution (PConv) and the pointwise convolution (PWConv) modules, replacing the original Bottleneck module in C2f. Under resource-constrained conditions, PConv performs convolution operations only on a portion of the input feature map, followed by PWConv, significantly reducing redundant computations and memory access while maintaining relatively high accuracy.

(3) The adaptive scale spatial fusion (ASSF) structure was designed to integrate information from multi-scale feature maps, addressing the feature fusion issue in low-resolution sidescan images. It processes and concatenates feature maps of different sizes, forming a unified dimensional composite view. Then, 3D convolution is applied to extract scale sequence features, enhancing target recognition.

(4) The detection head incorporates an improved multi-head self-attention (MHSA) mechanism. This method not only fully leverages contextual information to improve

detection accuracy but also ensures detection efficiency by using the attention mechanism without the feed-forward network (FFN) layer.

2. Data

The side-scan sonar dataset (SSSD) used in this study consists of two parts: one part is sourced from the publicly available sonar common target detection dataset (SCTD) [24], and the other part was collected by our team using a USV equipped with a multi-beam and SSS collection system, gathered in Weifang, China.

2.1. Data Collection

The proprietary data were collected using the SS900F(Qingdao Hydro-tech Marine Technology Co. Ltd., Qingdao, China) SSS device, which is mounted on the unmanned surface vessel, as shown in Figure 1. The imaging principle of this type of sonar is to transmit sound waves through a transmitting transducer, forming multiple narrow beams distributed vertically along the navigation direction within a certain space, and recording the echoes to obtain multiple channel information.

Figure 1. Unmanned surface vessel and SS900F SSS equipment.

The technical specifications of the SS900F SSS are also provided in Table 1. The data collection area is an operational wind farm, where the seafloor terrain is relatively flat with minor undulations. The majority of the surveyed sea area has a depth of less than 10 m. The seafloor features gentle undulations, with slope gradients ranging from 0.60‰ to 0.85‰ within the 0–5 m depth range, and approximately 0.23‰ within the 5–10 m depth range.

Table 1. The technical specifications of the SS900F SSS.

Technical Specifications	Parameters
Operating Frequency	900 kHz
Maximum Range	75 m @ 900 kHz
Horizontal Beam Width	0.2° @ 900 kHz
Vertical Beam Width	50°
Along Track Resolution	0.07 m @ 20 m;0.17 m @ 50 m;0.26 m @ 75 m;

We used the specialized sonar processing software SonarWiz (v7.09.03) to process the raw data from each collected SSS line. First, the SSS data are imported under the correct coordinate system. Then, the data are filtered, adjusted for gain, and color adjusted to achieve the desired visual effect. When filtering data, a low-pass filter is used to remove background noise and high-frequency interference generated by the sensor itself by selecting the "Filters" function in the SonarWiz menu. For gain adjustment, selecting the "Automatic Gain Control" function of "Gain Control" allows SonarWiz to automatically make gain adjustments to optimize image quality. For color adjustment, select "Color Palette" and use the default color settings to enhance image visualization. Two types of target objects were selected for target detection and classification: the submerged portions of the wind turbine pile foundation and the underwater artificial reef. The processed SSS images, as shown in Figure 2, were cropped to a size of 640 × 640 pixels for each image containing a target object.

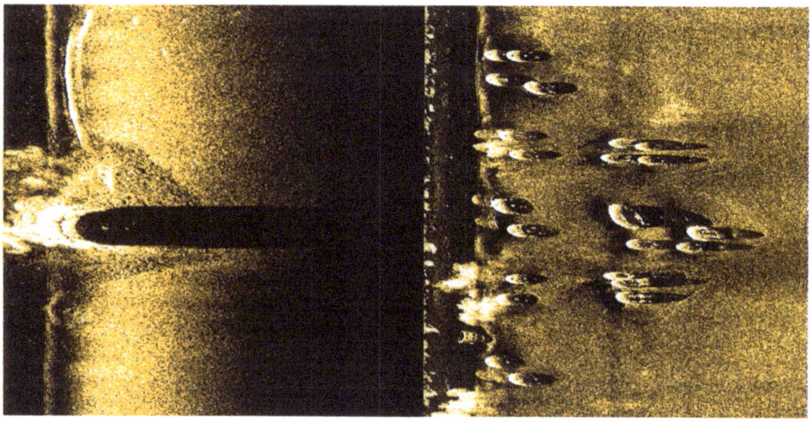

Figure 2. Processed SSS images of the wind turbine pile foundation and underwater artificial reef (the image on the left is of the wind turbine pile foundation, and the image on the right is of the underwater artificial reef).

In this study, the processed SSS images of the wind turbine pile foundation and underwater artificial reef are combined with data from the publicly available SCTD to form a new dataset, named SSSD. SCTD contains three categories of data samples, with a total of 363 samples, including 57 aircraft, 34 humans, and 271 ships. The detection targets in the SSSD are ultimately divided into the following five categories: aircraft, human, ship, foundation, and reefs. The LabelImg image annotation tool was used to annotate the category information of the target objects' locations in the images.

2.2. Data Post-Processing

To avoid longtail effects during model training, we ensured relative balance in the number of images for each target class, when merging and forming the new dataset. The newly constructed dataset specifically addresses the issue of insufficient small sample sizes in the original dataset, resulting in a more balanced distribution of samples across classes. This adjustment leads to a more uniform sample size distribution, which is shown in Figure 3. The original SCTD has only 357 images. After incorporating our experimental data, the total number of images in the SSSD was 682. The dataset was then split into three subsets at random: the training, validation, and testing sets. To ensure that each subset of the training, validation, and test sets accurately reflects the overall class distribution of the dataset, we employed a stratified sampling method. In total, 80% of the images are used for training in the training set, and another 15% of the images are provided to the validation set. The last 5% of the images are provided to the test set. This approach ensures that the class distribution in each subset is consistent with the distribution in the entire dataset, thus avoiding underrepresentation of any class in the subsets. Specifically, during the stratified sampling process, the data samples are randomly assigned to each subset, without considering the specific state or characteristics of the images, but solely based on the class distribution. This ensures adequate representation of each class in all subsets. The training set is used for model training to adjust the model weights by optimization algorithms such as back propagation and gradient descent. The validation set is used to evaluate the model performance and tune the hyperparameters. At the end of each training cycle, the model is evaluated on the validation set to determine whether there is overfitting or underfitting and to select the best hyperparameters. The test set is used for the final evaluation of the model's generalization ability, which are data that the model has not seen during the training and validation phases and is used to simulate the model's performance in real-world applications.

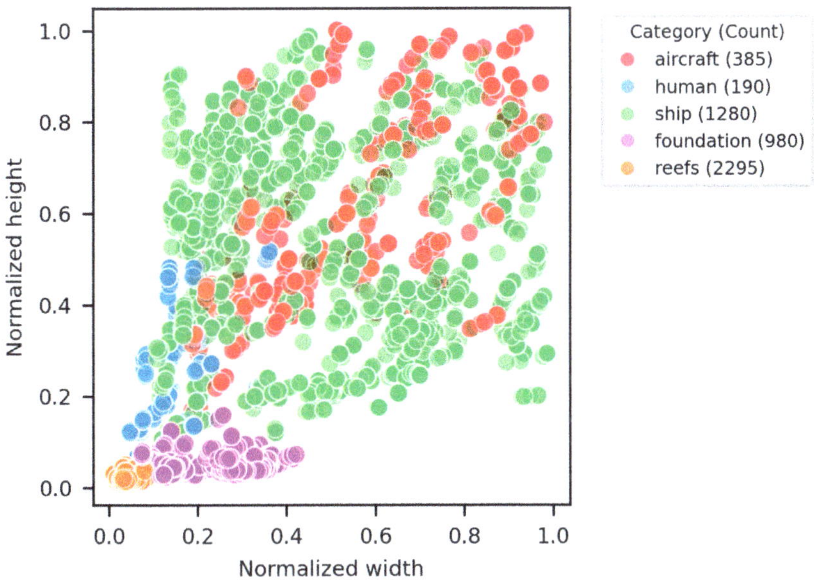

Figure 3. Distribution of the normalized width and heights of the objects in the combined dataset.

Given that the sample size of this training set is still small and may lead to overfitting problems, we performed data augmentation to expand the dataset. By using image augmentation techniques, we increased the diversity and quantity of the training data. This is crucial when dealing with limited datasets, as it allows the model to learn from a wider range of variations and conditions, thereby improving its generalization and accuracy in real-world scenarios. In this study, we applied various data augmentation techniques, such as adding random noise, adjusting image brightness, cutout, random rotation, cropping, translation, and mirroring to synchronously augment the original images and annotations. Each image was augmented by at least one method, and every original image was expanded into five images. We counted the total number of data samples in the training set after data augmentation. Since each image may contain multiple objects, the total number of data samples in the training set is greater than the number of images, with the total number of data samples being 5130. The sample quantity and size distribution are shown in Figure 3. Among them, the dataset includes 385 samples for aircraft, 190 for humans, 1280 for ships, 980 for foundations, and 2295 for reefs.

3. Method

To deploy the sonar target detection algorithm on our self-developed autonomous vessel for real-time detection, this research focuses on achieving model light weighting without compromising performance. The aim is to reduce the cost of model training and deployment, enabling the integration of the model on mobile and embedded devices. To effectively address this challenge, we selected the single-stage YOLOv8 model as the core of the algorithm and enhanced it specifically for underwater target detection.

3.1. The Network Structure

To meet various application objectives, YOLOv8 offers several network scales, including n, s, m, l, and x [25]. By modifying network scaling factors (hyperparameters such as widen_factor, deep_factor, and ratio), several different scales can be created. In particular, the widen_factor modifies the number of channels in each layer to alter the network's width, and the deep_factor modifies the number of repeats of specific structures inside layers to alter the network's depth. At the end of the network, the ratio is used to modify the number of channels in the feature maps [26]. Higher scaling factors can lead to deeper, wider networks that can learn more complicated characteristics, but they also increase the computational load. Considering that the SSS target detection task involves fewer categories, a smaller-scale network is sufficient to meet accuracy requirements while achieving faster speeds and lower memory access costs. Therefore, the model is improved based on YOLOv8n, by setting the hyperparameters widen_factor to 0.25, deep_factor to 0.33, and ratio to 2.0.

3.2. Improved Algorithm

Due to the limitation of the hydroacoustic communication bandwidth, it is relatively difficult to transmit large-volume data over long distances between the USV and the shore base station. Thus, the USV needs to independently complete the data processing and target detection tasks during operation and transmit the real-time detection results to the shore operator. In order to achieve efficient underwater target detection on the USV platform, it is necessary to lighten the detection model to improve the detection speed and minimize the consumption of computational resources while ensuring high detection accuracy. Although the YOLOv8 algorithm is already highly efficient, its performance for low-resolution, long-range SSS target recognition is limited by its feature extraction capability [27]. Therefore, it is necessary to design lightweight detection models that can efficiently extract structural features. This will ensure that the models can accurately and quickly complete target detection tasks with limited computational resources, meeting the needs of AUVs to independently perform underwater detection.

Based on these considerations, we improved the original YOLOv8 network, as illustrated in Figure 4. First, to address the complexity of the convolution operations in the original C2f module and the repetitive design of the Bottleneck module, we introduced the partial convolution(PConv) [28] structure from FasterNet, which reduces computation and memory access by processing only part of the input channels. Second, we integrated an adaptive scale spatial fusion (ASSF) module into the neck of the model, which preserves the structural features of the targets to the greatest extent, addressing the problem of information loss and degradation during the intermediate propagation of SSS images, thereby improving detection accuracy. Finally, we incorporated an improved multi-head self-attention (MHSA) attention mechanism into the detection head. This approach not only fully leverages contextual information to enhance detection precision, but also ensures detection efficiency by using an attention mechanism without an FFN layer.

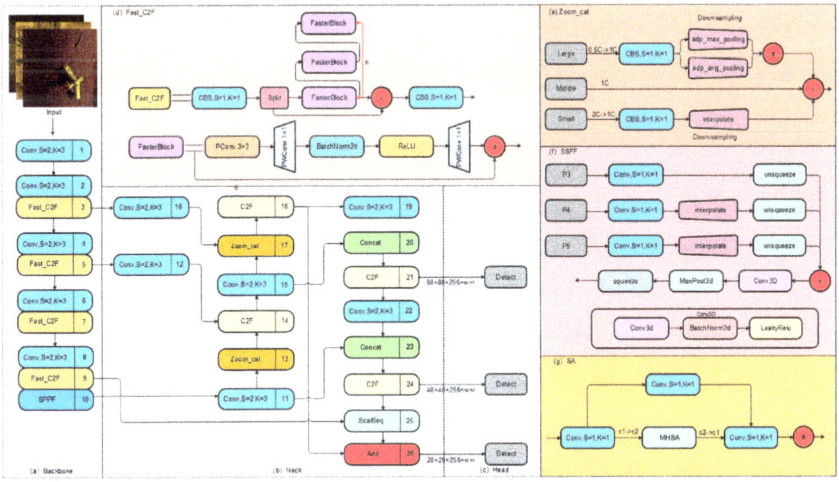

Figure 4. The network architecture diagram of SS-YOLO ((**a–c**) represent the Backbone, Neck, and Head parts of the YOLO network, respectively. (**d–g**) represent the proposed improved module structures within the overall network).

3.2.1. Fast-C2f

The C2f module is a key component of the YOLOv8 network backbone. Through operations such as feature transformation, branch processing, and feature fusion, it extracts and transforms the input data's features, generating output with stronger representational ability. This helps to improve the performance and representational capabilities of the network, allowing it to be better adapted to complex data tasks. These operations rely on a combination of multiple Bottleneck layers and complex convolutional operations, although this helps to improve the feature extraction capability of the network, allowing it to better adapt to complex data tasks, but they come at the cost of significantly increasing the number of parameters and computational complexity, leading to a slower inference time and higher computational overhead. This becomes especially problematic in contexts where real-time detection is critical, such as when deploying YOLOv8 on devices with limited resources, like edge devices or mobile platforms. Therefore, inspired by the FasterNet network design pattern, we introduced the partial convolution (PConv) and the pointwise convolution (PWConv) convolution modules to build a new FasterBlock module [28]. The new FasterBlock module is embedded into the original C2f structure of YOLOv8, replacing the original Bottleneck module and forming the new Fast-C2f structure. This approach reduces the number of parameters and speeds up the computation by reducing unnecessary convolution operations. Figure 4d depicts the Fast-C2f structure.

As shown in Figure 4d, the new PConv layer and PWConv layer act as primary operators that form the new FasterBlock module. Specifically, the PConv layer is followed by two PWConv layers to fuse information from different channels. The first PWConv layer fuses all channel functions to double the number of channels, while the second PWConv layer restores the number of channels to the original. Between these two PWConv layers, batch normalization and activation layers are applied, along with shortcut connections for efficiently reusing the input feature maps.

PConv optimizes memory access costs and saves computational resources by reducing feature map redundancy. Due to the high degree of similarity between feature maps across channels, PConv only processes a subset of the input channels using traditional spatial feature extraction, leaving the other channels unprocessed. For continuous or regular memory access, let the total number of channels be c. The first or last continuous c_p channels are selected to represent the entire feature map for spatial feature extraction. The design of PConv and its combination with PWConv are shown in Figure 5.

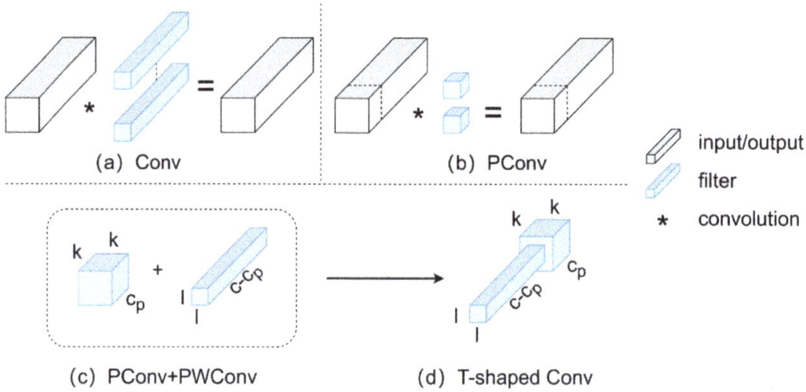

Figure 5. The convolutional structure design of PConv and the combination of PConv and PWConv form a T-shaped convolutional structure.

When the input and output feature map channels are identical, PConv's FLOPs are

$$h \times w \times k^2 \times c_p^2 \tag{1}$$

The memory storage of PConv is

$$h \times w \times 2c_p + k^2 \times c_p^2 \approx h \times w \times 2c_p \tag{2}$$

In Equation (1), when the ratio $r = c_p/c = 1/4$, the FLOPs of PConv is only $1/16$ of regular Conv, and the memory storage of PConv is only $1/4$ of regular Conv. PConv is followed by PWConv, and at this point, the effective receptive field on the input feature map forms a T-shaped Conv structure [29]. Compared to directly implementing a T-shaped convolution structure, this decomposed convolution fully exploits the redundancy between filters, further reducing computational demands. The FLOPs of the combined T-shaped Conv are calculated as follows, making full use of the information from the remaining channels.

$$h \times w \times \left(k^2 \times c \times c_p + c \times (c - c_p)\right) \tag{3}$$

3.2.2. Adaptive Scale Spatial Fusion

According to scale space theory, an image's scale axis—which denotes the range of possible scales for an object, is used to build the scale space. Rather than simply modifying the image size, images of different scales are generated based on Gaussian filtering with varying degrees of blurring of the original image. As a result, the larger the scale value, the blurrier the image generated [30]. The target's structural characteristics in the image, however, remain unchanged despite the scale shift. Due to the complexity of the underwater environment, the acoustic signal propagates in the water with problems such as deformation and noise pollution, which leads to problems such as low image illumination, blurred details, and feature loss, which will increase the difficulty of target detection.

Therefore, making full use of the invariance of the target's structural features and fusing image features of different scales is particularly important to improve detection accuracy.

In low-resolution SSS images, although blurry targets may lose some detail, the structural characteristics of the targets remain unchanged. However, when using the original classical FPN for multi-scale feature fusion, the high-level features, during interaction and propagation with the low-level features, often experience information loss or degradation after passing through multiple intermediate scales, reducing the effectiveness of the feature fusion between non-adjacent layers [31,32].

Therefore, modifying the feature pyramid structure to fully utilize the high-level features with low resolution and high semantic information is crucial for target recognition in low-resolution SSS images. Drawing on the design concepts of the feature fusion module by Kang et al. [33], we introduced an improved adaptive scale spatial fusion module (ASSF). This module consists of two parts: the Triple Feature Encoding module (TFE) and the Scale Sequence Feature Fusion module (SSFF).

Figure 4e shows the structure of the TFE module. First, convolution operation is used to adjust the number of feature channels, making the channel counts of the three different-sized feature maps the same. Let the original middle-size channel count be 1C; after convolution, the channel count of the large-size feature map is adjusted from the original 0.5C to 1C, and the channel count of the small-size feature map is adjusted from the original 2C to 1C.

Then, the large-size and small-size feature maps were downsampled and upsampled, respectively. For downsampling, a hybrid structure consisting of maximum pooling and average pooling is employed, which aids in reducing the number of parameters and computational load on the network. For feature maps of a small size, the nearest neighbor interpolation method is used for upsampling. The upsampling method can enhance low-resolution SSS images to high resolution, which allows the model to better learn and predict complex features in the data. Finally, the three feature maps of large, middle, and small sizes with the same dimensions are spliced in the channel dimension; the splicing method is as follows:

$$F_{TFE} = Concat(F_l, F_m, F_s) \tag{4}$$

And the feature maps of the TFE module output are indicated by F_{TFE}. The symbols F_l, F_m, and F_s stand for large, middle, and small feature maps, in that order. Concatenation of F_l, F_m, and F_s yields F_{TFE}. F_{TFE} has three times the channel number of F_m and the same spatial dimensions.

The original YOLOv8 structure predicts image content by constructing multi-scale feature maps, creating three scales: P3, P4, and P5. However, simply using summation or concatenation methods to fuse pyramid features cannot effectively leverage the correlations between feature maps of different scales. Therefore, to better merge the multi-scale feature maps, the Scale Sequence Feature Fusion module (SSFF) is proposed.

As shown in Figure 4f, the feature maps P3, P4, and P5 are first convolved with a series of Gaussian kernels of increasing standard deviation [34,35], so that the number of channels across the three feature maps is unified, as follows:

$$F_\sigma(i,j) = \sum_\mu \sum_\nu f(i-\mu, i-\nu) \times G_\sigma(\mu,\nu) \tag{5}$$

$$G_\sigma(x,y) = \frac{1}{2\pi\sigma^2} e^{-(x^2+y^2)/2\sigma^2} \tag{6}$$

Where f represents a two-dimensional (2D) feature map and F_σ is generated by smoothing with a series of convolutions using a 2D Gaussian filter with increasing standard deviation σ. G_σ is the 2D Gaussian filter that is used.

Next, the closest neighbor interpolation approach is employed to pair P4, P5, and P3 to the same resolution as P3 because the output feature maps from the above Gaussian smoothing have different resolutions. Each feature map is given the level dimension using the unsqueeze function, converting it from a 3D tensor [height, width, channel] to a 4D tensor [depth, height, width, channel] in order to provide a general view. P3 and P4, as well as P5 feature maps, are finally concatenated.

The 3D convolution block, which consists of 3D convolution, 3D batch normalization, and the leaky reLU activation function, receives the concatenated general view as an input. A $3 \times 3 \times 3$ (depth, height, width) kernel size with suitable padding and a stride of 1 is used in the 3D convolution procedure. Spatial feature information of the SSS images can be efficiently used by adding channels to convolution to extract the scale sequence features of general views. Thus, the deformed SSS targets can be better recognized for identification. Compared with the method of adding attention, this method of directly performing convolution operations to fuse spatial scale features is more lightweight and faster in calculation.

3.2.3. DetectSA

In the traditional YOLOv8 network structure, the design of the detection head is relatively complicated, and its number of parameters occupies almost half of the number of YOLOv8 parameters. This is because YOLOv8 uses a decoupling head to achieve target identification. However, since underwater SSS target detection task only requires the identification of fewer types of targets, a more complicated detection head is not required to guarantee the multiclassification target detection task. Therefore, we restructured the detection head of YOLOv8, abandoned the original means of target detection through multiple convolutional operations, and improved the real-time performance of the model by introducing a lightweight self-attention mechanism while ensuring the detection accuracy.

In order to help the neural network understand which locations and material require more attention, attention methods are frequently used in object detection. In the area of natural language processing, the transformer self-attention mechanism has acquired a lot of traction and shown competitive outcomes [13,36,37]. In fact, when dealing with image-related tasks, each pixel can be viewed as a three-dimensional vector, where the number of image channels represents the dimensions. Therefore, an image can be considered as a collection of vectors fed into the model. When self-attention mechanisms are introduced into the model, the model can independently determine the shape and type of the receptive field, making it more effective in capturing critical information.

However, in traditional attention mechanisms, since the relationship between each element and all other elements needs to be calculated, it results in a significant increase in computational cost and memory requirements. Therefore, we introduced a self-attention mechanism without the fully connected FFN to capture contextual information [38]. As shown in Figure 6, we employed 1×1 convolutions both before and after the (MHSA) [39] to reduce and then increase the dimensions, thereby reducing the memory consumption of the self-attention layer.

Specifically, the 1×1 convolution layer before the MHSA is used to reduce the feature dimension from c1 to c2, thereby reducing the computational load of the attention mechanism. MHSA is employed to capture long-range dependencies within the features, enhancing the ability to detect targets. The 1×1 convolution layer after MHSA is used to restore the original feature dimension by increasing the feature dimension from c2 back to c1. Additionally, a 1×1 convolution is added as a bypass to add the output of the self-attention module to the initial input, forming a residual connection [40].

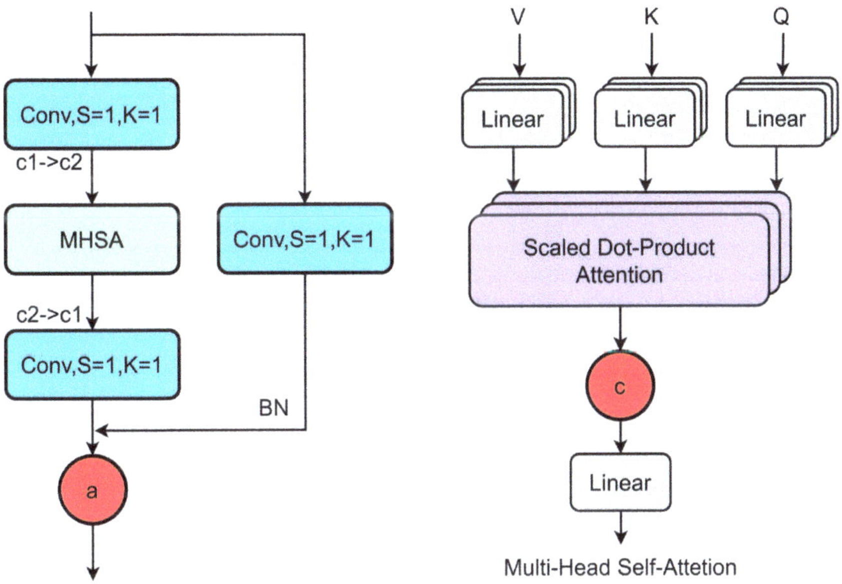

Figure 6. Memory-efficient self-attention mechanism.

4. Experiments and Analysis

In order to verify the superiority of our proposed algorithm and the completeness of the dataset, we first conducted ablation experiments on the SSSD for the different improvements, and then analyzed the comparative experiments for the different models on this dataset. The ablation experiments aim to confirm the effectiveness of our model improvement through qualitative analysis; while the comparison experiments aim to show the superiority of our proposed model in terms of both real-time and accuracy by comparing our proposed SS-YOLO model with other target detection models.

4.1. Evaluation Criteria

We employ a variety of standardized model detection evaluation metrics to compare the performance of various detection models, such as precision (P), recall rate (R), the average precision (AP), and the mean average precision (mAP). Below are the precise definitions and calculations for these indicators:

(1) Precision is used to assess the model's ability to accurately identify a target, and it represents the proportion of all detections that are judged by the model to be positive samples that are actually correctly identified as targets. In other words, the precision rate reflects how many of the results predicted to be in the positive category are correct when the model recognizes a target. A higher precision rate means that the model produces fewer false positives, i.e., fewer non-target instances are incorrectly identified as targets. Recall, on the other hand, is used to assess the model's ability to recognize all target classes, and denotes the proportion of all actual target samples that the model is able to correctly identify. Recall emphasizes the model's ability to provide complete coverage of targets, i.e., whether it is able to identify all actually existing targets as much as possible. A higher recall means that the model misses less and is able to find more real targets.

Thus, precision and recall focus on the trade-off between the accuracy and completeness of the model's detection results, respectively. A high precision rate indicates that the model more reliably avoids false positives, while a high recall rate indicates that the model is able to capture the target more comprehensively.

$$P = \frac{TP}{TP + FP} \tag{7}$$

$$R = \frac{TP}{TP + FN} \tag{8}$$

The real positives that are accurately predicted are known as true positives (*TP*). Actual negatives that are accurately anticipated negatives are known as true negatives (*TN*). Actual negatives that were mistakenly anticipated to be positives are known as false positives (*FP*). Actual positives that were mispredicted as negatives are known as false negatives (*FN*).

(2) The area under the precision-recall curve for a particular target category is typically used to determine the average detection precision, or *AP* (Average Precision). Both the precision and recall of the detection findings are taken into consideration when evaluating the model's detection performance in that category. On the other hand, the model's overall detection performance over the whole dataset is assessed using *mAP* (Mean Average Precision), which is the average of the APs of several target categories. The model's recognition ability across many categories can be fully reflected by *mAP*, and a larger *mAP* value means that the model performs better in each area. Therefore, *AP* is used to assess the detection ability of a single category, while *mAP* is more global and can comprehensively measure the performance of the model in a multi-category target detection task.

$$AP = \int_0^1 P(r)d_r \tag{9}$$

$$mAP = \sum_{i=1}^{N} \frac{AP_i}{N} \tag{10}$$

N is the number of detected categories.

4.2. Experiment Setup

The deep learning model training, validation and testing were performed using a computer system using the Windows 10 operating system, with two GPUs (NVIDIA Tesla T4 16G), running CUDA version 12.1, python version 3.10.3, deep learning framwork, and pytorch version 2.1.2.

The training of the deep learning model starts from scratch and the hyperparameters involved in the training process have been given in Table 2.

Table 2. Hyperparameters of model training.

Parameters	Configuration
image size	(960, 960)
batch size	16
epochs	200
initial learning rate	0.01
final learning rate	0.1
weight decay	0.0005
SGD momentum	0.937

4.3. SSSD Result

In order to illustrate the crucial significance that various alterations play, we first conduct ablation experiments on the SSSD. Figure 7 shows the confusion matrix for the experimental results of the various models. In the confusion matrix, the true category is represented by each column, while the predicted category to which the data belong is represented by each row. By analyzing the elements of the confusion matrix, it is possible to determine in which categories the model is prone to misclassification. From Figure 7a, it can be seen that the original YOLOv8 model demonstrates relatively low classification performance in the "aircraft" and "reefs" categories, with accuracy rates of 0.57 and 0.83, respectively. Particularly for these two categories, a high proportion of background is misclassified as targets. This indicates significant deficiencies in the model's ability to distinguish these categories, especially in separating them from complex backgrounds. Figure 7b–d, respectively, demonstrate the improvements after introducing different structures into the original model. With the optimization of the model, the classification accuracy for the "aircraft" and "reefs" categories shows significant improvement. In the SS-YOLO model shown in Figure 7d, the classification coefficient for "aircraft" increased by 0.14, while "reefs" improved by 0.07. Furthermore, the proportion of background being misclassified as "aircraft" and "reefs" dropped significantly. This indicates that the robustness of the improved model against complex backgrounds has been significantly enhanced.

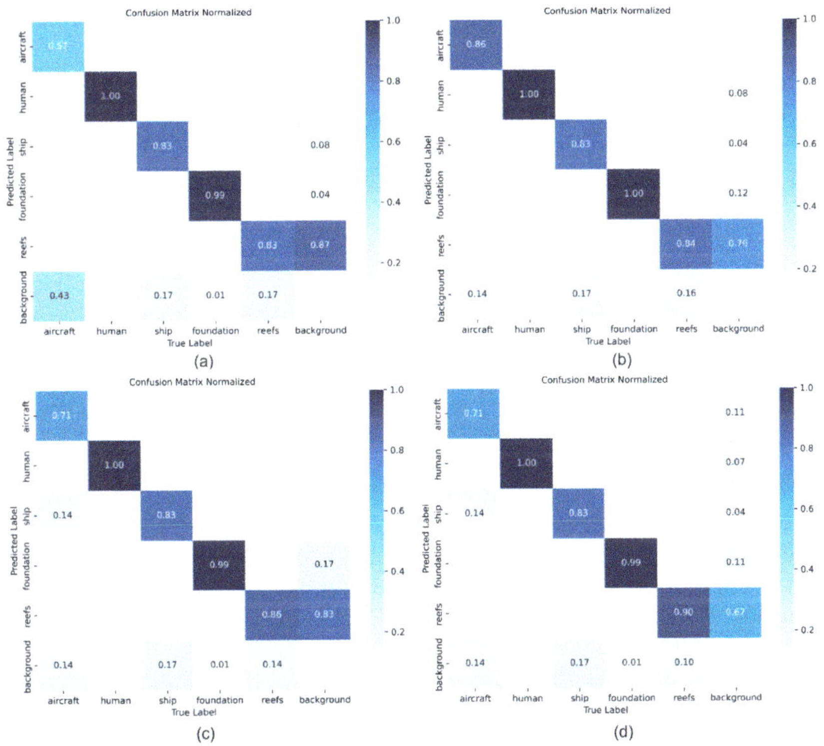

Figure 7. The confusion matrix obtained from ablation experiments based on the SSSD. (**a**) YOLOv8 (**b**) YOLOv8 + Fast-C2f (**c**) YOLOv8 + Fast-C2f + ASSF (**d**) YOLOv8 + Fast-C2f + ASSF + Detect.

Through the above analysis, it is evident that the MHSA attention mechanism enhances the model's ability to extract global features, enabling it to better capture the global characteristics of targets such as "aircraft" and "reefs" and avoid being misled by locally complex backgrounds. The ASSF module effectively integrates features at different scales, addressing the shortcomings of the original model in multi-scale object detection, particularly significantly improving the detection performance for small targets (such as "reefs"). Figure 7d shows a significant reduction in the proportion of background misclassified as targets, indicating that the improved SS-YOLO model exhibits higher robustness in distinguishing between target and non-target regions, thereby improving detection reliability. The SS-YOLO structure demonstrates greater consistency across multiple categories, reducing the mutual confusion between "reefs" and "foundation", which suggests that the model has clearer decision boundaries between different categories.

Furthermore, Figure 8 shows the changes in the loss functions, precision, recall, and map metrics for the SS-YOLO model on the training set and the validation set as the number of training epochs increases. According to the loss function charts, all three loss functions on the training and validation sets are covered to a stable state. Recall shows a consistent rising trend and coverages to about 0.85, whereas precision displays notable oscillations at first and stabilizes at 0.82 after 60 epochs. The model's mAP value eventually stabilizes around 0.93.

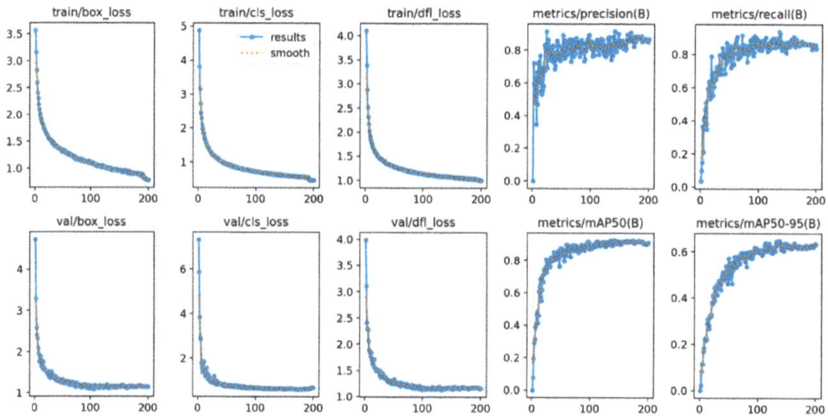

Figure 8. The result on our SSSD. The results and variations in the loss function, precision, recall and mapping evaluation metrics are included in the result images.

4.3.1. Ablation Experiments

We conducted a series of ablation experiments, making sure that the training settings and hyperparameters remained constant, to assess the contribution of each modification to the network. The YOLOv8n model served as the baseline in these studies, and the suggested improvements were implemented one after the other for comparison.

We evaluated all three enhanced detection networks on the same test dataset, with the detection results presented in Figure 9. For large targets with distinctive features, in Figure 9a, the YOLOv8 model in the first row exhibits a significant under-detection problem. This is likely attributed to the model's insufficient ability to adapt to complex backgrounds. While in Figure 9c, the YOLOv8 model in the first row has a significantly larger detection range and a duplicate detection problem, which suggests that the target has been misrecognized as multiple independent targets, resulting in the overlapping detection frames. For the small targets in Figure 9d, both the YOLOv8 model in the first row and the model integrating the Fast-C2f structure in the second row show lower confidence and more serious leakage detection, which is particularly underperforming in small object detection. These analyses show that the original YOLOv8 model has significant performance bottlenecks when processing SSS images in complex backgrounds, especially in the small target detection task, and is difficult to meet the demand for high-precision detection.

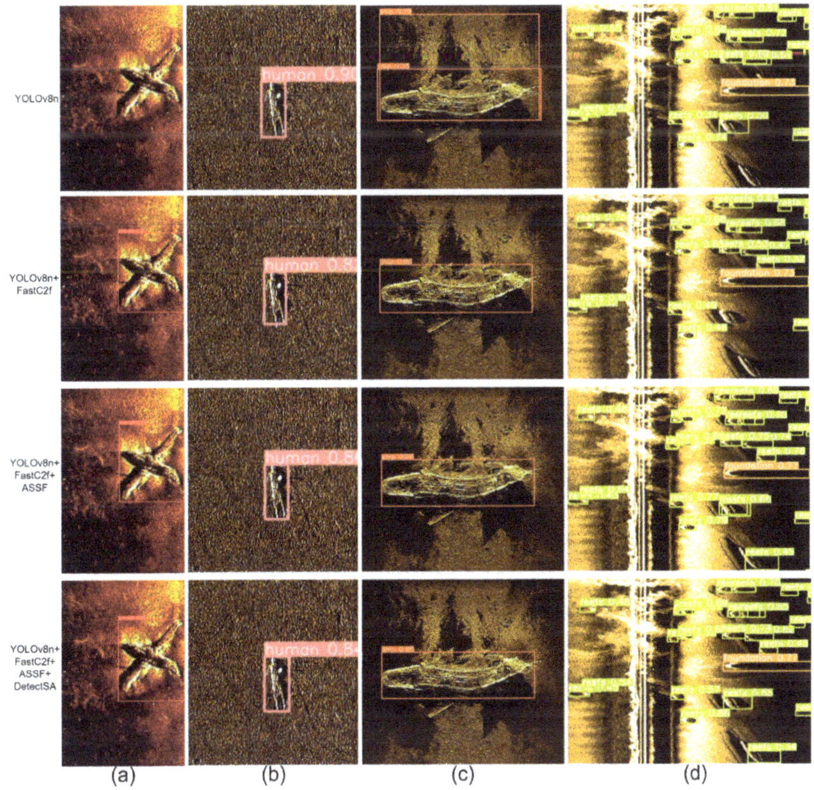

Figure 9. Visual analysis of different effects obtained from ablation experiments based on the SSSD. (**a**) YOLOv8 (**b**) YOLOv8 + Fast-C2f (**c**) YOLOv8 + Fast-C2f + ASSF (**d**) YOLOv8 + Fast-C2f + ASSF + Detect.

Based on the above, in order to improve the detection accuracy of the model, the ASSF structure is introduced, and the detection effect is shown in the third row in Figure 9. The detection effect in the third row shows a significant improvement compared to the model in the first and second rows. By introducing the ASSF structure, the model is able to better extract target features from information at different scales, which improves the recognition of multi-scale targets and reduces the effect of background interference on target detection. Especially for smaller targets, the model is able to capture the detailed information of the target area more accurately, which significantly improves the recognition confidence and reduces the leakage detection problem.

As shown in the fourth row of Figure 9, despite the reduction in the number of parameters in the detection head, the model's ability to perceive the target area remains unaffected. In fact, the attentional mechanism not only optimizes the recognition of small targets but also suppresses the interference from background noise. Compared with the first and second rows, the model in the fourth row can locate the target more accurately, and it maintains a similar detection effect with the third row while being lightweight. Especially in the complex background, the boundary of the target becomes clearer, and the missed detection is significantly reduced. The introduction of the attention mechanism allows the model to further improve the detection accuracy by focusing on key regions without increasing the computational overhead. At the same time, the model is able to adaptively allocate the attention so that even in low resolution images, the model can still effectively extract the target features, ensuring better detection effect and accuracy.

The quantitative evaluation scores of the ablation experimental outcomes are detailed in Table 3. Based on the data presented in Table 3, integrating the Fast-C2f structure into the YOLOv8n model reduces its parameters from 3.15 million to 2.45 million, and the GFLOPs from 8.9 to 7.1. This reduction in both parameters and computational complexity indicates a more lightweight model, which is crucial for deployment on resource-constrained edge devices. The decrease in parameters directly contributes to lower memory usage and faster inference times, which is essential for real-time target detection. This improvement in the model's architecture enables it to better meet the demands of real-time detection without sacrificing accuracy. Considering the impact of parameter reduction on model accuracy, the ASSF structure and DetectSA structure are addicted, respectively. The addition of the ASSF structure further improves the mAP 0.5 and mAP 0.5:0.95 scores of the YOLOv8n model by 1.8% and 2.5%. In summary, the proposed enhanced model exhibits a reduction in parameters by 0.49 million and a decrease in GFLOPs by 2.5, compared to the original model. Additionally, it achieves notable performance improvements, with an increase of 4.4% in mAP@0.5 and 3% in mAP@0.5:0.95, indicating a significant boost in detection accuracy.

Table 3. Quantitative results of the ablation experiments (Bolded and underlined data are the best data results for each parameter).

Method	Params	FLOPs	mAP@0.5	mAP@[0.5, 0.95]
YOLOv8n	3.15 M	8.9 G	0.88	0.617
YOLOv8n + Fast-C2f	**2.45 M**	7.1 G	0.874	0.634
YOLOv8n + Fast-C2f + ASSF	2.69 M	7.6 G	0.898	0.642
YOLOv8n + Fast-C2f + ASSF + DetectSA	2.55 M	**6.4 G**	**0.924**	**0.647**

4.3.2. Contrast Experiments

We tested two popular target detection algorithms as well as other algorithms in the YOLO family separately on the SSSD. The detection results are then compared with our proposed SS-YOLO algorithm in order to evaluate the detection performance of our algorithm in more detail. Table 4 provides a detailed presentation of the comparative experiments' quantitative data.

Table 4. Quantitative results of the contrast experiments (Bolded and underlined data are the best data results for each parameter).

Method	Params	FLOPs	P	R	mAP@0.5	mAP@[0.5,0.95]
SSD [41]	24.98 M	137.94 G	0.608	0.841	0.827	0.419
Faster R-CNN [42]	41.22 M	91.1 G	0.836	0.915	0.88	0.534
YOLOv5s [43]	7.2 M	16.5 G	**0.905**	0.902	0.896	0.628
YOLOv7t [44]	6.2 M	13.9 G	0.896	0.787	0.868	0.560
YOLOv8n [45]	3.15 M	8.9 G	0.877	0.835	0.88	0.617
YOLOv9s [46]	7.1 M	26.4 G	0.889	**0.921**	0.933	**0.698**
SS-YOLO	**2.55 M**	**6.4 G**	0.821	0.857	**0.924**	0.647

It can be observed that the proposed SS-YOLO network demonstrates superior performance across multiple key metrics. Compared with both the traditional single-stage object detection algorithm SSD [41] and the two-stage detection algorithm Faster R-CNN [42], SS-YOLO shows significantly lower parameters and higher accuracy. SS-YOLO's Params and FLOPs are only 2.55 and 6.4, respectively, while SSD's Params and FLOPs are 24.98

and 137.94. As a single-stage object detection algorithm, it is clear that SSD is slower than our proposed network. Furthermore, in terms of accuracy, SS-YOLO not only offers faster detection speed but also exhibits superior detection precision. SSD's mAP is only 0.827, and the lower accuracy is due to its simpler architecture compared to the two-stage detection algorithm Faster R-CNN and our SS YOLO, which sacrifices detection precision for higher speed [47]. Faster R-CNN's mAP is 0.88, slightly higher than SSD, but still significantly lower than our proposed network. Additionally, from the perspective of parameter count and inference time, the computationally intensive Faster R-CNN does not meet the requirements for deployment on edge devices.

Another point worth noting is that compared to other algorithms in the YOLO series, our network also demonstrates significant performance advantages. SS-YOLO not only surpasses other models in detection speed but also maintains relatively high accuracy. Compared to YOLOv5s [43] and YOLOv7t [44], SS-YOLO has significantly lower parameters and FLOPs. YOLOv5s has 7.2 M Params and 16.5G FLOPs, while YOLOv7t has 6.2 M Params and 13.9G FLOPs, both more than double those of SS-YOLO. In terms of mAP, SS-YOLO maintains a mAP of 0.924, while YOLOv5s and YOLOv7t have mAP values of 0.896 and 0.868, respectively, both noticeably lower than the improved SS-YOLO. Although the YOLOv9s [46] network achieves slightly higher accuracy with a mAP of 0.933, its Params and FLOPs are 7.1 M and 26.4 G, respectively, which indicates it sacrifices speed to gain higher accuracy. Overall, SS-YOLO achieves a balance between accuracy and speed, offering clear advantages over other models.

5. Conclusions

This paper introduces a lightweight network designed for deployment on unmanned vessels to detect objects in SSS images. To achieve network lightweighting, we first reduced the complexity of the convolution process by combining PConv and PWConv to form a new FasterBlock module, replacing the original convolution block in C2f. This achieved a balance between detection accuracy and speed. To fully utilize the scale information of objects and address the deformation issue of targets in SSS images, we optimized the ASSF structure. After concatenating feature maps of different sizes, we extracted scale-sequential features based on 3D convolutions. To address the high complexity of the original YOLOv8 model's detection head, we incorporated an MHSA attention mechanism without the FFN layer into the detection head. This increased the model's sensitivity to features while reducing the parameters of the detection head, thus improving detection speed. We constructed a new SSSD and conducted both ablation and comparative experiments on it. The results show that our proposed SS-YOLO model achieves an excellent balance between detection accuracy and speed, demonstrating superior performance. In the future, we will focus on the deployment of the model on autonomous vessels to realize the real-time detection of underwater targets by unmanned vessels equipped with SSS equipment.

Author Contributions: Conceptualization, N.Y., Z.W. and G.L.; methodology, N.Y. and S.W.; software, N.Y.; validation, N.Y.; formal analysis, N.Y.; investigation, N.Y.; resources, N.Y.; data curation, Z.W., H.R. and N.Y.; writing—original draft preparation, N.Y.; writing—review and editing, Z.W., G.L., X.Z. and Y.P.; visualization, Y.P.; supervision, Z.W.; project administration, X.Z. and Y.P; funding acquisition, Z.W. All authors have read and agreed to the published version of the manuscript.

Funding: This research was funded by Key R&D Program of Shandong Province, China (No. 2023CXPT054).

Data Availability Statement: Data are available on request due to restrictions, e.g., privacy or ethical. The data presented in this study are available on request from the corresponding author.

Acknowledgments: The first author would like to thank all the students and teachers who contributed to the data collection and processing for this study, and expresses gratitude to the corresponding author, Zhengrong Wei, and Kai Liu of the First Institute of Oceanography of Ministry of Natural Resources for their suggestions provided in revising this manuscript.

Conflicts of Interest: Author Mr. Guoyu Li was employed by the company Qingdao Xiushan Mobile Mapping Co. The remaining authors declare that the research was conducted in the absence of any commercial or financial relationships that could be construed as a potential conflict of interest.

References

1. Li, L.; Li, Y.; Yue, C.; Xu, G.; Wang, H.; Feng, X. Real-Time Underwater Target Detection for AUV Using Side Scan Sonar Images Based on Deep Learning. *Appl. Ocean Res.* **2023**, *138*, 103630. [CrossRef]
2. Grządziel, A. The Impact of Side-Scan Sonar Resolution and Acoustic Shadow Phenomenon on the Quality of Sonar Imagery and Data Interpretation Capabilities. *Remote Sens.* **2023**, *15*, 5599. [CrossRef]
3. Zhou, X.; Zhou, Z.; Wang, M.; Ning, B.; Wang, Y.; Zhu, P. Multi-Level Feature Enhancement Network for Object Detection in Sonar Images. *J. Vis. Commun. Image Represent.* **2024**, *100*, 104147. [CrossRef]
4. Yu, H.; Li, Z.; Li, D.; Shen, T. Bottom Detection Method of Side-Scan Sonar Image for AUV Missions. *Complexity* **2020**, *2020*, 8890410. [CrossRef]
5. Abu, A.; Diamant, R. A Statistically-Based Method for the Detection of Underwater Objects in Sonar Imagery. *IEEE Sens. J.* **2019**, *19*, 6858–6871. [CrossRef]
6. Febriawan, H.K.; Helmholz, P.; Parnum, I.M. Support Vector Machine and Decision Tree Based Classification of Side-Scan Sonar Mosaics Using Textural Features. *Int. Arch. Photogramm. Remote Sens. Spat. Inf. Sci.* **2019**, *XLII-2/W13*, 27–34. [CrossRef]
7. Azimi-Sadjadi, M.R.; Klausner, N.; Kopacz, J. Detection of Underwater Targets Using a Subspace-Based Method with Learning. *IEEE J. Ocean. Eng.* **2017**, *42*, 869–879. [CrossRef]
8. He, J.; Chen, J.; Xu, H.; Ayub, M.S. Small Target Detection Method Based on Low-Rank Sparse Matrix Factorization for Side-Scan Sonar Images. *Remote Sens.* **2023**, *15*, 2054. [CrossRef]
9. Fakiris, E.; Papatheodorou, G.; Geraga, M.; Ferentinos, G. An Automatic Target Detection Algorithm for Swath Sonar Backscatter Imagery, Using Image Texture and Independent Component Analysis. *Remote Sens.* **2016**, *8*, 373. [CrossRef]
10. Zhu, B.; Wang, X.; Chu, Z.; Yang, Y.; Shi, J. Active Learning for Recognition of Shipwreck Target in Side-Scan Sonar Image. *Remote Sens.* **2019**, *11*, 243. [CrossRef]
11. Wang, Z.; Zhang, S.; Huang, W.; Guo, J.; Zeng, L. Sonar Image Target Detection Based on Adaptive Global Feature Enhancement Network. *IEEE Sens. J.* **2022**, *22*, 1509–1530. [CrossRef]
12. Kong, W.; Hong, J.; Jia, M.; Yao, J.; Cong, W.; Hu, H.; Zhang, H. YOLOv3-DPFIN: A Dual-Path Feature Fusion Neural Network for Robust Real-Time Sonar Target Detection. *IEEE Sens. J.* **2020**, *20*, 3745–3756. [CrossRef]
13. Zhang, F.; Zhang, W.; Cheng, C.; Hou, X.; Cao, C. Detection of Small Objects in Side-Scan Sonar Images Using an Enhanced YOLOv7-Based Approach. *J. Mar. Sci. Eng.* **2023**, *11*, 2155. [CrossRef]
14. Wen, X.; Zhang, F. Underwater Target Detection by Side-Scan Sonar Based on Yolov7-Attention. In Proceedings of the 2023 7th Asian Conference on Artificial Intelligence Technology (ACAIT), Quzhou, China, 3–5 November 2023; pp. 1536–1542.
15. Wen, X.; Wang, J.; Cheng, C.; Zhang, F.; Pan, G. Underwater Side-Scan Sonar Target Detection: YOLOv7 Model Combined with Attention Mechanism and Scaling Factor. *Remote Sens.* **2024**, *16*, 2492. [CrossRef]
16. Mittal, P. A Comprehensive Survey of Deep Learning-Based Lightweight Object Detection Models for Edge Devices. *Artif Intell Rev* **2024**, *57*, 242. [CrossRef]
17. Liu, G.; Hu, Y.; Chen, Z.; Guo, J.; Ni, P. Lightweight Object Detection Algorithm for Robots with Improved YOLOv5. *Eng. Appl. Artif. Intell.* **2023**, *123*, 106217. [CrossRef]
18. Huyan, L.; Bai, Y.; Li, Y.; Jiang, D.; Zhang, Y.; Zhou, Q.; Wei, J.; Liu, J.; Zhang, Y.; Cui, T. A Lightweight Object Detection Framework for Remote Sensing Images. *Remote Sens.* **2021**, *13*, 683. [CrossRef]
19. Zhang, Y.; Zhang, T.; Wu, C.; Tao, R. Multi-Scale Spatiotemporal Feature Fusion Network for Video Saliency Prediction. *IEEE Trans. Multimed.* **2024**, *26*, 4183–4193. [CrossRef]
20. Tang, Y.; Wang, L.; Jin, S.; Zhao, J.; Huang, C.; Yu, Y. AUV-Based Side-Scan Sonar Real-Time Method for Underwater-Target Detection. *J. Mar. Sci. Eng.* **2023**, *11*, 690. [CrossRef]
21. Jocher, G.; Qiu, J.; Chaurasia, A. Ultralytics YOLO 2023. Available online: https://github.com/ultralytics/ultralytics (accessed on 17 November 2024).
22. Varghese, R.; Sambath, M. YOLOv8: A Novel Object Detection Algorithm with Enhanced Performance and Robustness. In Proceedings of the 2024 International Conference on Advances in Data Engineering and Intelligent Computing Systems (ADICS), Chennai, India, 18–19 April 2024; pp. 1–6.
23. Hao, C.-Y.; Chen, Y.-C.; Chen, T.-T.; Lai, T.-H.; Chou, T.-Y.; Ning, F.-S.; Chen, M.-H. Synthetic Data-Driven Real-Time Detection Transformer Object Detection in Raining Weather Conditions. *Appl. Sci.* **2024**, *14*, 4910. [CrossRef]
24. Zhang, P.; Tang, J.; Zhong, H.; Ning, M.; Liu, D.; Wu, K. Self-Trained Target Detection of Radar and Sonar Images Using Automatic Deep Learning. *IEEE Trans. Geosci. Remote Sens.* **2022**, *60*, 1–14. [CrossRef]

25. GitHub—Ultralytics/Ultralytics: NEW—YOLOv8 🚀 in PyTorch > ONNX > OpenVINO > CoreML > TFLite. Available online: https://github.com/ultralytics/ultralytics (accessed on 27 September 2024).
26. Liu, Z.; Rasika, D.; Abeyrathna, R.M.; Mulya Sampurno, R.; Massaki Nakaguchi, V.; Ahamed, T. Faster-YOLO-AP: A Lightweight Apple Detection Algorithm Based on Improved YOLOv8 with a New Efficient PDWConv in Orchard. *Comput. Electron. Agric.* **2024**, *223*, 109118. [CrossRef]
27. Wang, H.; Liu, C.; Cai, Y.; Chen, L.; Li, Y. YOLOv8-QSD: An Improved Small Object Detection Algorithm for Autonomous Vehicles Based on YOLOv8. *IEEE Trans. Instrum. Meas.* **2024**, *73*, 1–16. [CrossRef]
28. Chen, J.; Kao, S.; He, H.; Zhuo, W.; Wen, S.; Lee, C.-H.; Chan, S.-H.G. Run, Don't Walk: Chasing Higher FLOPS for Faster Neural Networks 2023. In Proceedings of the IEEE/CVF Conference on Computer Vision and Pattern Recognition 2023, Vancouver, BC, Canada, 17–24 June 2023.
29. Zhang, L.; Li, P.; Liu, X.; Yu, J.; Hu, G.; Yu, T. Dy-GNet: A Lightweight and Efficient 1DCNN-Based Network for Leakage Aperture Identification. *Meas. Sci. Technol.* **2024**, *35*, 056109. [CrossRef]
30. Park, H. J.; Kang, J. W.; Kim, B. G. ssFPN: Scale Sequence (S2) Feature Based Feature Pyramid Network for Object Detection. *Sensors* **2023**, *23*, 4432. [CrossRef] [PubMed]
31. Zhao, Z.; Pan, Y.; Guo, G.; Zhai, Y.; Liu, G. YOLO-AFPN: Marrying YOLO and AFPN for External Damage Detection of Transmission Lines. *IET Gener. Transm. Distrib.* **2024**, *18*, 1935–1946. [CrossRef]
32. Zhou, P.; Chen, J.; Tang, P.; Gan, J.; Zhang, H. A Multi-Scale Fusion Strategy for Side Scan Sonar Image Correction to Improve Low Contrast and Noise Interference. *Remote Sens.* **2024**, *16*, 1752. [CrossRef]
33. Kang, M.; Ting, C.-M.; Ting, F.F.; Phan, R.C.-W. ASF-YOLO: A Novel YOLO Model with Attentional Scale Sequence Fusion for Cell Instance Segmentation. *Image Vis. Comput.* **2024**, *147*, 105057. [CrossRef]
34. Liu, J.; Fan, X.; Jiang, J.; Liu, R.; Luo, Z. Learning a Deep Multi-Scale Feature Ensemble and an Edge-Attention Guidance for Image Fusion. *IEEE Trans. Circuits Syst. Video Technol.* **2022**, *32*, 105–119. [CrossRef]
35. Lindeberg, T. *Scale-Space Theory in Computer Vision*; Springer Science & Business Media: Berlin/Heidelberg, Germany, 2013; ISBN 978-1-4757-6465-9.
36. Vaswani, A.; Shazeer, N.; Parmar, N.; Uszkoreit, J.; Jones, L.; Gomez, A.N.; Kaiser, L.; Polosukhin, I. Attention Is All You Need. *arXiv* **2023**, arXiv:1706.03762.
37. Yang, Z.; Yang, D.; Dyer, C.; He, X.; Smola, A.; Hovy, E. Hierarchical Attention Networks for Document Classification. In Proceedings of the 2016 Conference of the North American Chapter of the Association for Computational Linguistics: Human Language Technologies, San Diego, CA, USA, 12–17 June 2016; pp. 1480–1489.
38. Zhang, H.; Goodfellow, I.; Metaxas, D.; Odena, A. Self-Attention Generative Adversarial Networks. In Proceedings of the 36th International Conference on Machine Learning, PMLR, Long Beach, CA, USA, 24 May 2019; pp. 7354–7363.
39. Cordonnier, J.-B.; Loukas, A.; Jaggi, M. On the Relationship Between Self-Attention and Convolutional Layers. Available online: https://arxiv.org/abs/1911.03584v2 (accessed on 7 September 2024).
40. Yu, H.; Wan, C.; Liu, M.; Chen, D.; Xiao, B.; Dai, X. Real-Time Image Segmentation via Hybrid Convolutional-Transformer Architecture Search. *arXiv* **2024**, arXiv:2403.10413.
41. Liu, W.; Anguelov, D.; Erhan, D.; Szegedy, C.; Reed, S.; Fu, C.-Y.; Berg, A.C. SSD: Single Shot MultiBox Detector. In Proceedings of the Computer Vision—ECCV 2016, Amsterdam, The Netherlands, 11–14 October 2016; Leibe, B., Matas, J., Sebe, N., Welling, M., Eds.; Springer International Publishing: Cham, Switzerland, 2016; pp. 21–37.
42. Ren, S.; He, K.; Girshick, R.; Sun, J. Faster R-CNN: Towards Real-Time Object Detection with Region Proposal Networks. *IEEE Trans. Pattern Anal. Mach. Intell.* **2016**, *39*, 1137–1149. [CrossRef] [PubMed]
43. Jocher, G. YOLOv5 by Ultralytics 2020. Available online: https://github.com/ultralytics/yolov5 (accessed on 17 November 2024).
44. Wang, C.-Y.; Bochkovskiy, A.; Liao, H.-Y.M. YOLOv7: Trainable Bag-of-Freebies Sets New State-of-the-Art for Real-Time Object Detectors. In Proceedings of the IEEE/CVF Conference on Computer Vision and Pattern Recognition, Vancouver, BC, Canada, 17–24 June 2023; pp. 7464–7475.
45. GitHub—Ultralytics/Ultralytics: Ultralytics YOLO11 🚀. Available online: https://github.com/ultralytics/ultralytics/tree/main (accessed on 17 November 2024).
46. Wang, C.-Y.; Yeh, I.-H.; Liao, H.-Y.M. YOLOv9: Learning What You Want to Learn Using Programmable Gradient Information. *arXiv* **2024**, arXiv:2402.13616.
47. Qu, S.; Cui, C.; Duan, J.; Lu, Y.; Pang, Z. Underwater Small Target Detection under YOLOv8-LA Model. *Sci. Rep.* **2024**, *14*, 16108. [CrossRef] [PubMed]

Journal of
Marine Science and Engineering

MDPI

Article

AquaYOLO: Enhancing YOLOv8 for Accurate Underwater Object Detection for Sonar Images

Yanyang Lu [1], Jingjing Zhang [1], Qinglang Chen [1], Chengjun Xu [1], Muhammad Irfan [2] and Zhe Chen [3,4,*]

1 Hangzhou Applied Acoustics Research Institute, Hangzhou 310023, China; cqjndys@sina.com (Y.L.); zjj03514@163.com (J.Z.); cql2024@163.com (Q.C.); 17730221837@163.com (C.X.)
2 School of Software, Northwestern Polytechnical University, Xi'an 710129, China; mirfan@mail.nwpu.edu.cn
3 School of Information and Communication, Guilin University of Electronic Technology, Guilin 541004, China
4 Cognitive Radio and Information Processing Key Laboratory Authorized by China's Ministry of Education Foundation, Guilin University of Electronic Technology, Guilin 541004, China
* Correspondence: chenzhe@mail.nwpu.edu.cn

Abstract: Object detection in underwater environments presents significant challenges due to the inherent limitations of sonar imaging, such as noise, low resolution, lack of texture, and color information. This paper introduces AquaYOLO, an enhanced YOLOv8 version specifically designed to improve object detection accuracy in underwater sonar images. AquaYOLO replaces traditional convolutional layers with a residual block in the backbone network to enhance feature extraction. In addition, we introduce Dynamic Selection Aggregation Module (DSAM) and Context-Aware Feature Selection (CAFS) in the neck network. These modifications allow AquaYOLO to capture intricate details better and reduce feature redundancy, leading to improved performance in underwater object detection tasks. The model is evaluated on two standard underwater sonar datasets, UATD and Marine Debris, demonstrating superior accuracy and robustness compared to baseline models.

Keywords: underwater object detection; underwater sonar images; tracking; underwater classification; marine detection; marine classification

1. Introduction

Conventional intelligent systems, designed for tasks like object detection, recognition, and segmentation, heavily rely on high-resolution data from optical sensors. Inspired by the human visual system, these sensors have undergone significant technological advancements, resulting in an abundance of high-quality optical data. Landmark datasets such as ImageNet [1], COCO [2], and PASCAL VOC [3] have been instrumental in advancing these systems, driving substantial improvements in their performance and capabilities in recent years.

Underwater environments present significant challenges for optical sensors due to the absorption and scattering of light in water, which severely degrades image quality and limits their effectiveness. To overcome these limitations, acoustic sensors, particularly sonar systems, have become the preferred solution for underwater data collection. Unlike optical sensors, sonar systems perform well in low-visibility and challenging underwater conditions, delivering superior performance and dependable data [4]. This makes sonar systems indispensable for underwater applications where conventional optical sensors are inadequate [5].

Interpreting acoustic images is challenging due to the harsh underwater environment, significant noise, reflections, interference, attenuation, reverberations, scattering, and shadows [6]. These factors severely degrade image quality, complicating the interpretation process. Additionally, sonar images lack texture and color information, leading to lower resolution and reduced detail compared to optical images. In addition, the inherent limitations of sonar technology, such as restricted range and resolution, exacerbate these challenges, making underwater object detection and recognition particularly difficult. Consequently, developing robust algorithms and techniques for underwater object detection from sonar images remains an active area of research. Moreover, factors such as depth, water transparency, operating frequency, and sensor parameters introduce variations in resolution and scale across underwater sonar images. These complexities are further amplified because traditional object detection methods rely heavily on texture and color cues.

The primary reason for the limited research in underwater sonar imagery is the lack of large, publicly available datasets, unlike those in the optical sensor domain. Acquiring underwater images requires specialized equipment, such as sonar systems, and these images differ significantly from high-resolution optical images due to the absence of texture and color information. Additionally, labeling sonar images is challenging as it demands specialized domain expertise, making the annotation process complex and time-consuming. Interpreting these images also requires significant technical knowledge, further complicating dataset creation. As a result, existing underwater sonar datasets [7–12] are sparse and limited to a few classes.

To address the challenge of object detection and recognition in underwater environments, significant progress has been made with advancements in sonar technology, data acquisition techniques, and algorithm development. Numerous research efforts are currently focused on designing robust deep learning architectures for this domain. In recent years, several object detection methods [13–21] have been proposed for underwater sonar data. However, most of these approaches address specific problems and evaluate performance using custom-tailored datasets. Despite the availability of some benchmark underwater datasets, the lack of standardized evaluation protocols makes it challenging to compare the overall progress of underwater object detection systems, unlike in the optical sensor domain.

Recently, many researchers have leveraged transfer learning to adapt existing models for underwater object detection. For example, Majchrowska et al. [22] evaluated EfficientDet, Mask R-CNN, and DETR on the Trashcan and TACO datasets, while Xue et al. [23] proposed a custom ResNet50-YOLOv3 model. Fulton et al. [9] trained SSD, Faster R-CNN, and Tiny YOLO on the Deep-Sea Image Dataset. Wang et al. [24] enhanced PP-YOLOv2 with noise filtering and attention mechanisms, achieving improved performance on the Marine Debris dataset. Similarly, Qin et al. [25] developed YOLOv7C, incorporating attention and Multi-GnBlock modules to enable high-order feature interactions, resulting in a 1.9% mAP improvement and 2.5× faster detection speed through redundancy pruning.

While transformer-based models like DETR [26] have demonstrated promising results in object detection, their high computational complexity and slower inference speeds make them impractical for real-time underwater tasks, especially on hardware with limited processing power, such as onboard sonar systems. Additionally, methods specifically designed for sonar imagery, such as SSD-based approaches [9] and custom ResNet-YOLO hybrids [23], often target narrow use cases or rely on custom datasets. This lack of standardized benchmarks limits direct comparisons and raises concerns about their generalization capabilities.

This paper introduces AquaYOLO, an enhanced version of YOLOv8 specifically tailored for subtle and complex underwater sonar images. The model is designed to capture the intricate details of sonar imagery, which are often challenging to detect. To achieve this,

AquaYOLO replaces the traditional Convolutional Neural Network (CNN) backbone with a residual block structure, enhancing the model's ability to learn complex features and fine-grained details critical for underwater sonar data. Furthermore, AquaYOLO incorporates a Dynamic Spatial Attention Module (DSAM) and Context-Aware Feature Selection (CAFS) in the neck network, replacing the conventional approach of simply concatenating multi-level features. These modules enable the model to effectively learn interrelationships between features across different levels and intelligently select relevant features based on contextual information, leading to improved feature representation. The proposed method delivers a balance of effectiveness, speed, and adaptability, making it highly suitable for real-time object detection tasks. AquaYOLO is particularly advantageous for underwater sonar data, where low-latency detection is essential for practical and operational applications. The key contributions of this work are as follows:

- Replacing traditional CNN layers with residual blocks in the backbone network improves the model's ability to learn complex features and capture fine details in noisy sonar images.
- Integrating DSAM in the neck network enables the dynamic aggregation of multi-level features, reducing redundancy and improving feature correlation.
- Introducing CAFS for intelligent feature selection based on contextual information further enhances feature representation and object localization accuracy.
- AquaYOLO is evaluated through four different variants on two standard benchmarks of underwater sonar data Underwater Acoustic Target Detection Dataset (UATD) [27] and the Marine Debris Dataset [12], demonstrating superior performance compared to existing methods despite the limited availability of sonar data.

2. Related Work

Research in underwater object detection has progressed significantly with the adoption of deep learning techniques. Due to the inherent challenges of sonar imaging such as noise, low resolution, lack of texture, and limited color information, traditional object detection models often under perform in these environments. Furthermore, the lack of large publicly available underwater sonar datasets hinders the development and training of robust models, making it difficult to achieve high accuracy and generalization.

To address these challenges, many researchers experimented with transfer learning on different state-of-the-art object detection models on underwater sonar images. Majchrowska et al. [22] evaluated the performance of EfficientDet, Mask R-CNN, and DETR on the Trashcan and Taco dataset. Xue et al. [23] proposed custom Resnet50-Yolov3 and Fulton et al. [9] trained SSD, Faster R-CNN, Tiny Yolo, and Yolov2 on the Deep-sea image dataset [28]. Chia et al. [29] performed transfer learning on RetinaNet, Faster R-CNN, Tiny Yolov3, Yolov4, and Yolov5, evaluating their performance on the Trashcan data set.

Other notable works include Watanabe et al. [30], who trained Yolov3 on custom-made underwater images of marine debris dataset. Valdenegro et al. [31,32] developed a CNN model with different optimizers and evaluated its performance on a custom-made dataset. Additionally, Fuchs et al. [33] evaluated performance of Resnet50 on the ARACATI dataset [34]. Recently, Saba et al. [35] evaluated performance of various scaled variants of YOLOv5 and YOLOv8 on the Marine Debris dataset [12] and a custom tailored dataset.

Furthermore, Wang et al. [24] proposed an improved object detection method based on PP-YOLOv2, incorporating image resegmentation, noise filtering, and attention mechanisms to enhance feature extraction, with a decoupled head optimizing classification. The method was evaluated on the Marine Debris dataset, showing improved performance in challenging sonar conditions. Qin et al. [25] proposed YOLOv7C, which enhances sonar object detection by integrating an attention mechanism for better feature extraction and

Multi-GnBlock modules for high-order feature interaction. The model achieves a 1.9% Mean Average Precision (mAP) improvement and 2.5× faster detection speed with a 47.5% reduction in model size through redundancy pruning.

While significant progress has been made in underwater object detection, most existing models, including PP-YOLOv2 [24], YOLOv7C [25], and other variants of YOLO, focus on improving detection accuracy in underwater sonar images but often face challenges with noise, low resolution, and limited texture and color information. These models, despite their improvements, tend to focus on feature extraction from general object detection tasks and often require large datasets for effective training. In contrast, the proposed AquaYOLO model specifically addresses these challenges by introducing residual blocks in the backbone to enhance feature extraction in low-resolution sonar images. Additionally, AquaYOLO incorporates DSAM and CAFS in the neck and head, which allows the model to better aggregate multi-scale features and reduce redundancy. These innovations not only improve the robustness of detection in sonar data but also make AquaYOLO more effective in handling the unique challenges of underwater environments. Experimental results demonstrate that AquaYOLO outperforms these existing methods in terms of accuracy and robustness, making a significant contribution to the field of underwater object detection.

YOLOv8, selected as the baseline model for this study, consists of three core components: the backbone, the neck, and the output layers. YOLOv8 introduces various improvements. It introduces the C2f module in place of the CSPLayer. The C2f incorporates a two-convolution cross-stage by combining high-level features with circumstantial data to enhance performance. Additionally, YOLOv8 employs an anchor-free architecture with decoupled heads. This task-specific specialization improves the model's performance for various computer vision tasks. The backbone module is tasked with mining key features from input data. The neck module consolidates and refines these features. Using a feature pyramid network (FPN) design, it combines features of different levels to generate a richer and more complete illustration of the input.

The prediction module is responsible for detecting and pinpointing input classes. It comprises various detectors of different sizes and levels, enabling it to identify input of varying dimensions. In the last layer, YOLOv8 employs the sigmoid activation to compute object scores, reflecting the likelihood of an object's existence within a bounding container. Softmax activation is used to illustrate the probability of each category, indicating the chances that an item fits into each possible category. To compute the loss value at the bounding box level, it integrates CIoU and DFL. In addition, it utilizes binary cross-entropy. These mechanisms are highly useful in enhancing performance, particularly for small-scale items.

The principles introduced in the Adaptive Selection and Interaction Aggregation (ASIA) module and the Dynamic Feature Selection (DFS) module, as proposed in [36], have been extended to enhance the neck and head network of YOLOv8. Specifically, the traditional concatenation of multi-level features is replaced with a more dynamic approach inspired by the ASIA and DFS modules introduced as DSAM and CAFS. These modules help the network learn interrelations between features at different levels, reducing feature redundancy and improving the model's capability to focus on crucial features.

Additionally, incorporating CAFS, a slight modification of the DFS module, further improves the selection and representation of features based on their contextual relevance. This modification allows for more intelligent feature selection, ensuring that the network emphasizes the most relevant features in the given context. This method enables more precise feature aggregation, leading to improved detection performance and overall effectiveness of YOLOv8 in complex underwater scenarios.

3. Proposed Methodology

Detecting objects in underwater environments using sonar images is a challenging task due to the inherent noise and low resolution of these images. While traditional object detection models like YOLOv8 have demonstrated remarkable effectiveness in various domains, they often struggle with the fine-grained details required for accurate detection in sonar imagery.

To address these limitations, we propose AquaYOLO, a model specifically designed for the complexities of underwater sonar image analysis. AquaYOLO is structured into three primary components: the backbone module, the neck module, and the prediction head module. The backbone module focuses on robust feature extraction, incorporating residual block layers instead of standard convolutional layers to enhance feature representation. The neck module is responsible for efficient feature aggregation and processing, where we introduce a novel Dynamic Spatial Attention Module (DSAM) as a replacement for traditional concatenation and up-sampling methods. This module enables the network to capture intricate spatial features more effectively. The prediction head module performs object identification and localization, leveraging the enriched feature maps from the previous stages. These enhancements improve AquaYOLO's ability to extract and process intricate features, significantly boosting its performance in detecting objects within noisy and low-resolution sonar images.

In AquaYOLO, we utilize residual blocks. Residual blocks introduce skip connections, which help mitigate the vanishing gradient problem and allow the training of much deeper networks as shown in Figure 1. In the residual network based layers, the output of the particular lth layer is calculated through the combining in and out of the previous layer $(l-1)$th as [37]

$$x_l = g(x_{l-1}, k_l) + x_{l-1} \tag{1}$$

where k_l is the kernel, x is the input. $g()$ is the variable that needs to be learned. A simple residual layer combination is shown in Figure 1. $g(x_{t-1})$ shows the output for input x_{t-1}. Output $g(x_{t-1})$ is combination with input x_{t-1} and after it, the activation function is utilized. The inclusion of residual block layers enables the model to learn more complex features and finer details from the sonar images. Residual block is particularly advantageous for processing sonar images due to their ability to retain information across many layers. This characteristic is essential for sonar imagery, where subtle differences in texture and structure can be critical for accurate object detection.

The integration of residual block into the backbone network significantly enhances the model's feature extraction capabilities. Residual block's ability to capture intricate details and maintain feature integrity across multiple layers is particularly beneficial for sonar images, where fine-grained features are crucial for accurate detection. This improvement allows the model to better differentiate between objects and background noise, leading to higher detection accuracy.

Underwater sonar images often contain subtle variations that are critical for identifying objects. Residual block enables the AquaYOLO model to learn these fine details more effectively than traditional convolutional layers. This capability is essential for distinguishing between objects of similar size and shape but different textures or materials. Sonar images are inherently noisy, which can hinder the performance of standard object detection models. In addition, Sonar images lack the fine edge details due to unavailability of color information. Traditional CNNs may suffer to maintain the edges information. The residual connections in the network enable the network to become more vigorous by facilitating the flow of gradients during training. This helps the network to preserve and carry edge information. The residual block architecture allows for the creation of deeper networks without the risk of vanishing gradients. Moreover, we also observed that incorporation of residual blocks results in effective training on smaller datasets compared to other models due to their capability to extract features from data with reduced dimensional representations. This capability is useful for underwater images datasets which suffer from scarcity. Above-mentioned reasons support that utilization of residual block enable AquaYOLO to perform well even in challenging underwater environments where noise levels are high.

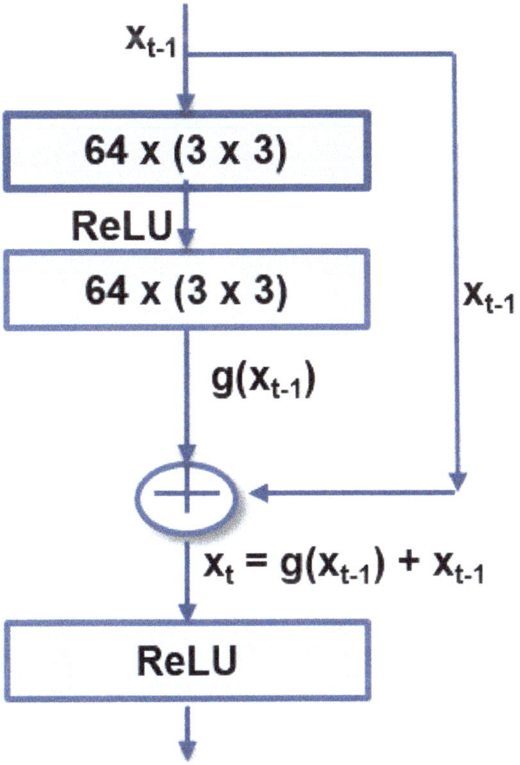

Figure 1. Residual block used in backbone of AquaYOLO.

This scalability means that the AquaYOLO can be expanded to include more layers if necessary, further improving its ability to learn complex features from sonar images. The deeper network can capture more detailed and abstract representations of the input data, enhancing overall detection performance. The output head leverages the enhanced features provided by the residual block backbone to make more accurate predictions.

The neck network of the proposed model introduces DSAM replacing concatenation and upsampling at different levels of the models. YOLOv8 integrates features learned at different levels by concatenation which causes information redundancy and also hinders localization. DSAM dynamically interacts and integrates multi-level features by selectively adjusting features based on contextual information. DSAM results in better feature correlation at different levels, reduces feature redundancy, and limits excessive variation in learned features at different scales. Figure 2 represents the detailed architecture of the DSAM module which starts with the Feature Alignment Unit (FAU) which processes input features Fa and Fb to align resolution and channel count with the current level feature Fa, hence replacing up sampling layers in YOLOv8 neck. FAU is a combination of an upsampling layer, a convolution layer, batch normalization, and finally ReLU, which means FAU fuses features from adjacent levels to address the semantic gap between them. Then, the input features Fa and Fb are fed to the CAFS network.

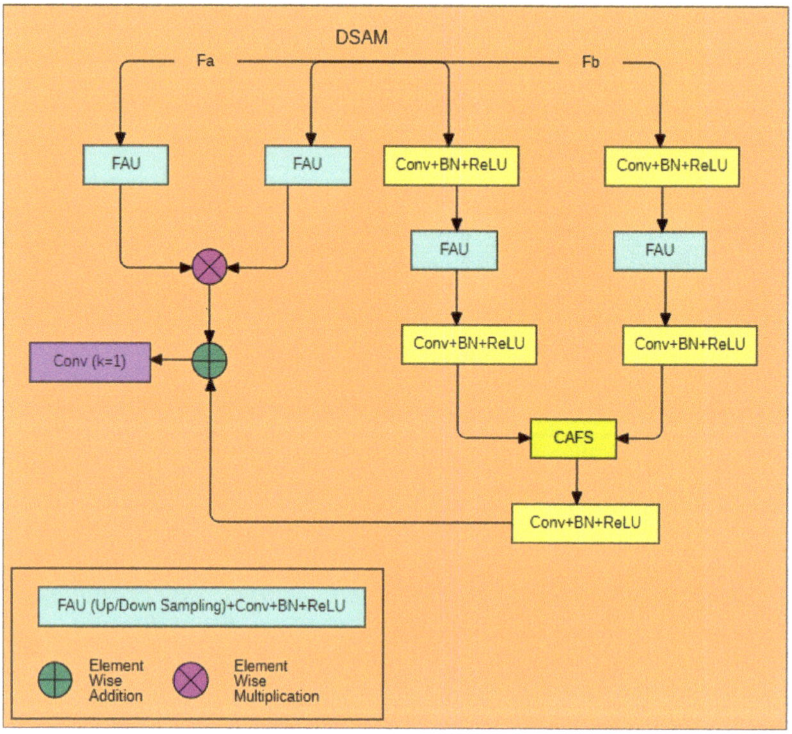

Figure 2. Detailed architecture of dynamic selection aggregation module (DSAM).

CAFS first concatenates two input features and then a series of convolution operations are performed to improve feature extraction as depicted in Figure 3, and then finally softmax is applied to produce feature weights Wa and Wb. These learned weights are used to select features at different levels by performing element-wise multiplication with input feature to suppress any interference or irrelevant information and then performing element-wise summation to obtain the final enhanced feature. To learn fused features, sigmoid and element-wise multiplication are applied to obtain Wf. Finally, the output feature is obtained by element-wise summation. CAFS captures multi-level feature interaction by using multiplicative fusion with addition and concatenation. In the back-propagation operation, gradients of concatenation remain constant. The results of features at one level do not impact features at other levels, leading to a lack of correlation between different levels. Conversely, in the multiplicative fusion method, the gradient of each level's features is influenced by features from other levels. This allows features at different levels to constrain and support each other, thereby significantly enhancing the localization accuracy of learned features, thus improving feature illustration. The detailed architecture of AquaYOLO is depicted in Figure 4.

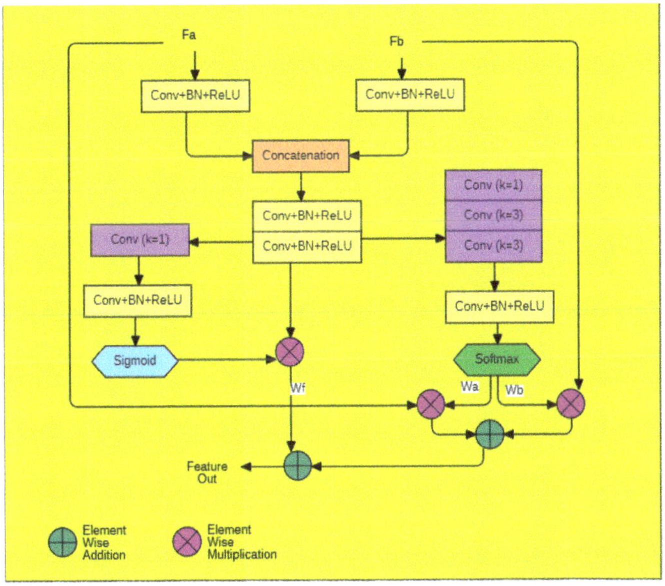

Figure 3. Detailed architecture of context-aware feature selection (CAFS).

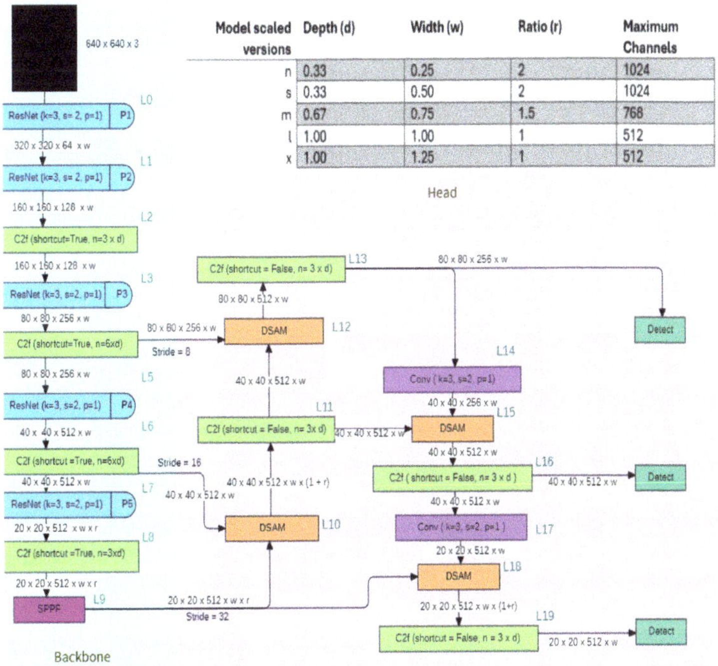

Figure 4. AquaYOLO: Detailed model architecture. In AquaYOLO we utilized ResNet Blocks denoted as P1, P2. . . P5. Model layers are denoted by L1, L2. . . L19.

4. Experimental Configuration

4.1. Datasets

AquaYOLO is evaluated on two publicly available standard benchmark datasets for underwater sonar images, the Underwater Acoustic Target Detection dataset (UATD) [27] and the Marine Debris dataset [12]. As AquaYOLO is an enhanced version of Ultralytics YOLOv8, for the implementation using Ultralytics YOLOv8 models, a different annotation format is required. Specifically, these models expect annotations in the form of .txt files. In this format, each row corresponds to a single object in an image and includes the following information: ['class index', 'center x', 'center y', 'width', 'height']. The class index starts from 0 and increments for each additional class. Moreover, the coordinates of the bounding boxes are normalized relative to the resolution of the image, ensuring that all values fall between 0 and 1.

UATD is a relatively large-scale dataset comprising 9000 sonar images. The dataset is captured using Multibeam Forward Looking Sonar (MFLS). Gemini 1200ik sonar manufactured by Tritech, Scotland, is used for recording sonar images which is a high-resolution sonar and operates at 720 kHz for long-range targets and 1200 kHz for shorter ranges. The good part of the dataset is that it is captured in the real marine environment. Some part of the dataset is captured on Golden Pebble beach at Dalian and the rest is captured in Haoxin lake at Moaming. It comprises sonar images captured in shallow water, 4 to 10 m in Golden Pebble Beach, and at-most 4 m in Haoxin lake. A total of ten object classes are used: Cube, Ball, Cylinder, Human Body, Plane, Circle Cage, Square Cage, Metal Bucket, Tyre, and ROV. Figure 5 shows some sample sonar images from the UATD dataset which illustrates the complexity of underwater sonar images.

Figure 5. Some sample pictures from UATD dataset [27].

The dataset is already divided into train, test, and validation sets comprising 7600, 800, and 800 sonar images, respectively. Annotations are saved in XML files in PASCAL VOC format. The PASCAL VOC format is a commonly used annotation format in computer vision. In this format, each annotation file contains detailed information about the objects within an image, encapsulated within the "object" tag. This tag comprises two essential child tags: "name" and "bndbox". The "name" tag indicates the class name of the object, providing a clear identification of what the object is, while the "bndbox" tag specifies the bounding box coordinates that localize the object within the image. This bounding box is further detailed with four tags: "xmin", "ymin", "xmax", and "ymax". These tags define the coordinates of the bounding box in terms of the top-left and bottom-right corners. Specifically, "xmin" and "ymin" represent the x and y coordinates of the top-left corner, while "xmax" and "ymax" correspond to the x and y coordinates of the bottom-right corner of the bounding box. A python script was developed and used to convert VOC standard format to YOLO format and later saved in txt format so that it can be used by AquaYOLO.

The Marine Debris dataset is an openly accessible collection of 1868 images, captured using the Forward Looking Sonar ARIS Explorer 3000 sensor desinged and manufactured by Sound Metrics, Bellevue, Washington. These images were taken within a controlled water tank environment with dimensions of 3 m by 2 m by 4 m. This dataset encompasses 11 distinct classes, each representing common marine debris items such as bottle, can, drink carton, hook, propeller, shampoo bottle, tire, chain, valve, hook, wall, and standing bottle. Figure 6 shows some sample sonar images from the Marine Debris Dataset which shows that this dataset is rather easy as it shows crisp boundaries of objects captured using Forward Looking Sonar. The accompanying dataset annotations are stored in the COCO XML format, a widely used standard in computer vision. Within each annotation file, the details for every object are encapsulated within the "object" tag, which includes two critical child tags. The "name" tag specifies the class name of the object, while the "bndbox" tag defines the bounding box coordinates. The "bndbox" tag further breaks down into "x", "y", "w", and "h" tags, where "x" and "y" represent the positions of the top-left position. Here, "w" and "h" depict the width and height. To facilitate the conversion from the COCO XML format to the YOLO .txt format, a Python script was employed. This script automates the translation of the bounding box coordinates and class labels into the appropriate format for YOLO, ensuring compatibility with the AquaYOLO. For training, testing, and validation of the models, the dataset was split into three subsets with an 80:10:10 ratio.

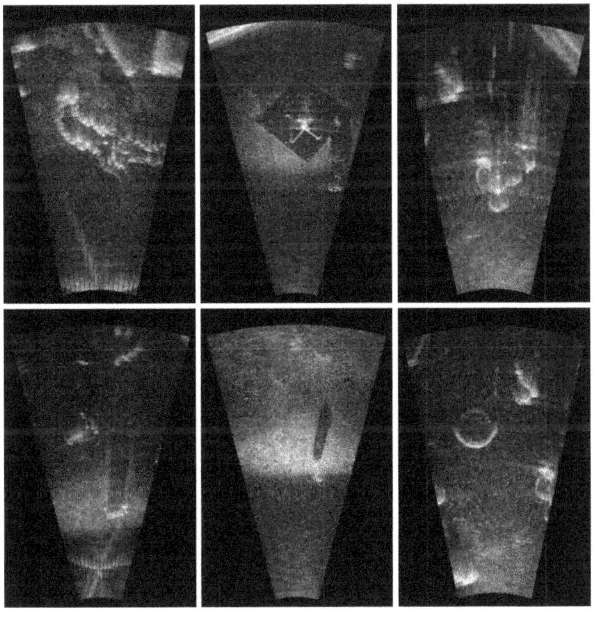

Figure 6. Sonar images from Marine Debris Dataset [12].

4.2. Environmental Setup

To ensure a fair comparison of performance, we use the hardware platform having Nvidia GeForce RTX 2070 GPU with 8 GB of memory manufactured by NVidia, United States. This setup provides a consistent hardware environment for evaluating each model's capabilities. The performance is evaluated based on mAP at a 50% Intersection over Union (IoU) threshold and mAP at 50–95% IOU. Additionally, inference speed, measured in milliseconds, is used to compare the efficiency of the models on the test data.

The experiments are performed on a computer with Windows 10. The deep learning models are implemented in torch version 2.4. This setup ensured compatibility and stability across all experiments. By using these standardized metrics and consistent hardware and software environments, we aim to provide a reliable and accurate comparison of the models' performance in detecting objects in underwater sonar images.

4.3. Evaluation Metrics

When assessing target detection tasks, various important metrics are frequently utilized: recall (R), precision (P), and average precision (AP). Recall represents the fraction of actual targets correctly identified, as outlined in Equation (2) [38]. Conversely, precision quantifies the proportion of correctly identified targets among all predicted targets, as explained in Equation (3) [38].

$$R = \frac{TP}{TP + FN} \tag{2}$$

$$P = \frac{TP}{TP + FP} \tag{3}$$

Here, TP stands for the count of true positives, or correctly identified targets, FP denotes false positives, and FN signifies false negatives, or targets that were missed.

Average precision (AP) measures the region beneath the precision–recall (P-R) graph, with recall represented on the horizontal axis and precision on the vertical axis. The AP is calculated using Equation (4) [39]:

$$AP = \int_{r=0}^{1} P(r)\, dr \tag{4}$$

The mean average precision (mAP) is determined by computing the average of the AP scores across all classes, as illustrated in Equation (5) [39]:

$$mAP = \frac{1}{K} \sum_{n=1}^{k} AP(n) \tag{5}$$

In this research, mAP is at 50%, employing a 50% IoU criterion, and mAP is at 50–95%, utilizing an IoU range of 50% to 95% as thresholds are utilized. For convenience, mAP at 50% is referred to as mAP50 and mAP at 50–95% is referred to as mAP50–95 throughout the paper.

5. Experimental Results

5.1. Qualitative Results

Table 1 presents performance comparison between various deep learning models in the literature with AquaYOLO, all trained on the UATD dataset [27]. Among all these models, AquaYOLO demonstrated superior performance, achieving best precsion, recall, and mAP50.

Table 1. Comparison of AquaYOLO with Various other Models on UATD Dataset.

Model	Precision	Recall	mAP50	mAP50–95
RetinaNet [40]	63.2	62.4	62.5	30.6
FLS Detector	82.3	75.7	75.6	
RT DETR L	82.6	80.7	77.3	
FSLD-Net	86.7	81.9	80.3	
FasterRCNN [41,42]	74.3	75.3	75.1	37.9
YOLOv3 [43]	85.4	82.1	79.1	-
SDDNet [44]	81.3	79.7	80.2	41.9
YOLO-DCN [43]	86.2	83.4	80.5	-
YOLOv3SPP [45]	91.1	93.0	92.2	45.7
YOLOv5-s [42]	92.8	92.2	93.9	53.8
YOLOv8	85.4	81	83.3	37.9
AquaYOLO (our)	94.7	94.5	94.9	48.1

Table 2 presents performance comparison between different variants of YOLOv5, YOLOv8, and AquaYOLO. All three were trained on the Marine Debris dataset [12]. Among these models, AquaYOLO demonstrated superior performance achieving an mAP50 score of 97.6%.

Table 2. Comparison of AquaYOLO with Various other Models on Marine Debris Dataset.

Model	mAP50	mAP50–95	Inference Speed (ms)
Tiny-YOLOv3	70.0	61	3.3
Faster R-CNN	80.8	64	54
RetinaNet	85.1	67	53
Tiny-YOLOv4	87.4	69	3.1
YOLOv5s	93.6	72	3.4
YOLOv5l	94.1	71.8	6.5
YOLOv5m [35]	95.3	72	3.2
YOLOv8s	94.6	73.2	4.4
YOLOv8m	95.1	73.4	5.4
YOLOv8l	94.9	73.6	8.2
YOLOv8n [35]	95.3	73.3	1.3
AquaYOLO	97.6	75.7	1.3

Table 3 shows the mAP50 scores per class for AquaYOLO, trained on the Marine Debris Dataset. Table 3 indicates that both the Shampoo-bottle and Valve classes received relatively low mAP scores, suggesting that they are frequently misclassified as other classes. These misclassifications can be attributed to the inherent challenges of sonar imaging. Objects like the Shampoo-bottle and Standing-bottle share similar shapes and sizes, while the Valve class resembles other metallic debris. The lack of texture and color information in sonar images makes it difficult for the model to differentiate between these classes. Additionally, the presence of noise and low resolution in sonar images further obscures subtle features, contributing to these errors. This analysis highlights the specific areas where AquaYOLO struggles, providing insight into the model's limitations in handling certain object classes.

To better understand the feature extraction process, feature visualization was performed using Eigen-CAM [46]. This technique allowed us to visualize the regions of the input image that the model focuses on at different layers of the network, providing a deeper understanding of how the model interprets the sonar images. By generating activation maps, Eigen-CAM highlights the most important areas that influence the decision-making process of the model, offering a clear view of the features learned by the network.

Table 3. Class-wise mAP50 score for AquaYOLO on Marine Debris dataset.

Class	mAP50
Bottle	98.7
Can	95.4
Chain	98.7
Drink-carton	98.8
Hook	99.3
Propeller	94.7
Shampoo-bottle	92.5
Standing-bottle	98.1
Tire	98.95
Valve	93.5
Wall	97.7

We performed a comparison of the features learned at different layers by YOLOv8n and AquaYOLO. The source image, randomly selected from the test set, contains three objects: Wall, Drink-carton, and Tire, and it was used to represent feature maps through Eigen-CAM. Figure 7 illustrates the feature visualization of the initial feature extraction layers, randomly chosen from the backbone of both YOLOv8 and AquaYOLO. As shown in Figure 7, AquaYOLO captures a larger number of fine-grained features in the early layers (L1 and L2) because of residual blocks, while later layers (L5 and L6) become more focused and specific to the objects of interest, demonstrating the model's ability to refine its attention and distinguish between different objects in the scene.

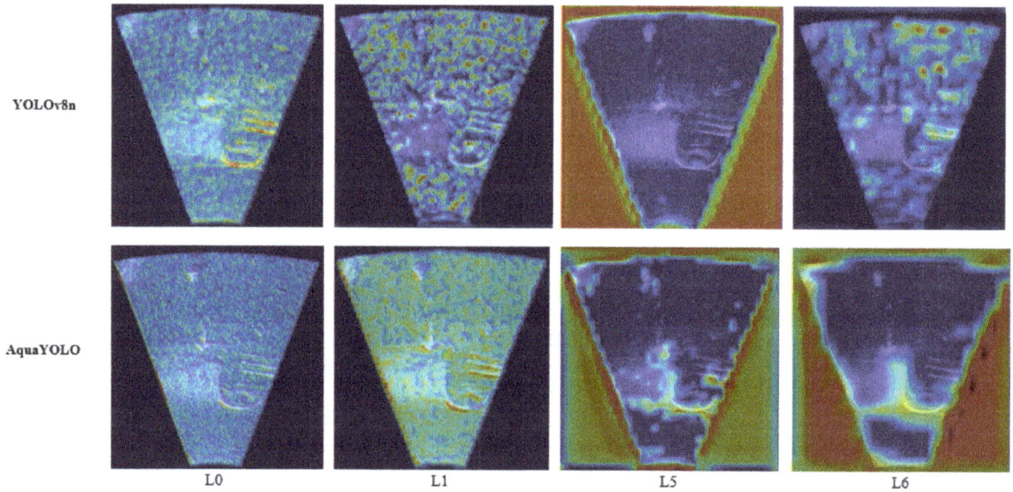

Figure 7. Feature visualization of backbone layers in YOLOv8n and AquaYOLO using Eigen-CAM on the Marine Debris Dataset. Red color show the high intensity areas which were mainly considered for decision making by the model.

Figure 8 shows the feature maps of the neck layers. Visualizing L7 and L9 reveals that the inclusion of residual blocks in AquaYOLO enables the model to focus on learning only the most relevant features, specifically the objects of interest. The visualizations in L10 and L12 demonstrate the improvement achieved by replacing simple concatenation with the DSAM, enhancing feature representation and object detection performance.

Figure 9 shows the feature visualization of the head layers of YOLOv8n and AquaYOLO. L13 and L15 highlight the high-intensity features learned by AquaYOLO near the objects of interest, while YOLOv8n tends to learn more features from the background.

A similar pattern was observed in the feature maps of L18. While both models successfully detected all three objects, the feature intensities in L19 of AquaYOLO were notably stronger than those in YOLOv8n, indicating higher confidence in detecting the objects of interest.

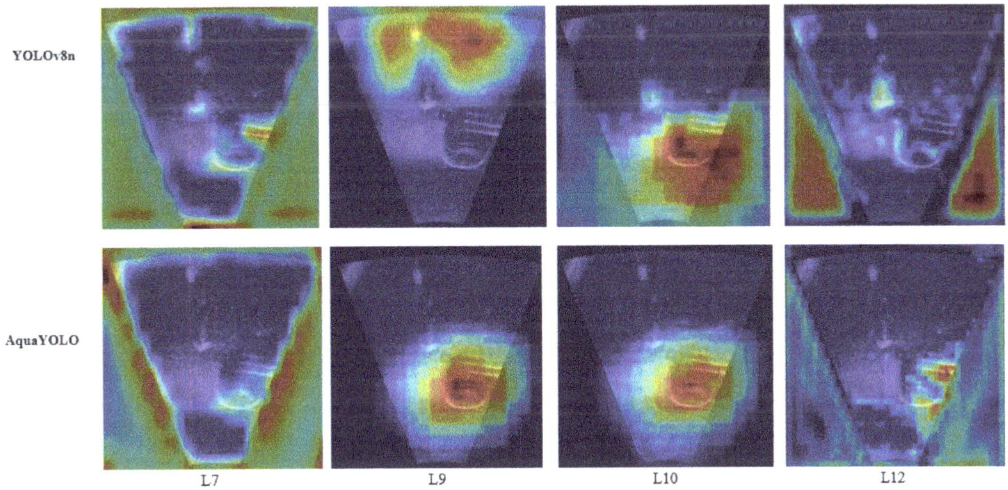

Figure 8. Feature visualization of neck layers in YOLOv8n and AquaYOLO using Eigen-CAM on the Marine Debris Dataset. Red color show the high intensity areas which were mainly considered for decision making by the model.

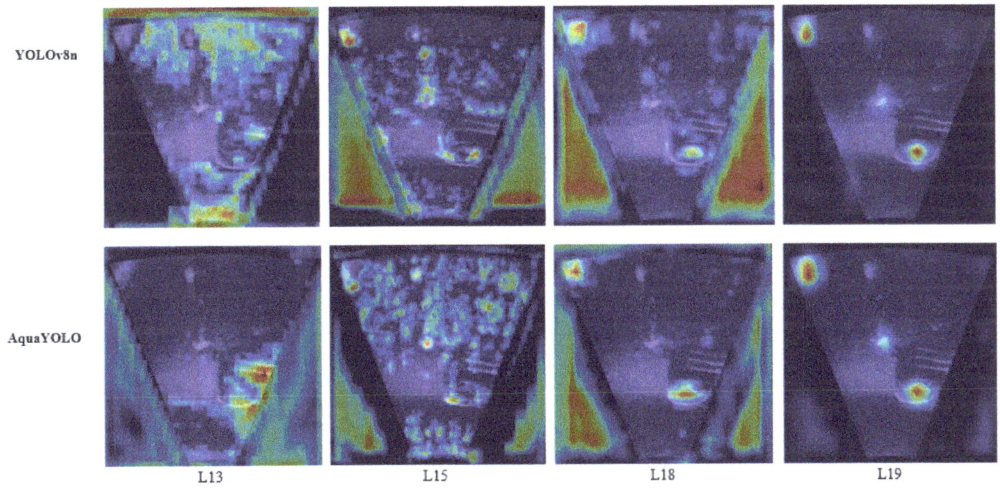

Figure 9. Feature visualization of head layers in YOLOv8n and AquaYOLO using Eigen-CAM on the Marine Debris Dataset. Red color show the high intensity areas which were mainly considered for decision making by the model.

5.2. Quantitative Results

Figure 10 illustrates some inference results produced by AquaYOLO on the test data, along with the ground truth annotations on the Marine Debris Dataset.

Figure 11 shows some inference results produced by AquaYOLO on test data along with ground truth values on the UATD Dataset.

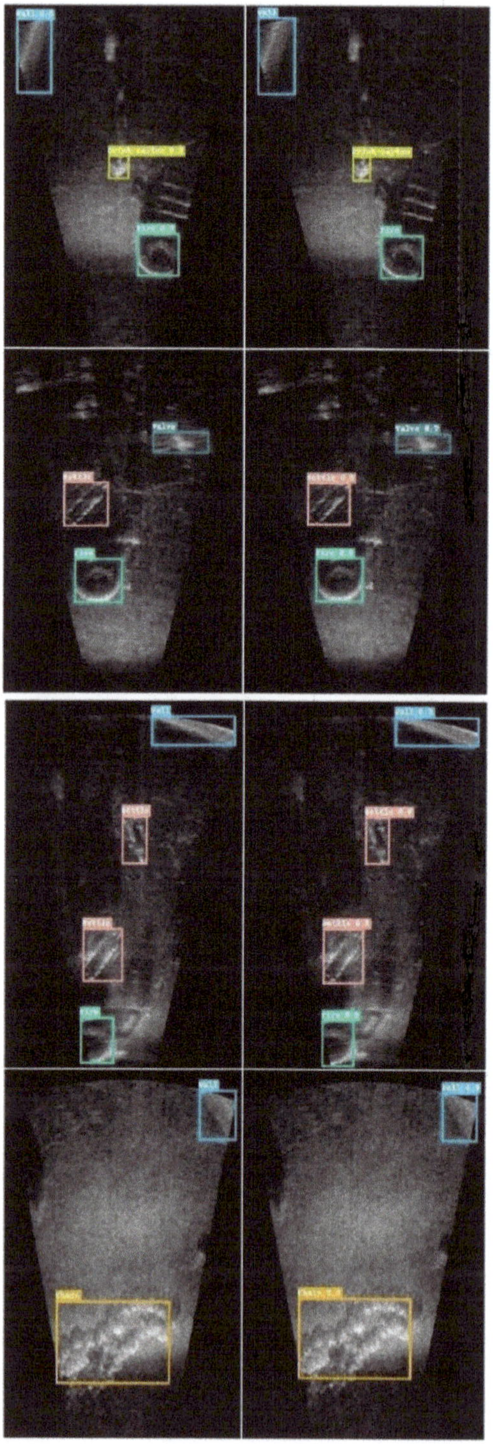

Figure 10. Inference Results of AquaYOLO, ground truth (**left**) and inference results (**right**) on Marine Debris Dataset.

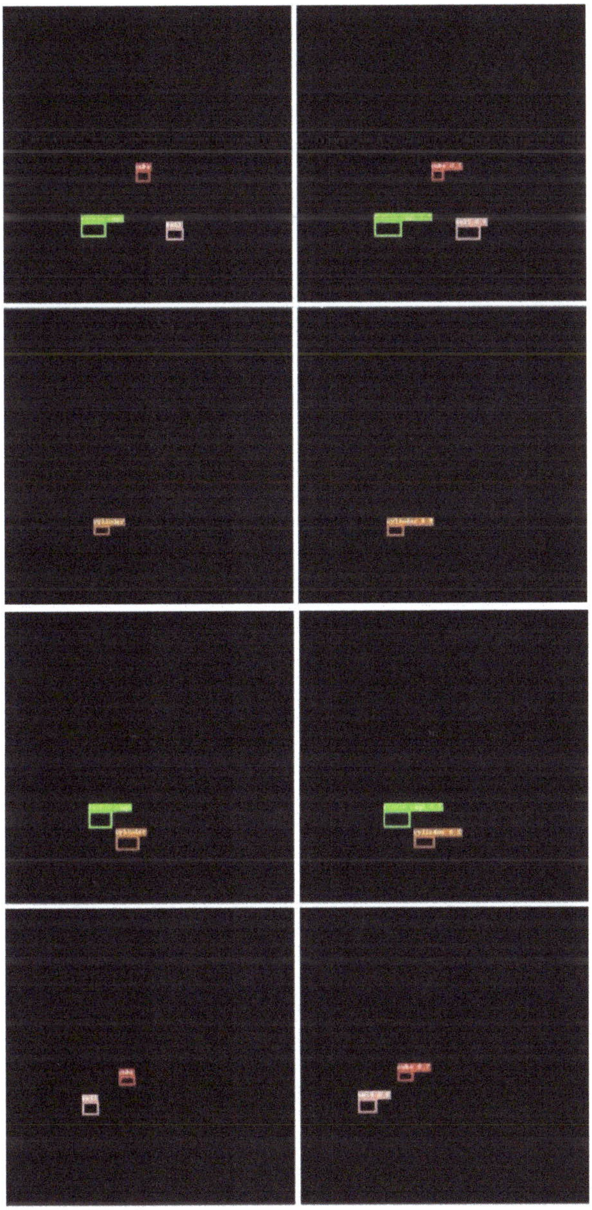

Figure 11. Inference results of AquaYOLO ground truth (**left**) along with inference (**right**) on UATD Dataset.

6. Ablation Study

Figure 12 represents the precision–confidence curve which plots the precision value at different confidence thresholds. It can be observed that as the model's confidence threshold increases, the precision typically increases as well. It can also be observed that the precision–confidence curve rises quickly and maintains high precision across most confidence levels.

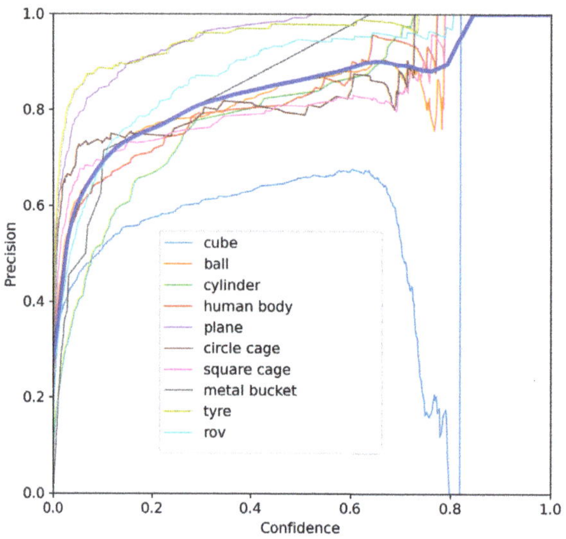

Figure 12. Precision confidence curve for UATD Dataset. Different color lines represent graph for different classes. Thick blue line represents the overall graph.

Figure 13 represents the precision–recall curve which plots precision values at different recall levels. It can be observed that as recall increases, precision is decreasing. This is because to increase recall, the model needs to predict more positives, which can result in more false positives, reducing precision.

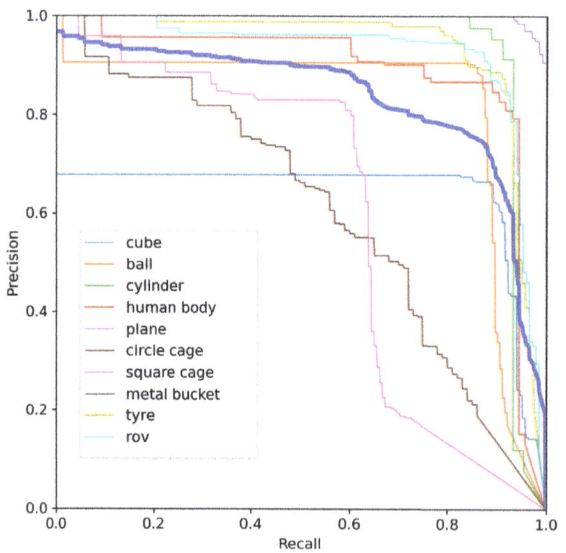

Figure 13. Precision recall curve for UATD Dataset. Different color lines represent graph for different classes. Thick blue line represents the overall graph.

Figure 14 represents the recall–confidence curve which plots the recall value at different confidence thresholds. It can be observed as the confidence threshold increases, recall decreases. This is because the model becomes more conservative, classifying fewer instances as positive, which may result in missing more true positives.

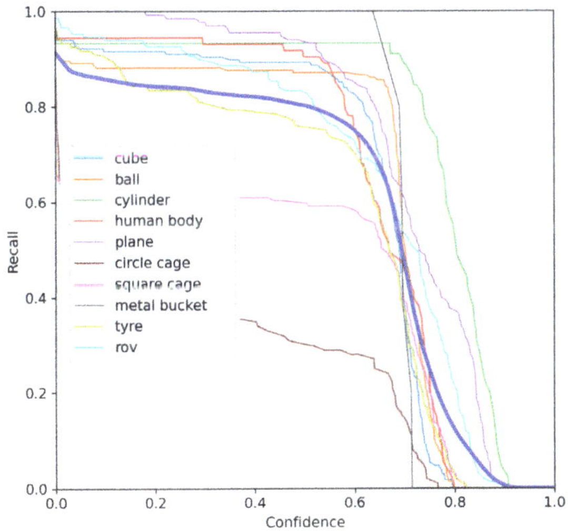

Figure 14. Recall confidence curve for UATD Dataset. Different color lines represent graph for different classes. Thick blue line represents the overall graph.

Figure 15 represents the confusion matrix for the UATD dataset by comparing the predicted labels with the actual labels. Although the overall performance of AquaYOLO remains strong, the confusion matrix highlights specific areas where the model struggles. Notably, the Square Cage class is frequently misclassified as a Cube due to their similar structural features in sonar imagery. Similarly, the Circle Cage class often overlaps with predictions for the Square Cage and the Ball, likely because of their circular outlines and the lack of distinguishing texture in sonar data. These misclassifications are influenced by factors such as reverberation artifacts, shadowing effects, and the inherent noise in sonar images. Such insights provide a clearer understanding of AquaYOLO's limitations in handling objects with similar geometric profiles.

Table 4 presents performance comparison when different modules are incorporated in the proposed AquaYOLO on the UATD dataset. These modules include the residual block, the DSAM module, and the CAFS module. However, here, we elaborate the impact of these modules. It can be inferred that our proposed modules improved performance in the object detection problem.

Table 4. Comparison of AquaYOLO with various other models on Marine Debris Dataset.

Residual	DSAM	CAFS	mAP50	mAP50–95
			95.1	73.3
yes			95.4	73.5
	yes		95.6	73.2
		yes	95.3	74.4
	yes	yes	97.2	74.4
yes	yes		96.7	75.2
yes		yes	97.1	74.3
yes	yes	yes	97.6	75.7

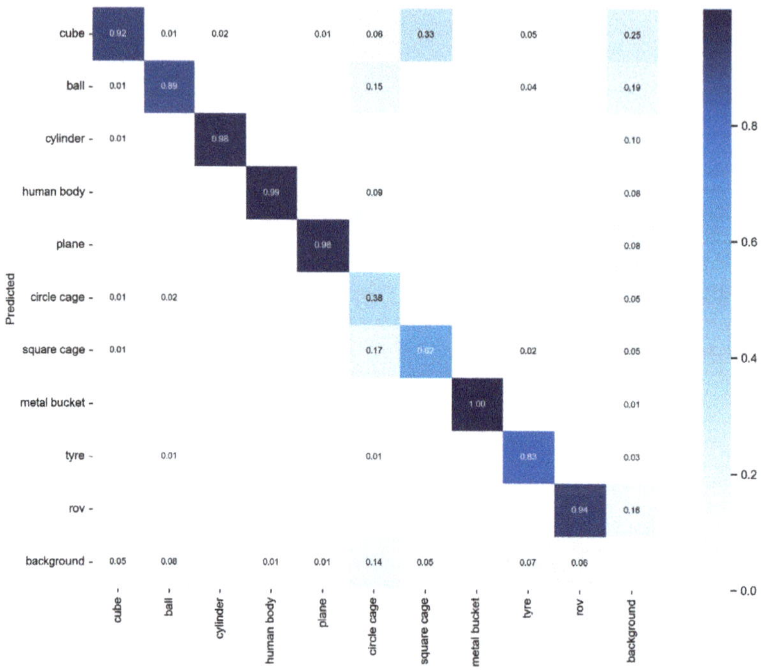

Figure 15. Confusion matrix results for UATD Dataset.

7. Conclusions

AquaYOLO represents a significant advancement in underwater object detection by effectively addressing the challenges of sonar imaging. Underwater object detection is inherently difficult due to factors such as noise, low resolution, and the lack of texture in sonar images. By incorporating residual block layers and introducing the Dynamic Spatial Attention Module (DSAM) and Context-Aware Feature Selection (CAFS) into the network architecture, AquaYOLO can capture subtle details in complex underwater sonar data and selectively extract the most relevant information from feature maps, thereby enhancing detection accuracy. Experimental results on the UATD and Marine Debris datasets demonstrate that AquaYOLO outperforms existing models in both accuracy and robustness. These results confirm that the proposed model excels at handling the unique challenges of underwater sonar imagery, establishing it as a promising solution for underwater object detection tasks. Future research could focus on enhancing the model's scalability and robustness across different sonar systems and environments.

Author Contributions: Conceptualization, Z.C.; Methodology, Y.L., M.I. and Z.C.; Validation, J.Z.; Investigation, Q.C., C.X. and M.I.; Data curation, Y.L., J.Z., M.I. and Z.C.; Writing—original draft, Y.L.; Writing— review & editing, Q.C., C.X., M.I. and Z.C.; Funding acquisition, Z.C. All authors have read and agreed to the published version of the manuscript.

Funding: This research was funded by Guangxi Science and Technology Base and Talent Project (No. GuikeAD21220098) and the 2021 Open Fund project of the Key Laboratory of Cognitive Radio and Information Processing of the Ministry of Education (No. CRKL210102).

Data Availability Statement: The datasets used in the paper can be downloaded here "UATD Dataset" at https://figshare.com/articles/dataset/UATD_Dataset/21331143/3 accessed on 18 June 2024 and "Marine Debris Dataset" at https://github.com/mvaldenegro/marine-debris-fls-datasets/ accessed on 10 January 2024.

Conflicts of Interest: The authors declare no conflicts of interest.

Abbreviations

The following abbreviations are used in this manuscript:

COCO	Common Objects in Context
C2f	Coarse to Fine
CIoU	Complete Intersection of Union
DETR	Detection Transformer
DFL	Distribution Focal Loss
RFB	Radial Basis Functions
SSD	Single Shot MultiBox Detector
VOC	Visual Object Classes
YOLO	You Only Look Once

References

1. Krizhevsky, A.; Sutskever, I.; Hinton, G.E. Imagenet classification with deep convolutional neural networks. In Proceedings of the Advances in Neural Information Processing Systems 25 (NIPS 2012), Lake Tahoe, NV, USA, 3–6 December 2012.
2. Lin, T.Y.; Maire, M.; Belongie, S.; Hays, J.; Perona, P.; Ramanan, D.; Dollár, P.; Zitnick, C.L. Microsoft COCO: Common objects in context. In *European Conference on Computer Vision*; Springer International Publishing: Cham, Switzerland, 2014; pp. 740–755.
3. Everingham, M.; Van Gool, L.; Williams, C.K.; Winn, J.; Zisserman, A. The pascal visual object classes (VOC) challenge. *Int. J. Comput. Vis.* **2010**, *88*, 303–338. [CrossRef]
4. Hurtós, N.; Palomeras, N.; Nagappa, S.; Salvi, J. Automatic detection of underwater chain links using a forward-looking sonar. In Proceedings of the 2013 MTS/IEEE OCEANS-Bergen, Bergen, Norway, 10–14 June 2013; IEEE: Piscataway, NJ, USA, 2013; pp. 1–7.
5. Aslam, M.A.; Zhang, L.; Liu, X.; Irfan, M.; Xu, Y.; Li, N.; Zhang, P.; Zheng, J.; Yaan, L. Underwater sound classification using learning based methods: A review. *Expert Syst. Appl.* **2024**, *255*, 124498. [CrossRef]
6. Irfan, M.; Jiangbin, Z.; Iqbal, M.; Arif, M.H. A novel lifelong learning model based on cross domain knowledge extraction and transfer to classify underwater images. *Inf. Sci.* **2021**, *552*, 80–101. [CrossRef]
7. Cormier, R.; Elliott, M. SMART marine goals, targets and management—Is SDG 14 operational or aspirational, is 'Life Below Water'sinking or swimming? *Mar. Pollut. Bull.* **2017**, *123*, 28–33. [CrossRef]
8. Hong, J.; Fulton, M.; Sattar, J. Trashcan: A semantically-segmented dataset towards visual detection of marine debris. *arXiv* **2020**, arXiv:2007.08097.
9. Fulton, M.; Hong, J.; Islam, M.J.; Sattar, J. Robotic detection of marine litter using deep visual detection models. In Proceedings of the 2019 International Conference on Robotics and Automation (ICRA), Montreal, QC, Canada, 20–24 May 2019; IEEE: Piscataway, NJ, USA, 2019; pp. 5752–5758.
10. Kikaki, K.; Kakogeorgiou, I.; Mikeli, P.; Raitsos, D.E.; Karantzalos, K. MARIDA: A benchmark for Marine Debris detection from Sentinel-2 remote sensing data. *PLoS ONE* **2022**, *17*, e0262247. [CrossRef]
11. Sánchez-Ferrer, A.; Gallego, A.J.; Valero-Mas, J.J.; Calvo-Zaragoza, J. The CleanSea set: A benchmark corpus for underwater debris detection and recognition. In Proceedings of the Iberian Conference on Pattern Recognition and Image Analysis, Aveiro, Portugal, 4–6 May 2022; Springer: Berlin/Heidelberg, Germany, 2022; pp. 616–628.
12. Singh, D.; Valdenegro-Toro, M. The marine debris dataset for forward-looking sonar semantic segmentation. In Proceedings of the IEEE/CVF International Conference on Computer Vision, Montreal, BC, Canada, 11–17 October 2021; pp. 3741–3749.
13. Valdenegro-Toro, M. End-to-end object detection and recognition in forward-looking sonar images with convolutional neural networks. In Proceedings of the 2016 IEEE/OES Autonomous Underwater Vehicles (AUV), Tokyo, Japan, 6–9 November 2016; IEEE: Piscataway, NJ, USA, 2016; pp. 144–150.
14. Valdenegro-Toro, M. Object recognition in forward-looking sonar images with convolutional neural networks. In Proceedings of the OCEANS 2016 MTS/IEEE Monterey, Monterey, CA, USA, 19–23 September 2016; IEEE: Piscataway, NJ, USA, 2016; pp. 1–6.

15. Valdenegro-Toro, M. Objectness scoring and detection proposals in forward-looking sonar images with convolutional neural networks. In Proceedings of the Artificial Neural Networks in Pattern Recognition: 7th IAPR TC3 Workshop, ANNPR 2016, Ulm, Germany, 28–30 September 2016; Proceedings 7; Springer: Berlin/Heidelberg, Germany, 2016; pp. 209–219.
16. Kim, J.; Yu, S.C. Convolutional neural network-based real-time ROV detection using forward-looking sonar image. In Proceedings of the 2016 IEEE/OES Autonomous Underwater Vehicles (AUV), Tokyo, Japan, 6–9 November 2016; IEEE: Piscataway, NJ, USA, 2016; pp. 396–400.
17. Valdenegro-Toro, M. Learning objectness from sonar images for class-independent object detection. In Proceedings of the 2019 European Conference on Mobile Robots (ECMR), Prague, Czech Republic, 4–6 September 2019; IEEE: Piscataway, NJ, USA, 2019; pp. 1–6.
18. Neves, G.; Cerqueira, R.; Albiez, J.; Oliveira, L. Rotation-invariant shipwreck recognition with forward-looking sonar. *arXiv* **2019**, arXiv:1910.05374.
19. Ribeiro, P.O.; dos Santos, M.M.; Drews, P.L.; Botelho, S.S. Forward looking sonar scene matching using deep learning. In Proceedings of the 2017 16th IEEE International Conference on Machine Learning and Applications (ICMLA), Cancun, Mexico, 18–21 December 2017; IEEE: Piscataway, NJ, USA, 2017; pp. 574–579.
20. Zacchini, L.; Franchi, M.; Manzari, V.; Pagliai, M.; Secciani, N.; Topini, A.; Stifani, M.; Ridolfi, A. Forward-looking sonar CNN-based automatic target recognition: An experimental campaign with FeelHippo AUV. In Proceedings of the 2020 IEEE/OES Autonomous Underwater Vehicles Symposium (AUV), St Johns, NL, Canada, 30 September–2 October 2020; IEEE: Piscataway, NJ, USA, 2020; pp. 1–6.
21. Kvasić, I.; Mišković, N.; Vukić, Z. Convolutional neural network architectures for sonar-based diver detection and tracking. In Proceedings of the OCEANS 2019-Marseille, Marseille, France, 17–20 June 2019; IEEE: Piscataway, NJ, USA, 2019; pp. 1–6.
22. Majchrowska, S.; Mikołajczyk, A.; Ferlin, M.; Klawikowska, Z.; Plantykow, M.A.; Kwasigroch, A.; Majek, K. Deep learning-based waste detection in natural and urban environments. *Waste Manag.* **2022**, *138*, 274–284. [CrossRef] [PubMed]
23. Xue, B.; Huang, B.; Wei, W.; Chen, G.; Li, H.; Zhao, N.; Zhang, H. An efficient deep-sea debris detection method using deep neural networks. *IEEE J. Sel. Top. Appl. Earth Obs. Remote Sens.* **2021**, *14*, 12348–12360. [CrossRef]
24. Wang, F.; Li, H.; Wang, K.; Su, L.; Li, J.; Zhang, L. An Improved Object Detection Method for Underwater Sonar Image Based on PP-YOLOv2. *J. Sens.* **2022**, *2022*, 5827499. [CrossRef]
25. Qin, K.S.; Liu, D.; Wang, F.; Zhou, J.; Yang, J.; Zhang, W. Improved YOLOv7 model for underwater sonar image object detection. *J. Vis. Commun. Image Represent.* **2024**, *100*, 104124. [CrossRef]
26. Carion, N.; Massa, F.; Synnaeve, G.; Usunier, N.; Kirillov, A.; Zagoruyko, S. End-to-End Object Detection with Transformers. *arXiv* **2020**, arXiv:2005.12872.
27. Xie, K.; Yang, J.; Qiu, K. A dataset with multibeam forward-looking sonar for underwater object detection. *Sci. Data* **2022**, *9*, 739. [CrossRef]
28. Sasaki, T.; Azuma, S.; Matsuda, S.; Nagayama, A.; Ogido, M.; Saito, H.; Hanafusa, Y. IN53C–1911: Jamstec e-library of deep-sea images (j-edi) realizes a virtual journey to the earth's unexplored deep ocean. In Proceedings of the AGU Fall Meeting Abstracts, San Francisco, CA, USA, 12–16 December 2016; Volume 2016.
29. Chia, K.Y.; Chin, C.S.; See, S. Deep Transfer Learning Application for Intelligent Marine Debris Detection. In Proceedings of the International Conference on Engineering Applications of Neural Networks, Cham, Switzerland, 14–17 June 2023; pp. 479–490.
30. Watanabe, J.I.; Shao, Y.; Miura, N. Underwater and airborne monitoring of marine ecosystems and debris. *J. Appl. Remote Sens.* **2019**, *13*, 044509. [CrossRef]
31. Valdenegro-Toro, M. Submerged marine debris detection with autonomous underwater vehicles. In Proceedings of the 2016 International Conference on Robotics and Automation for Humanitarian Applications (RAHA), Amritapuri, India, 18–20 December 2016; IEEE: Piscataway, NJ, USA, 2016; pp. 1–7.
32. Valdenegro-Toro, M. Best practices in convolutional networks for forward-looking sonar image recognition. In Proceedings of the OCEANS 2017, Aberdeen, UK, 19–22 June 2017; IEEE: Piscataway, NJ, USA, 2017; pp. 1–9.
33. Fuchs, L.R.; Gällström, A.; Folkesson, J. Object recognition in forward looking sonar images using transfer learning. In Proceedings of the 2018 IEEE/OES Autonomous Underwater Vehicle Workshop (AUV), Porto, Portugal, 6–9 November 2018; IEEE: Piscataway, NJ, USA, 2018; pp. 1–6.
34. Dos Santos, M.M.; De Giacomo, G.G.; Drews, P.L., Jr.; Botelho, S.S. Cross-view and cross-domain underwater localization based on optical aerial and acoustic underwater images. *IEEE Robot. Autom. Lett.* **2022**, *7*, 4969–4974. [CrossRef]
35. Mehmood, S.; Irfan, M.; Hamid, U.; Ali, S. Underwater Object Detection from Sonar Images Using Transfer Learning. In Proceedings of the 2024 21st International Bhurban Conference on Applied Sciences & Technology (IBCAST), Murree, Pakistan, 20–23 August 2024; IEEE: Piscataway, NJ, USA, 2024.
36. Xu, Y.; Ren, J.; Irfan, F.A.; Muhammad, Z.J. Adaptive feature coupling and multi-level supervision for image forgery localization. 2024, *submitted*.

37. Irfan, M.; Jiangbin, Z.; Iqbal, M.; Masood, Z.; Arif, M.H.; ul Hassan, S.R. Brain inspired lifelong learning model based on neural based learning classifier system for underwater data classification. *Expert Syst. Appl.* **2021**, *186*, 115798. [CrossRef]
38. Irfan, M.; Jiangbin, Z.; Ali, S.; Iqbal, M.; Masood, Z.; Hamid, U. DeepShip: An underwater acoustic benchmark dataset and a separable convolution based autoencoder for classification. *Expert Syst. Appl.* **2021**, *183*, 115270. [CrossRef]
39. Liu, Y.; Sun, P.; Wergeles, N.; Shang, Y. A survey and performance evaluation of deep learning methods for small object detection. *Expert Syst. Appl.* **2021**, *172*, 114602. [CrossRef]
40. Wang, Z.; Guo, J.; Zeng, L.; Zhang, C.; Wang, B. MLFFNet: Multilevel feature fusion network for object detection in sonar images. *IEEE Trans. Geosci. Remote Sens.* **2022**, *60*, 5119119. [CrossRef]
41. Luo, S.; Yu, J.; Xi, Y.; Liao, X. Aircraft target detection in remote sensing images based on improved YOLOv5. *IEEE Access* **2022**, *10*, 5184–5192. [CrossRef]
42. Wu, W.; Luo, X. Sonar Object Detection Based on Global Context Feature Fusion and Extraction. In Proceedings of the 2024 12th International Conference on Intelligent Control and Information Processing (ICICIP), Nanjing, China, 8–10 March 2024; IEEE: Piscataway, NJ, USA, 2024; pp. 195–202.
43. Hou, J. Underwater Detection using Forward-Looking Sonar Images based on Deformable Convolution YOLOv3. In Proceedings of the 2024 4th International Conference on Neural Networks, Information and Communication (NNICE), Guangzhou, China, 19–21 January 2024; IEEE: Piscataway, NJ, USA, 2024; pp. 490–493.
44. Lin, T.Y.; Goyal, P.; Girshick, R.; He, K.; Dollár, P. Focal loss for dense object detection. In Proceedings of the IEEE International Conference on Computer Vision, Venice, Italy, 22–29 October 2017; pp. 2980–2988.
45. Pebrianto, W.; Mudjirahardjo, P.; Pramono, S.H.; Setyawan, R.A. YOLOv3 with Spatial Pyramid Pooling for Object Detection with Unmanned Aerial Vehicles. *arXiv* **2023**, arXiv:2305.12344.
46. Muhammad, M.B.; Yeasin, M. Eigen-CAM: Class Activation Map using Principal Components. *arXiv* **2020**, arXiv:2008.00299.

Journal of
Marine Science and Engineering

MDPI

Article

Dual-CycleGANs with Dynamic Guidance for Robust Underwater Image Restoration

Yu-Yang Lin [1], Wan-Jen Huang [1] and Chia-Hung Yeh [2,3,*]

[1] Institute of Communications Engineering, National Sun Yat-Sen University, Kaohsiung 80404, Taiwan; wjhuang@faculty.nsysu.edu.tw (W.-J.H.)
[2] Department of Electrical Engineering, National Taiwan Normal University, Taipei 10610, Taiwan
[3] Department of Electrical Engineering, National Sun Yat-Sen University, Kaohsiung 80404, Taiwan
* Correspondence: chyeh@ntnu.edu.tw

Abstract: The field of underwater image processing has gained significant attention recently, offering great potential for enhanced exploration of underwater environments, including applications such as underwater terrain scanning and autonomous underwater vehicles. However, underwater images frequently face challenges such as light attenuation, color distortion, and noise introduced by artificial light sources. These degradations not only affect image quality but also hinder the effectiveness of related application tasks. To address these issues, this paper presents a novel deep network model for single under-water image restoration. Our model does not rely on paired training images and incorporates two cycle-consistent generative adversarial network (CycleGAN) structures, forming a dual-CycleGAN architecture. This enables the simultaneous conversion of an underwater image to its in-air (atmospheric) counterpart while learning a light field image to guide the underwater image towards its in-air version. Experimental results indicate that the proposed method provides superior (or at least comparable) image restoration performance, both in terms of quantitative measures and visual quality, when compared to existing state-of-the-art techniques. Our model significantly reduces computational complexity, resulting in a more efficient approach that maintains superior restoration capabilities, ensuring faster processing times and lower memory usage, making it highly suitable for real-world applications.

Keywords: underwater image restoration; deep learning; unsupervised learning; generative adversarial networks

1. Introduction

In recent years, the exploration of underwater environments has become increasingly important because of the increasing exhaustion of natural resources and the advancement of the global economy. Many applications in ocean engineering and research now rely heavily on underwater imagery captured by autonomous underwater vehicles (AUVs). One of the primary functions of AUVs is to be able to capture underwater imagery that is essential to explore, understand, and interact with the marine environment. Much research regarding underwater image processing has been proposed for scientific exploration of deep-sea environments [1,2]. However, underwater imaging faces more challenges than atmospheric imaging. Underwater images frequently experience degradation caused by attenuation, color distortion, and noise from artificial lighting. Specifically, scattering and absorption caused by particles in the water, such as microscopic phytoplankton or non-algae particles, can attenuate direct transmission and produce ambient scattered light. The diminished direct transmission lowers scene intensity and causes color distortion, while scattered ambient light further alters the scene's appearance. These degradations complicate the restoration of underwater image quality, seriously affecting related tasks in underwater exploration, such as target detection, pattern recognition, and scene understanding.

Image restoration is fundamentally an ill-posed problem used to obtain high-quality images from degraded input images. Quality degradation may occur due to the capture process (such as noise and lens blur), post-processing (such as compression), or under non-ideal conditions (such as haze and fog). Various image restoration techniques have been proposed in the literature, which rely on prior knowledge, assumptions, and learning strategies. With the rapid advancement of deep learning algorithms, more and more image restoration technologies based on deep learning have emerged [3,4]. The success of deep learning-based methods often depends on sufficient and effective training datasets. Considering the scarcity of paired training image samples of underwater images and their corresponding ground truth (or clean) versions, this poses a significant challenge to training deep models for single under-water image restoration. Although several underwater image datasets are synthesized through physical-based models, there is still a lack of publicly accessible collections. Furthermore, most underwater image synthesis methods do not intend to reproduce atmospheric scenes, resulting in incomplete enhancement and difficulty in approaching actual underwater conditions.

In recent years, underwater image restoration methods based on single images have attracted attention due to their effectiveness and flexibility. The restoration of underwater images generally falls into two primary categories: conventional techniques and deep learning-based approaches. Chiang et al. focused on improving the quality of individual underwater images by applying image dehazing techniques, addressing attenuation discrepancies along the propagation path [5]. Li et al. employed dehazing through blue-green channels and corrected the red channel for restoring single underwater images [6]. Ancuti et al. proposed a fusion-based haze removal framework for single underwater images, which integrates two images produced by applying color compensation and white balancing to the input image [7]. For example, Li et al. developed a convolutional neural network (CNN) specifically designed for the enhancement of underwater images and relied on underwater scene priors to generate synthetic training data [8]. Additionally, Dudhane et al. introduced a deep learning network trained on this synthetic dataset [9]. Another notable contribution is [10], which presented a large-scale benchmark for underwater image enhancement. This benchmark includes reference images generated through 12 selected enhancement methods, with the optimal result for each underwater image determined by a voting process. The simultaneous enhancement and super-resolution method called Deep SESR is a generative model based on a residual network, designed to enhance image quality and improve spatial resolution during restoration [11]. Naik et al. proposed a Shallow-UWnet based on a shallow neural network model to maintain performance with fewer parameters [12]. Zhou et al. introduced a method for underwater image restoration that involves estimating depth maps and employing backscatter reduction [13]. A lightweight multi-level network called Lit-Net is proposed by Pramanik et al., focusing on multi-resolution and multi-scale image analysis for recovering underwater images [14].

Training end-to-end CNN models on paired training data is challenging due to the difficulty in acquiring a data set consisting of pairs of underwater images and their corresponding ground truth (clean) images. To address this issue, generative adversarial network (GAN) architectures [15], including the cycle-consistent adversarial network called CycleGAN [16], have been employed in the restoration of underwater images. The WaterGAN framework [17] uses GAN to generate realistic underwater images from in-air (atmospheric) images and depth information to facilitate color correction of monocular underwater images. proposed a model based on a conditional generative adversarial network for real-time enhancement of underwater images [18]. Guo et al. introduced a multi-scale dense GAN designed to enhance underwater images [19]. Cong et al. introduced PUGAN, a GAN model guided by physical models, for enhancing underwater image processing [20].

Existing methods for underwater image restoration often face two major challenges including inadequate color correction and insufficient detail reconstruction. To tackle these challenges, this paper presents a solution by proposing a dual-CycleGAN model specifically designed for single underwater image restoration, which restores an underwater image with dynamically learning guidance. In more detail, our framework utilizes one CycleGAN to learn a light field guidance image, which is generated from the target image to improve color accuracy. The second CycleGAN focuses on training a generator specifically for underwater image restoration. Both CycleGANs are trained concurrently, with the guidance image dynamically steering the output of the restoration generator, leading to more effective and efficient restoration of underwater images. Two CycleGANs work together, with one focusing on extracting useful color features from the underwater images, while the other focuses on restoration and reconstruction. This division of labor helps improve the final image quality and consistency but also ensures stability throughout the training process.

The paper is structured as follows. In Section 2, we present a background review of light field generation techniques, which are utilized in the design of our guidance image. Section 3 describes the proposed deep learning network for single underwater image restoration and outlines the problem that this study aims to solve. Section 4 presents the experimental results, and Section 5 concludes with final remarks.

2. Related Work

2.1. Light Filed Map

The light characteristics of underwater scenes are distinct from those of aerial images due to particle scattering being highly random, making it difficult to accurately simulate them with traditional physics methods. A light field preservation approach is introduced to capture and integrate the diverse underwater light field information into the target image. The background light field map is generated by the multi-scale filtering process, and this is achieved by applying a Gaussian blur at different levels of intensity [21]. This multi-scale approach helps address the limitations of using a single level of filtering, offering a more precise depiction of the background light. The method is inspired by the multi-scale Retinex technique [22]. The preserved features in underwater light field images emphasize the natural stylistic elements of various underwater scenes while omitting the detailed and structured information found in the original underwater images. In our proposed algorithm, we use the light field map as a learning guide image to iteratively guide the output of a generator to effectively enhance the color presentation of underwater image restoration. Figure 1 illustrates the light field map derived from several underwater images.

Figure 1. Illustration of the output from the light field module.

2.2. CycleGAN

Cycle-GAN, extended from GAN [15], which was originally proposed by Zhu et al. [16], aims at enhancing GAN for unsupervised image style transfer by utilizing the cycle consistency principle. Cycle-GAN has been adapted and extended to various other image generation tasks, yielding impressive results in each case [23]. CycleGAN learns a generator that generates images in one domain under given images in another domain. Many possible mappings can be inferred without using matching information, which demonstrates outstanding performance in image processing tasks. Deep learning has been shown to perform well in a variety of underwater tasks, but large datasets are difficult to obtain in underwater environments. GAN-based methods are suitable for underwater image restoration. Although generative adversarial networks have excellent results, there are also unavoidable problems [24–26]. One challenge lies in the difficulty of training GANs and the complexity of objectively assessing the generated output. Another issue is the potential for model collapse, which can occur if the training penalties are not properly adjusted. Our proposed method uses light field maps as guide images to stabilize model training and achieve better restoration performance.

3. Proposed Framework

The overall structure of the proposed dual-CycleGAN that includes two CycleGANs is shown in Figure 2. As shown in Figure 2, the upper section referred to as the Light Field CycleGAN, abbreviated as "LFCycleGAN" is mainly designed for guidance learning to guide an input underwater image toward its in-air version by producing the light field image of the input image, while the bottom part of Figure 2, referred to as the Restoration CycleGAN, abbreviated as "RCycleGAN", is the major CycleGAN designed for producing the finally restored image. The two models will be jointly trained until the network converges to obtain the final generator G_R (Figure 3).

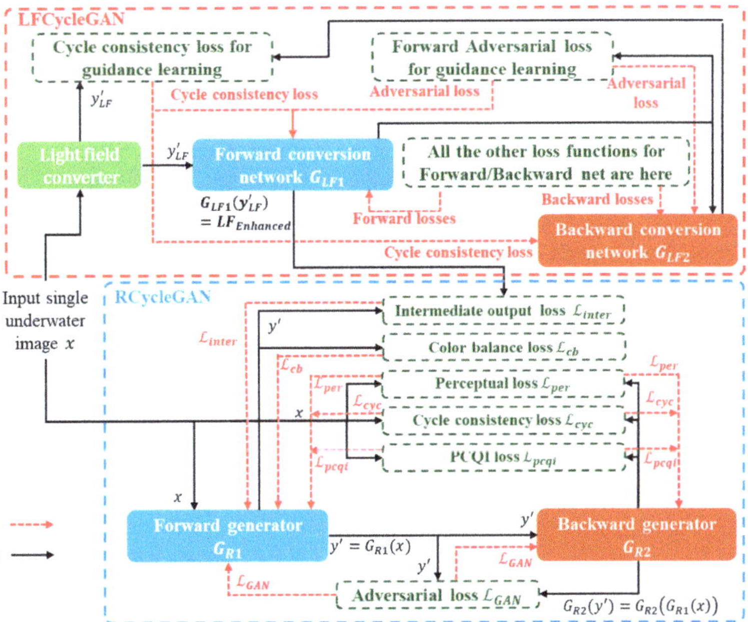

Figure 2. The proposed unsupervised adversarial learning framework consisting of dual-CycleGAN with unpaired training images.

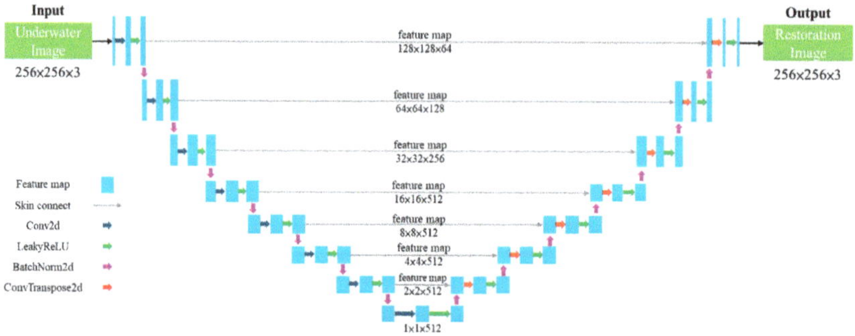

Figure 3. The architecture of the generator in the proposed deep single underwater image restoration network.

3.1. Dual-CycleGAN

The primary objective of this paper is to design a deep neural network that transforms a single underwater image $x \in X$ into its corresponding in-air image, $y \in Y$. X and Y denote the collections of underwater images (the source domain) and in-air images (the target domain), respectively. To design an effective generator model, we base it on an autoencoding architecture, the image is first passed through an encoder, then fed into an 8-layer U-Net to extract features, and finally processed by a decoder. The encoder is made up of 8 convolutional layers, each with a 4×4 kernel size, using a stride of 2, and both padding and output padding values set to 1. The feature maps have sizes of 64, 128, 256, and 512, respectively. The decoder is made up of 8 transposed convolutional (deconvolution) layers, each with a 4×4 kernel size and a stride of 2, and both padding and output padding set to 1. The feature maps for these layers are 512, 256, 128, and 3, respectively. Finally, we can obtain an image with the same shape ($H \times W \times 3$) as the input.

In the RCycleGAN architecture, the process begins with the input of a degraded underwater image x into the discriminator. The goal of the RCycleGAN is to transform this degraded underwater image into its in-air version. This is done through the first generator G_{R1}, which takes the underwater image x as input and produces an intermediate in-air version $G_{R1(x)}$. However, since the aim is to recover the original degraded input, the next step involves converting the in-air version back into a degraded underwater image. This is accomplished using the second generator G_{R2}, which takes the in-air image $G_{R1(x)}$ as input and outputs the degraded underwater version $G_{R2}\left(G_{R1(x)}\right)$. The process in RCycleGAN essentially cycles the image between these two transformations: from degraded underwater to in-air and back to degraded underwater.

In contrast, the LFCycleGAN introduces an additional layer of complexity by incorporating a guidance mechanism for improved restoration. In this architecture, an additional generator G_{LF1} is used to generate a light field image that serves as a guidance map. This light field image captures the illumination and other environmental characteristics of the scene, which are then used to help guide the RCycleGAN process. Specifically, the guidance light field image produced by G_{LF1} is applied to the input underwater image, assisting the RCycleGAN's first generator G_{R1} in transforming the degraded underwater image into a more accurate in-air version. This guided approach improves the overall restoration performance by providing additional information to inform the generation of the in-air image. Since underwater images often exhibit color distortion compared to corresponding in-air images, we believe that the inherent light field image of the input underwater image should be consistent with its restored image. Our main idea is to use the light field map obtained by G_{LF1} in LFCycleGAN of the input underwater image to guide the restoration process in RCycleGAN to produce its in-air version. As shown in

Figure 2, for the generated in-air image $y' = G_R(x)$ of the input underwater image x in the proposed RCycleGAN, we directly get its "baseline" light field version y'_{LF}. To obtain an "enhanced" light field version of the input underwater image, it is needed to exclude possible noises, blurring effects, or color distortions within the image while preserving its inherent image color representation. To achieve this purpose, we propose a generator denoted by G_{LF1} in our LFCycleGAN (circled by the red dotted line region in Figure 2) for transforming an underwater image to the corresponding enhanced light field image. To train G_{LF}, the LFCycleGAN mainly consists of the forward generator G_{LF1} and the backward generator G_{LF2}.

According to the Retinex theory, an image S can be decomposed into the light field information L and the reflectance R as their product

$$S = L \cdot R, \tag{1}$$

where L is the light field information and R is the reflectance. This equation explains how light field information influences the appearance of the image. In the LFCycleGAN, the input image X is processed through the light field module to get the light field information y'_{LF} by Equation (1). The LFCycleGAN takes the original image X and generates the enhanced light field $LF_{enhanced}$. This process can be expressed as a functional relationship as

$$LF_{enhanced} = G_{LF1}(y'_{LF}). \tag{2}$$

We hypothesize that if the enhanced light field information $LF_{enhanced}$ accurately reflects the light conditions of the real scene, then the smaller $C(y', LF_{enhanced})$ is, the closer the light distribution of the restored image y' is to that of the real scene. Therefore, the RCycleGAN can leverage this comparison mechanism to improve the restoration quality of the image. This can be further expressed as:

$$Q(y') = g(C(y', LF_{enhanced})), \tag{3}$$

where $Q(y')$ is the quality assessment function of the image, and g is a function that describes the impact of the comparison result $C(y', LF_{enhanced})$ on the restoration quality. The ultimate goal is to maximize $Q(y')$, i.e., improve the restoration quality of the image. This can be achieved by minimizing the comparison function $C(y', LF_{enhanced})$

$$\max_{G_{R1}} Q(y') \text{ subject to } \max_{G_{R1}} C(y', LF_{enhanced}) \tag{4}$$

This mathematical analysis describes how LFCycleGAN uses the light field information generated by the RCycleGAN as a comparison baseline for image restoration. By minimizing the difference between the restored image and the light field information, it enhances the quality of the restored image.

3.2. Loss Design in RCycleGAN and LFCycleGAN

To achieve the goal of underwater image restoration in RCycleGAN and mimic features of light field from LFCycleGAN, we include seven kinds of formulate loss functions in the proposed dual-CycleGAN and they are adversarial loss ($\mathcal{L}_{GAN}$), cycle consistency loss ($\mathcal{L}_{cyc}$), identity loss ($\mathcal{L}_{id}$), perceptual loss ($\mathcal{L}_{per}$), patch-based contrast quality index(PCQI) [27] loss ($\mathcal{L}_{pcqi}$), color balance loss ($\mathcal{L}_{cb}$), and intermediate output (Inter) loss ($\mathcal{L}_{inter}$).

Adversarial loss $\mathcal{L}_{GAN}$ is used to determine the similarity between the produced data and the actual data distribution. Different from the adversarial loss of the original CycleGAN, in our RCycleGAN and LFCycleGAN, we not only optimize the equations for converting underwater image into in-air image, G_{R1} and on-road image into underwater images, G_{R2}, but also, we optimize the equations for converting underwater light field into in-air light field G_{LF1} and equations for converting in-air light field into underwater light field. We also incorporate four discriminators: D_R, comprising D_{R1} and D_{R2}, and D_{LF}, comprising D_{LF1} and D_{LF2}, to distinguish the translated samples from those of the real samples. Therefore, the proposed adversarial loss $\mathcal{L}_{GAN}$ is expressed as:

$$\begin{aligned}\mathcal{L}_{GAN}(G_{R1}, G_{LF1}, D_{R1}, D_{LF1}, X, Y) \\ = \mathbb{E}_{y\sim p_{data}(y)}\left[{\log(D_{R1}(y)) \atop +\log(D_{LF1}(y))}\right] \\ +\mathbb{E}_{x\sim p_{data}(x)}\left[{\log(1-D_{R1}(G_{R1}(x))) \atop +\log(1-D_{LF1}(G_{LF1}(x)))}\right],\end{aligned} \tag{5}$$

where $x \sim p_{data}(x)$ and $y \sim p_{data}(y)$ denote the data distributions of the samples $\{x_i\}$ and $\{y_i\}$, respectively. The expectation $\mathbb{E}_{x\sim p_{data}(x)}$ represents the average value of a function over the distribution $p_{data}(x)$, meaning it calculates the expected value of some operation across all possible samples x drawn from the real data distribution. Similarly, $\mathbb{E}_{y\sim p_{data}(y)}$ represents the expected value over the distribution $p_{data}(y)$, calculated across all possible samples y. G_{R1} and G_{LF1} generate images that aim to resemble domain Y, starting from domain X (i.e., $X \to Y$). D_{R1} and D_{LF1} are the discriminators responsible for distinguishing the RGB and light field characteristics of the generated images from the real ones. Similarly, we define an adversarial loss for the reverse mapping ($Y \to X$), using generators G_{R2}, G_{LF2} and discriminators D_{R2}, D_{LF2}.

Cycle consistency loss, L_{cyc} is used to ensure that the image converted back retains the complete information and characteristics of the original image. Different from the cycle consistency loss of the original CycleGAN, we introduce the respective optimizations in underwater images and underwater light fields in the proposed cycle consistency loss, L_{cyc}, defined as:

$$\begin{aligned}\mathcal{L}_{cyc}(G_{R1}, G_{R2}, G_{LF1}, G_{LF2}) \\ = \mathbb{E}_{x\sim p_{data}(x)}\left[{\|G_{R2}(G_{R1}(x))-x\|_1 \atop +\|G_{LF2}(G_{LF1}(x))-x\|_1}\right] \\ +\mathbb{E}_{y\sim p_{data}(y)}\left[{\|G_{R1}(G_{R2}(y))-y\|_1 \atop +\|G_{LF1}(G_{LF2}(y))-y\|_1}\right].\end{aligned} \tag{6}$$

We utilize the L1 norm (i.e., Manhattan distance) as part of the cycle consistency loss to ensure that the mapping from one domain to another is reversible. Compared to the L2 norm (i.e., Euclidean distance), the L1 norm does not impose excessive penalties on small errors, allowing the generator to capture the features of real images more stably during the training process. Additionally, the sparsity characteristics of the L1 norm help produce images that are more interpretable and visually appealing. By focusing on the similarity between the generated and original images, our model effectively preserves the semantic information of the input images, thereby achieving higher-quality image translation.

Identity loss, $\mathcal{L}_{id}$, ensures that when the input image belongs to the target domain, the generator does not alter it and outputs the same image as the input. This avoids unnecessary changes and maintains consistency in the content of the image. Different from the identity loss of the original CycleGAN, we introduce the respective optimizations in RGB and light field components in the proposed identity loss $\mathcal{L}_{id}$, is defined as:

$$\begin{aligned}\mathcal{L}_{id}(G_{R1}, G_{R2}, G_{LF1}, G_{LF2}) \\ = \mathbb{E}_{y\sim p_{data}(y)}\begin{bmatrix}\|G_{R1}(y)-y\|_1 \\ +\|G_{LF1}(y)-y\|_1\end{bmatrix} + \mathbb{E}_{x\sim p_{data}(x)}\begin{bmatrix}\|G_{R2}(x)-x\|_1 \\ +\|G_{LF2}(x)-x\|_1\end{bmatrix}.\end{aligned} \tag{7}$$

Perceptual loss focuses more on the perceived quality of the image, aligning it better with how a human observer perceives image quality. To address the blurring issue in images generated by GAN, we incorporate perceptual loss, L_{per}, which is based on the VGG16 network, is defined as:

$$\begin{aligned}\mathcal{L}_{per}(G_R, G_{LF}) = \mathbb{E}_{x\sim p_{data}(x)}&\begin{bmatrix}\sum_i \lambda_i\|\phi_i(x) - \phi_i(G_R(x))\|_1 \\ +\sum_i \lambda_i\|]\phi_i(x) -]\phi_i(G_{LF}(x))\|_1\end{bmatrix} \\ +\mathbb{E}_{y\sim p_{data}(y)}&\begin{bmatrix}\sum_i \lambda_i\|\phi_i(y) - \phi_i(G_R(y))\|_1 \\ +\sum_i \lambda_i\|\phi_i(y) -]\phi_i(G_{LF}(y))\|_1\end{bmatrix},\end{aligned} \quad (8)$$

where ϕ_i and λ_i represents the feature extraction function and weight assigned from the i-th layer of the VGG16, respectively.

To refine the color quality of the restored image $G_R(x)$ from the input x in accordance with human visual perception, we incorporate the PCQI loss function, which is defined as:

$$\mathcal{L}_{pcqi}(G_{R2}(G_{R1}(x)), x) = e^{-1\times PCQI(G_{R2}(G_{R1}(x)),x)}, \quad (9)$$

where the PCQI function is defined in [27].

In addition, the color balance loss function $\mathcal{L}_{cb}$ is used to adjust and optimize the color distribution in images to achieve a more balanced or desired color representation and is defined as:

$$\mathcal{L}_{cb}(y') = \sum_{C\in\{R,\ G,\ B\}}\left|\frac{1}{H\times W}\sum_{i,j=0}^{H\times W} y'_C(i,j) - M_{y'}\right|, \quad (10)$$

where y' is the output map of G_{R1}, y'_C denotes the color component of $C \in \{R, G, B\}$, H and W are the rows and hights in y', $M_{y'}$ is the pixel mean across the three color channels of y', and i and j are row and column coordinates of the y' image. Finally, we use intermedia output loss $\mathcal{L}_{inter}$ to assess the color similarity between the light field image produced by $G_{LF1}(x)$ and the image generated by $G_{R1}(x)$, as defined by

$$\mathcal{L}_{inter}\ (x) = \sum_{i,j=0}^{H\times W}\left|G_{R1}(x)_{(i,j)} - G_{LF1}(x)_{(i,j)}\right|. \quad (11)$$

The loss functions of RCycleGAN, $\mathcal{L}_{LOSS}^{RCycleGAN}$, and LFCycleGAN, $\mathcal{L}_{LOSS}^{LFCycleGAN}$ are defined as:

$$\begin{aligned}\mathcal{L}_{LOSS}^{RCycleGAN} = &\mathcal{L}_{GAN}^{RCycleGAN} + \mathcal{L}_{cyc}^{RCycleGAN} + \mathcal{L}_{id}^{RCycleGAN} + \mathcal{L}_{per}^{RCycleGAN} + \mathcal{L}_{pcqi}^{RCycleGAN} \\ &+\mathcal{L}_{cb}^{RCycleGAN} + \mathcal{L}_{inter},\end{aligned} \quad (12)$$

$$\mathcal{L}_{LOSS}^{LFCycleGAN} = \mathcal{L}_{GAN}^{LFCycleGAN} + \mathcal{L}_{cyc}^{LFCycleGAN} + \mathcal{L}_{id}^{LFCycleGAN} + \mathcal{L}_{per}^{LFCycleGAN}. \quad (13)$$

In the overall training process for jointly training our RCycleGAN and LFCycleGAN, the complete loss function employed to train the proposed model dual-CycleGAN model is defined as:

$$\mathcal{L}_{Loss}^{Totall} = \mathcal{L}_{Loss}^{RGBCycleGAN} + \mathcal{L}_{Loss}^{GrayCycleGAN}. \quad (14)$$

In the training process, the perceptual loss is iteratively calculated by $\mathcal{L}_{per}(y', y'_{LF})$, as depicted in Figure 2. That is, minimizing this perceptual loss can be viewed as the linkage between the two CycleGANs. During the overall network training process, the iteratively updated light field image y'_{LF} is used as the guidance to guide the restoration of the input underwater image.

4. Experimental Results

To train the dual-CycleGAN for underwater image restoration with unpaired images, four datasets of different domains are used. Three of these datasets are widely recognized for underwater image augmentation, including UFO-120 [11], EUVP [28], and UIEB [10]. The DIV2K (DIVerse 2K) dataset [29] is used as the in-air domain dataset. The proposed method was implemented using PyTorch version 2.0.1 with the Python programming language. The model optimization was performed using the Adam optimizer [30], with the initial learning rate configured at 0.00002. The training input patch size was set to 256 × 256, and the model was trained for four hundred epochs. This section is organized under subheadings, offering a succinct and clear explanation of the experimental results, their analysis, and the conclusions derived from the experiments.

4.1. Quantitative Comparisons

To assess the performance of our method quantitatively, three well-known quantitative metrics are used and there are PSNR (peak signal-to-noise ratio), SSIM (structural similarity index measure), and UIQM (underwater image quality measurement) [31], which leverages multiple factors influencing underwater image quality. It integrates three components: underwater image color measurement (UICM), underwater image sharpness measurement (UISM), and underwater image contrast measurement (UIConM) to provide a comprehensive evaluation. Typically, a higher UIQM value indicates better image quality. To assess the performance, we compare our method against five state-of-the-art deep learning-based underwater image enhancement techniques, which are Deep SESR [11], Shallow-UWnet [12], UGAN [18], WaterNet [17], and PUGAN [20]. Among the five comparison methods, the first two are end-to-end model architectures that require paired training data, while the last three are GAN based and do not require paired training data. As revealed by Table 1, our method exhibits better or comparable quantitative performances. Our proposed method achieves a PSNR that is only 1.89 lower than Deep SESR on the UFO-120 dataset, while our GFLOPs are just 12.4% of those of Deep SESR. Additionally, the Deep SESR method requires paired data for training. Compared to PUGAN, our method demonstrates superior performance compared to others, achieving higher PSNR and SSIM values while achieving comparable UIQM. Also, our method has only 25% of the computational cost of PUGAN, which represents a significant improvement in computational efficiency.

Table 1. Quantitative performance assessments on UFO-120, EUVP, and UIEB datasets.

Method	**UFO-120**			**EUVP**			**UIEB**		
	PSNR	**SSIM**	**UIQM**	**PSNR**	**SSIM**	**UIQM**	**PSNR**	**SSIM**	**UIQM**
Deep SESR	27.15	0.84	3.13	25.25	0.75	2.98	19.26	0.73	2.97
Shallow-UWnet	25.20	0.73	2.85	27.39	0.83	2.98	18.99	0.67	2.77
UGAN	23.45	0.80	3.04	23.67	0.67	2.70	20.68	0.84	3.17
WaterNet	22.46	0.79	2.83	20.14	0.68	2.55	19.11	0.80	3.04
PUGAN	23.70	0.82	2.85	24.05	0.74	2.94	21.67	0.78	3.28
Dual-CycleGAN	25.23	0.84	3.06	27.39	0.91	2.97	22.12	0.85	3.26

Table 2 provides a detailed comparison of the GFLOPs (giga floating point operations) and the number of network parameters required by the evaluated methods. Our model stands out by demonstrating a significant reduction in both GFLOPs and network parameters when compared to the other methods. This results in a more computationally efficient model that retains or even surpasses the performance of state-of-the-art methods in terms of underwater image restoration quality. By achieving a lighter model complexity without compromising performance, our approach not only addresses computational resource limitations but also demonstrates scalability, making it highly suitable for practical applications with constrained resources. This efficiency, alongside its superior restoration capability, is one of the most notable strengths of our proposed method, as it ensures faster processing times and lower memory usage, thus making it a highly practical solution for real-world deployment.

Table 2. Complexity evaluations for difference methods.

Method	FLOPs	Parameters
Deep SESR	146.10 G	2.46 M
Shallow-UWnet	21.63 G	0.22 M
UGAN	38.97 G	57.17 M
WaterNet	193.70 G	24.81 M
PUGAN	72.05 G	95.66 M
Dual-CycleGAN	18.15 G	54.41 M

4.2. Qualitative Comparisons

As illustrated in Figure 4, our proposed method significantly outperforms state-of-the-art approaches, particularly in preserving image details and enhancing color representation. Moreover, benefited from the proposed guidance learning, our method recovers better color representation than those recovered by other methods.

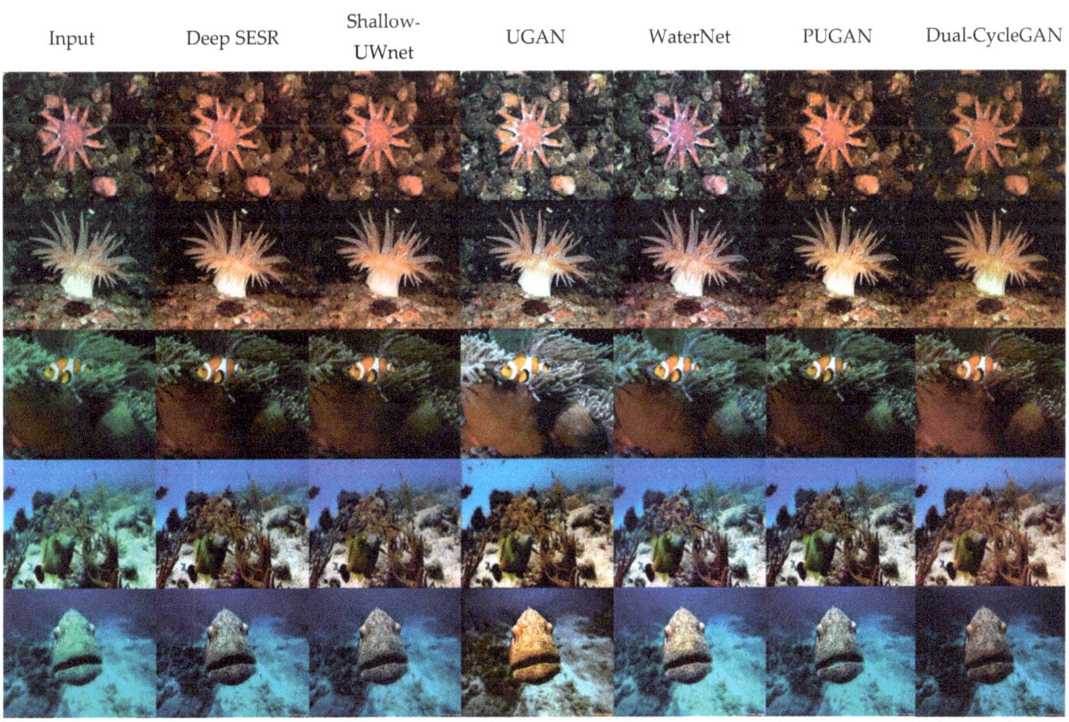

Figure 4. Qualitative evaluation results on the UFO-120 dataset.

4.3. Ablation Study

To assess the contribution of each component in the proposed model, we performed the ablation studies, shown as follows. We remove the guidance learning component and loss function from the proposed framework to evaluate the effectiveness of the learned light field guidance, $\mathcal{L}_{inter}$, $\mathcal{L}_{pcqi}$, and $\mathcal{L}_{cb}$. The results of the ablation study are presented in Table 3. From the findings, it is evident that removing LFCycleGAN from the proposed method results in a decline in image quality, including blur, loss of structure, and color distortion. As shown in Table 3, the complete method demonstrates the best performance in both quantitative and qualitative restoration. All components, including the loss functions and the guidance learning component, are essential for optimal underwater image restoration. Removing any of these components resulted in a significant degradation in image quality, highlighting their necessity in achieving the best restoration performance.

Table 3. Quantitative results of ablation studies.

	UIEB		
	PSNR	SSIM	UIQM
Complete Dual-CycleGAN	22.12	0.85	3.26
$(w/o)\ \mathcal{L}_{inter}$	20.81	0.82	2.99
$(w/o)\ \mathcal{L}_{pcqi}$	20.75	0.84	2.96
$(w/o)\ \mathcal{L}_{cb}$	20.8	0.84	2.96
$(w/o)\ \mathcal{L}_{pcqi}$ & $\mathcal{L}_{cb}$	20.88	0.84	2.93

4.4. Discussion

Compared with the existing literature, the proposed method outperforms GAN-based models, such as WaterNet and PUGAN, in quantitative metrics like PSNR, SSIM, and UIQM, while maintaining lower computational complexity. Importantly, the model's reliance on unpaired training data overcomes a major challenge in applying underwater restoration models to real-world scenarios, highlighting its methodological flexibility and applicability. Compared to traditional physics-based approaches or shallow networks like Shallow-UWnet, the dual-CycleGAN leverages light field guidance to enhance color authenticity and detail restoration. This improvement likely stems from the light field module's ability to capture additional information related to light propagation, effectively mitigating the scattering and absorption effects prevalent in underwater environments. Additionally, the CycleGAN's adversarial learning framework resolves the data pairing challenges inherent in underwater scenarios, offering greater stability compared to conventional GAN models.

However, several limitations of this study warrant further investigation. First, while the light field guidance module significantly improves restoration, its reliability in highly turbid or dynamic water bodies requires further validation. Second, the model's generalizability across diverse underwater scenarios may benefit from additional real-world training data. Furthermore, although the framework's computational demands are relatively low, its real-time application in resource-constrained environments, such as underwater drones, remains an area for optimization. Future research directions include designing more efficient and generalizable light field generation methods to handle diverse underwater conditions. Incorporating advanced deep learning techniques, such as self-attention mechanisms or variational autoencoders, could further enhance restoration detail and stability. Additionally, developing multi-frame image restoration techniques for dynamic underwater scenarios presents another promising avenue.

5. Conclusions

This paper proposes a dual-CycleGAN architecture, extending and refining concepts established in previous works on underwater image enhancement and restoration. The model employs a dynamic guided learning approach, utilizing one CycleGAN to generate light field information as a reference for the second CycleGAN, which performs the actual restoration. This innovative dual-CycleGAN structure addresses persistent challenges in underwater image processing, such as color attenuation, contrast loss, and noise, which are often exacerbated by the unique optical properties of water. Drawing from existing literature, most of the methods often rely on paired datasets or heuristic algorithms; our approach circumvents the need for paired training data, thus enhancing flexibility and applicability across diverse underwater environments. By transforming underwater images to their in-air counterparts, our model achieves substantial improvements in color correction, detail preservation, and texture recovery, as validated by comprehensive experimental results. The proposed model not only outperforms the existing state-of-the-art methods in both quantitative metrics and qualitative visual assessments but also demonstrates comparable or superior performance in recovering crucial image features. Moreover, its lower computational complexity facilitates faster processing times, making it highly scalable and suitable for real-time applications, such as AUVs and remote sensing systems. This advancement in computational efficiency positions our method as a promising solution for large-scale deployment in practical scenarios where computational resources may be limited.

Author Contributions: Conceptualization, C.-H.Y.; Methodology, C.-H.Y.; Software, Y.-Y.L.; Validation, Y.-Y.L.; Formal analysis, C.-H.Y.; Investigation, Y.-Y.L.; Data curation, Y.-Y.L.; Writing—original draft, Y.-Y.L.; Writing—review & editing, C.-H.Y. and W.-J.H.; Visualization, Y.-Y.L.; Supervision, C.-H.Y. and W.-J.H.; Project administration, C.-H.Y.; Funding acquisition, W.-J.H. All authors have read and agreed to the published version of the manuscript.

Funding: This research was funded by National Science and Technology Council, grant number MOST 110-2221-E-003-005-MY3, MOST 111-2221-E-003-007-MY3, and NSTC 113-2221-E-003-018-MY3.

Institutional Review Board Statement: Not applicable.

Informed Consent Statement: Not applicable.

Data Availability Statement: The data presented in this study are available in [10,11,28,29].

Conflicts of Interest: The authors declare no conflict of interest.

References

1. Peng, L.; Zhu, C.; Bian, L. U-Shape transformer for underwater image enhancement. *IEEE Trans. Image Process.* **2023**, *32*, 3066–3079. [CrossRef]
2. Liu, X.; Chen, Z.; Xu, Z.; Zheng, Z.; Ma, F.; Wang, Y. Enhancement of underwater images through parallel fusion of transformer and CNN. *J. Mar. Sci. Eng.* **2024**, *12*, 1467. [CrossRef]
3. Yeh, C.-H.; Lin, C.-H.; Lin, M.-H.; Kang, L.-W.; Huang, C.-H.; Chen, M.-J. Deep learning-based compressed image artifacts reduction based on multi-scale image fusion. *Inf. Fusion* **2021**, *67*, 195–207. [CrossRef]
4. Yeh, C.-H.; Lai, Y.-W.; Lin, Y.-Y.; Chen, M.-J.; Wang, C.-C. Underwater image enhancement based on light field guided rendering network. *J. Mar. Sci. Eng.* **2024**, *12*, 1217. [CrossRef]
5. Chiang, Y.-W.; Chen, Y.-C. Underwater image enhancement by wavelength compensation and dehazing. *IEEE Trans. Image Process.* **2012**, *21*, 1756–1769. [CrossRef] [PubMed]
6. Li, C.; Guo, J.; Pang, Y.; Chen, S.; Wang, J. Single underwater image restoration by blue-green channels dehazing and red channel correction. In Proceedings of the IEEE International Conference on Acoustics, Speech and Signal Processing, Shanghai, China, 20–25 March 2016; pp. 1731–1735.
7. Ancuti, C.O.; Ancuti, C.; De Vleeschouwer, C.; Bekaert, P. Color balance and fusion for underwater image enhancement. *IEEE Trans. Image Process.* **2018**, *27*, 379–393. [CrossRef] [PubMed]

8. Li, C.; Anwar, S.; Porikli, F. Underwater scene prior inspired deep underwater image and video enhancement. *Pattern Recognit.* **2020**, *98*, 107038. [CrossRef]
9. Dudhane, A.; Hambarde, P.; Patil, P.W.; Murala, S. Deep underwater image restoration and beyond. *IEEE Signal Process. Lett.* **2020**, *27*, 675–679. [CrossRef]
10. Li, C.; Guo, C.; Ren, W.; Cong, R.; Hou, J.; Kwong, S.; Tao, D. An underwater image enhancement benchmark dataset and beyond. *IEEE Trans. Image Process.* **2019**, *29*, 4376–4389. [CrossRef] [PubMed]
11. Islam, M.J.; Luo, P.; Sattar, J. Simultaneous enhancement and super-resolution of underwater imagery for improved visual perception. *arXiv* **2020**, arXiv:2002.01155.
12. Naik, A.; Swarnakar, A.; Mittal, K. Shallow-uwnet: Compressed model for underwater image enhancement (student abstract). In Proceedings of the AAAI Conference on Artificial Intelligence, Virtual, 19–21 May 2021.
13. Zhou, J.; Liu, Q.; Jiang, Q.; Ren, W.; Lam, K.-M.; Zhang, W. Underwater camera: Improving visual perception via adaptive dark pixel prior and color correction. *Int. J. Comput. Vis.* **2023**, 1–19. [CrossRef]
14. Pramanick, A.; Sur, A.; Saradhi, V.V. Harnessing multi-resolution and multi-scale attention for underwater image restoration. *arXiv* **2024**, arXiv:2408.09912.
15. Goodfellow, I.; Pouget-Abadie, J.; Mirza, M.; Xu, B.; Warde-Farley, D.; Ozair, S.; Bengio, Y. Generative adversarial nets. In Proceedings of the Annual Conference on Neural Information Processing Systems, Montreal, QC, Canada, 8–13 December 2014; pp. 2672–2680.
16. Zhu, J.Y.; Park, T.; Isola, P.; Efros, A.A. Unpaired image-to-image translation using cycle-consistent adversarial networks. In Proceedings of the IEEE International Conference on Computer Vision, Venice, Italy, 22–29 October 2017; pp. 2242–2251.
17. Li, J.; Skinner, K.A.; Eustice, R.M.; Johnson-Roberson, M. WaterGAN: Unsupervised generative network to enable real-time color correction of monocular underwater images. *IEEE Robot. Autom. Lett.* **2018**, *3*, 387–394. [CrossRef]
18. Fabbri, C.; Islam, M.J.; Sattar, J. Enhancing underwater imagery using generative adversarial network. In Proceedings of the IEEE International Conference on Robotics and Automation, Brisbane, Australia, 21–25 May 2018; pp. 7159–7165.
19. Guo, Y.; Li, H.; Zhuang, P. Underwater image enhancement using a multiscale dense generative adversarial network. *IEEE J. Ocean. Eng.* **2020**, *45*, 862–870. [CrossRef]
20. Cong, R.; Yang, W.; Zhang, W.; Li, C.; Guo, C.-L.; Huang, Q.; Kwong, S. PUGAN: Physical model-guided underwater im-age enhancement using GAN with dual-discriminators. *IEEE Trans. Image Process* **2023**, *32*, 4472–4485. [CrossRef] [PubMed]
21. Ye, T.; Chen, S.; Liu, Y.; Ye, Y.; Chen, E.; Li, Y. Underwater light field retention: Neural rendering for underwater imaging. In Proceedings of the IEEE/CVF Conference on Computer Vision and Pattern Recognition, New Orleans, LA, USA, 18–24 June 2022; pp. 488–497.
22. Rahman, Z.; Jobson, D.J.; Woodell, G.A. Multi-scale retinex for color image enhancement. In Proceedings of the IEEE International Conference on Image Processing, Lausanne, Switzerland, 16–19 September 1996; Volume 3, pp. 1003–1006.
23. Chang, B.; Zhang, Q.; Pan, S.; Meng, L. Generating handwritten chinese characters using cyclegan. In Proceedings of the 2018 IEEE Winter Conference on Applications of Computer Vision (WACV), Lake Tahoe, NV, USA, 12–15 March 2018; pp. 199–207.
24. Arjovsky, M.; Bottou, L. Towards principled methods for training generative adversarial networks. *arXiv* **2017**, arXiv:1701.04862.
25. Srivastava, A.; Valkov, L.; Russell, C.; Gutmann, M.-U.; Sutton, C. Veegan: Reducing mode collapse in gans using implicit variational learning. In *Advances in Neural Information Processing Systems 30*; Long Beach Convention & Entertainment Center: Long Beach, CA, USA, 2017.
26. Li, W.; Fan, L.; Wang, Z.; Ma, C.; Cui, X. Tackling mode collapse in multi-generator gans with orthogonal vectors. *Pattern Recognit.* **2021**, *110*, 107646. [CrossRef]
27. Wang, S.; Ma, K.; Yeganeh, H.; Wang, Z.; Li, W. A patch-structure representation method for quality assessment of contrast changed images. *IEEE Signal Process. Lett.* **2015**, *22*, 2387–2390. [CrossRef]
28. Islam, M.J.; Xia, Y.; Sattar, J. Fast underwater image enhancement for improved visual perception. *IEEE Robot. Autom. Lett.* **2020**, *5*, 3227–3234. [CrossRef]
29. Agustsson, E.; Timofte, R. NTIRE 2017 Challenge on single image super-resolution: Dataset and study. In Proceeding of the 2017 IEEE Conference on Computer Vision and Pattern Recognition Workshops, Honolulu, HI, USA, 21–26 July 2017; pp. 126–135.
30. Kinga, D.; Adam, J.B. A method for stochastic optimization. *arXiv* **2017**, arXiv:1412.6980.
31. Panetta, K.; Gao, C.; Agaian, S. Human-visual-system-inspired underwater image quality measures. *IEEE J. Ocean. Eng.* **2015**, *41*, 541–551. [CrossRef]

Journal of
*Marine Science
and Engineering*

MDPI

Review

Comparative Analysis of Traditional and Deep Learning Approaches for Underwater Remote Sensing Image Enhancement: A Quantitative Study

Yunsheng Ma [1,2,†], Yanan Cheng [3,†] and Dapeng Zhang [1,*]

1 Ship and Maritime College, Guangdong Ocean University, Zhanjiang 524005, China; yitcycyic@163.com
2 School of Electronics and Information Engineering, Guangdong Ocean University, Zhanjiang 524088, China
3 Taizhou Institute of Science & Technology, College of Business, NJUST, Taizhou 225300, China
* Correspondence: zhangdapeng@gdou.edu.cn
† These authors contributed equally to this work.

Abstract: Underwater remote sensing image enhancement is complicated by low illumination, color bias, and blurriness, affecting deep-sea monitoring and marine resource development. This study compares a multi-scale fusion-enhanced physical model and deep learning algorithms to optimize intelligent processing. The physical model, based on the Jaffe–McGlamery model, integrates multi-scale histogram equalization, wavelength compensation, and Laplacian sharpening, using cluster analysis to target enhancements. It performs well in shallow, stable waters (turbidity < 20 NTU, depth < 10 m, PSNR = 12.2) but struggles in complex environments (turbidity > 30 NTU). Deep learning models, including water-net, UWCNN, UWCycleGAN, and U-shape Transformer, excel in dynamic conditions, achieving UIQM = 0.24, though requiring GPU support for real-time use. Evaluated on the UIEB dataset (890 images), the physical model suits specific scenarios, while deep learning adapts better to variable underwater settings. These findings offer a theoretical and technical basis for underwater image enhancement and support sustainable marine resource use.

Keywords: underwater remote sensing; deep learning algorithms; multi-scale fusion-enhanced physical model; underwater image enhancement

1. Introduction

The oceans are abundant in resources [1–3], and with the continuous advancement of science and technology, humanity is increasingly exploring their depths [4,5]. Marine fishery farming, as one of the most important industries in the marine sector, has faced challenges in recent years due to offshore water pollution and other factors [6,7]. As a result, there has been a gradual shift from traditional offshore net cage farming to the exploration and development of deep-sea marine ranching [8–10].

Underwater imagery is one of the most important sources of marine information, playing a crucial role in the monitoring and management of deep-sea marine ranches [11,12]. However, the complex marine environment often hampers the acquisition of high-quality underwater images, leading to issues such as color distortion, reduced contrast, uneven illumination, and noise. These challenges arise from the absorption and scattering effects of light as it propagates through water, resulting in images with decreased clarity, contrast, and color fidelity. This directly impacts the accuracy of visual perception and the subsequent extraction and analysis of important information [13–15]. Figure 1a shows three underwater degradation images. Figure 1b shows a schematic diagram of the imaging process for an underwater image. Figure 1c shows the absorption behavior of light underwater.

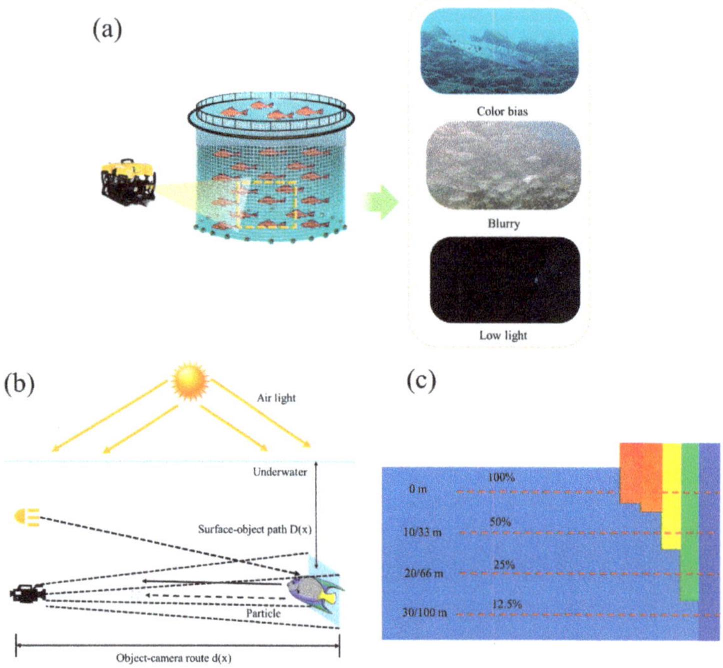

Figure 1. (**a**) Three types of underwater degradation images. (**b**) Schematic diagram of underwater imaging process. (**c**) Absorption properties of light in water.

Zhou et al. proposed a multi-feature fusion method (MFFM) to enhance underwater images affected by color distortion and low contrast. MFFM combines color correction and contrast enhancement techniques, generating a fusion weight map based on the features of the color-corrected and contrast-enhanced images. This method effectively balances color and contrast while preserving the natural characteristics of the image [16].

Li et al. introduced the Underwater Image Enhancement Benchmark (UIEB), a dataset of 950 real-world underwater images designed to evaluate and advance underwater image enhancement algorithms. They also presented water-net, a convolutional neural network model trained on this dataset to improve image quality. Water-net is a fully convolutional network that integrates inputs with predicted confidence maps. A feature transformation unit optimizes the input before fusion, enhancing the overall results [17].

A new underwater image enhancement method was proposed by Ancuti et al., which introduces a multi-scale fusion technique that combines inputs from white-balanced images. This method applies gamma correction and edge sharpening to improve visibility. By using normalized weight mapping, the technique efficiently blends the inputs to produce artifact-free images [18].

Garg et al. proposed a novel method for enhancing underwater images that combines contrast-limited adaptive histogram equalization (CLAHE) with a percentile approach to improve image clarity and visibility [19].

Yang et al. present a solution for underwater image enhancement using a deep residual framework, which involves generating synthetic training data and employing a VDSR model for super-resolution tasks via CycleGAN. They introduce the underwater reset (uresNet) model, which improves the loss function by incorporating a multinomial approach that includes Edge Difference Loss (EDL) and Mean Error Loss, enhancing image quality [20].

Hou et al. present a large synthetic underwater image dataset (SUID) designed to enhance the evaluation of underwater image enhancement and restoration algorithms. The SUID consists of 900 images generated using the Underwater Image Synthesis Algorithm (UISA), featuring various degradation types and turbidity levels [21].

Zhang et al. present a contrast enhancement method that involves color channel attenuation correction while preserving image details. The approach utilizes a specially designed attenuation matrix to address poor-quality color channels and employs a bi-histogram-based technique for both global and local contrast enhancement [22].

Guo et al. present a novel approach to underwater image enhancement using a multi-scale dense generative adversarial network (MSDB-GAN) that integrates residual learning and dense connectivity techniques. The method employs a multinomial loss function to generate visually appealing enhancement results [23].

Underwater image enhancement methodologies have evolved along two distinct trajectories: physics-based traditional approaches and data-driven deep learning paradigms. While traditional methods (e.g., Jaffe–McGlamery model [24]) provide interpretable solutions grounded in optical propagation principles, their performance often degrades in complex scenarios due to oversimplified assumptions about water turbidity and heterogeneous lighting conditions. Conversely, deep learning techniques demonstrate remarkable adaptability through learned feature representations (e.g., ResNet architectures [25]), yet remain constrained by substantial data requirements [26] and limited physical interpretability [27]. This methodological dichotomy creates critical knowledge gaps in three aspects:

(1) Absence of empirical evidence quantifying performance boundaries between physics-driven and data-driven approaches across marine environments [28];

(2) Underexplored synergies combining physical priors with neural network architectures (e.g., RD-Unet [29] or MetaUE [30]);

(3) Lack of standardized evaluation protocols addressing both full-reference and non-reference metrics (PSNR/UIQM/UCIQE) and operational feasibility (real-time processing) [31,32].

Our comparative analysis reveals the differences under different approaches for underwater imagery by systematically evaluating classical algorithms against four state-of-the-art depth models in 890 images from the UIEB dataset, but also develops a practical guide for selecting enhancement methods based on depth gradient and turbidity levels—a decision support framework urgently needed in marine rangeland monitoring, where equipment limitations and environmental variability coexist.

2. Traditional Methodologies

2.1. Preprocessing Framework for Degradation Characterization

To enhance underwater images, the first step is to effectively classify the images in the dataset. This requires the selection of appropriate recognition and classification methods based on the characteristics of each degradation type [17]. To this end, we performed a preliminary analysis of the dataset. Images categorized as "fuzzy" exhibited blurred object edges, leading to a loss of detail [33]. This made it difficult to recognize complex textures or fine features. These images resemble out-of-focus photographs, with an overall lack of clarity [34]. Low-light images, on the other hand, are characterized by overall dimness and a lack of contrast between highlights and shadows, making the outlines of objects hard to distinguish [35]. Additionally, these images typically contain a high level of noise [36,37]. Color-biased images primarily exhibit a green or blue hue [38,39], resulting in unnatural color contrast [40] and a loss of the original color fidelity [41]. Red and yellow objects, in particular, are heavily affected, appearing either severely distorted or nearly invisible [42].

2.1.1. Grayscale Conversion and Luminance Analysis

Color images contain a vast amount of information, which can complicate computational processes and reduce efficiency [43]. To address this, RGB three-channel images are converted into single-channel grayscale images [44]. There are three common methods for grayscale conversion: the maximum value method, the average value method, and the weighted average method [45–47]. In this study, the weighted average method is employed, where the R, G, and B components are averaged based on specific weights [48].

$$I_{greyscale}(i,j) = 0.299R(i,j) + 0.587G(i,j) + 0.114B(i,j) \tag{1}$$

where $I_{greyscale}$ represents the pixel value of the image after grayscale conversion, while R, G, and B denote the pixel values of the red, green, and blue channels of the original image, respectively.

In digital image processing, Digital Average Brightness (DA) is a metric that quantifies the overall brightness level of an image. It is calculated as the average of pixel values, providing a straightforward measure of the image's overall luminance.

$$DA = \frac{1}{M \times N} \sum_{i=1}^{M} \sum_{j=1}^{N} I(i,j) \tag{2}$$

Here, M and N represent the height and width of the image, respectively, defining its pixel dimensions. $I(i,j)$ denotes the luminance value at the pixel located at position (i,j).

The absolute value of the luminance deviation is defined as the average of the ab-solute values of the pixel luminance values that deviate from the mean value.

$$D = \frac{1}{M \times N} \sum_{i=1}^{M} \sum_{j=1}^{N} |I(i,j) - L_{mean}| \tag{3}$$

where L_{mean} is the average brightness of the image. According to the intermediate value of the grey scale value, L_{mean} takes the value of 128. The larger the value of D, the greater the fluctuation in the distribution of luminance values within the image, resulting in a more pronounced contrast between bright and dark areas. Conversely, the smaller the value of D, the more uniform the luminance distribution becomes.

The weighted calculation utilizes information about the luminance distribution of the image. The grayscale histogram, Hist[i], represents the frequency of pixels with a luminance value of i in the image, indicating how often each luminance value appears. The range of luminance values in the image spans from 0 to 255.

The luminance anomaly parameter k is a parameter that measures the relative relationship between the overall deviation in image luminance and the deviation in luminance distribution.

$$k = \frac{D}{M} \tag{4}$$

When the value of k is less than 1, it indicates that the overall luminance deviation of the image is small relative to the deviation in the luminance distribution. This means the luminance distribution is more uniform, with no significant luminance anomalies. Conversely, when k is greater than 1, it suggests that the overall brightness deviation of the image is larger than the deviation in the brightness distribution. This often results in noticeable brightness abnormalities, such as brightness being concentrated within a certain range or exhibiting significant unevenness.

2.1.2. Laplacian Sharpness Detection

Sharpness degradation, characterized by blurred edges and the loss of fine details, is a common issue in underwater imaging [49]. The Laplacian variance-based method provides a robust quantitative measure for evaluating image sharpness, enabling the targeted enhancement of degraded regions [50].

The Laplacian variance-based method is a classic algorithm for evaluating image sharpness. It assesses sharpness by calculating the Laplacian operator of the image, which highlights the high-frequency components that correspond to image details [51]. The Laplacian response is shown in Figure 2.

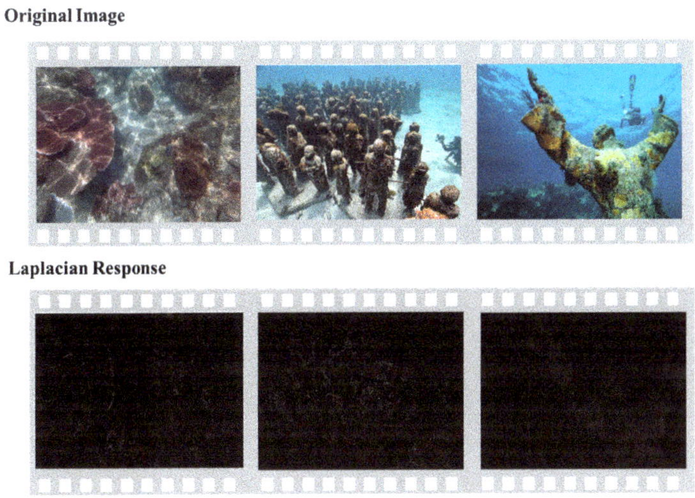

Figure 2. Laplacian response.

The Laplace operator is a second-order derivative operator, which is a scalar form.

$$\nabla^2 f(x,y) = \frac{\partial^2 f(x,y)}{\partial x^2} + \frac{\partial^2 f(x,y)}{\partial y^2} \tag{5}$$

$$\mu_L = \frac{1}{N}\sum_{i=1}^{N} \nabla^2 f(x,y)_i \tag{6}$$

$f(x,y)_i$ is the i-th pixel value of the Laplace operator result. N is the total number of pixels. μ is the mean value of the Laplace operator result. The higher response value of the Laplace operator indicates that there are more high-frequency components in the image with significant edges and details. The pixel values in its output image fluctuate and have high variance. By comparing the variance with the threshold value, it can be judged whether the image is clear or not. In this section, the value of T is taken as 12.

$$Decision = \begin{cases} Sharp, \sigma^2 > T \\ Blurry, \sigma^2 \leq T \end{cases} \tag{7}$$

2.1.3. LAB Color Space Analysis

We have selected a color deviation detection technique in the LAB color space to identify color-distorted images. This method leverages the separation of luminance and chromatic information in the LAB space, where L represents luminance, A corresponds to the green-to-red chromaticity channel, and B represents the blue-to-yellow chromaticity channel [52,53]. The advantage of using the LAB color space is its close alignment with human visual perception, offering enhanced adaptability to color variations under different lighting conditions [54,55]. Figure 3 illustrates the color shifts.

Figure 3. Color cast.

The color space conversion from RGB to LAB involves three main steps: RGB → XYZ → LAB. This is because the LAB color space is based on the CIE XYZ color model, with XYZ being one of the standard color spaces used for color representation.

Convert Linear RGB to XYZ: the RGB image is first converted to XYZ color space by the CIE 1931 standard matrix and subsequently, nonlinear mapping is applied to generate LAB components [56].

Convert Linear RGB to LAB:

$$\begin{cases} L* = 116f(\frac{Y}{Y_n}) - 16 \\ a* = 500[f(\frac{X}{X_n}) - f(\frac{Y}{Y_n})] \\ b* = 200[f(\frac{Y}{Y_n}) - f(\frac{Z}{Z_n})] \end{cases} \tag{8}$$

$$f(t) = \begin{cases} t^{\frac{1}{3}}, t > (\frac{6}{29})^3 \\ \frac{1}{3}(\frac{29}{6})^2 t + \frac{4}{29}, otherwise \end{cases} \tag{9}$$

where $L*$, $a*$, and $b*$ represent the values of the three channels in the final LAB color space, respectively. X, Y, and Z are the calculated values after conversion from RGB to XYZ.

The design of the LAB color space ensures that each channel value directly corresponds to a specific color property. The A and B channel values for each pixel represent the red–green and blue–yellow attributes of that pixel's color, respectively. This distinction enables the average values of the A and B channels to effectively describe the overall color tendency of an image, indicating shifts towards red–green or blue–yellow tones.

$$\begin{cases} d_a = \dfrac{\sum_{i=1}^{M}\sum_{j=1}^{N} A(i,j)}{M \times N} - L_{mean} \\ d_b = \dfrac{\sum_{i=1}^{M}\sum_{j=1}^{N} B(i,j)}{M \times N} - L_{mean} \end{cases} \tag{10}$$

If $|d_a|$ is large, the image deviates more significantly in the red–green direction. If $|d_a|$ is large, the image deviates more significantly in the blue–yellow direction.

By analyzing the standardized deviation of the B channel, the breadth or dispersion of the color distribution in the image can be quantitatively described.

$$\begin{cases} m_{sqA} = \dfrac{\sum_{y=0}^{225} |y - L_{mean} - d_a| \cdot HistA[y]}{M \times N} \\ m_{sqB} = \dfrac{\sum_{y=0}^{225} |y - L_{mean} - d_b| \cdot HistB[y]}{M \times N} \end{cases} \quad (11)$$

Proportion of chromatic aberration:

$$R = \frac{\sqrt{d_a^2 + d_b^2}}{\sqrt{m_{sqA}^2 + m_{sqB}^2}} \quad (12)$$

As in the case of fuzzy recognition, here again, a threshold is required, T. In this section, the value of T is taken as four.

$$Decision = \begin{cases} Color\_bias, R > T \\ Normal, R \leq T \end{cases} \quad (13)$$

2.2. *Jaffe–McGlamery Model*

The Jaffe–McGlamery model is a classical framework commonly applied in underwater imaging and image enhancement [57,58]. To investigate the degradation characteristics of underwater images and validate the accuracy of the degradation model, this paper develops a degradation model based on the underwater optical transmission model introduced by Jaffe and McGlamery.

2.2.1. Core Imaging Principle

The imaging principle of the Jaffe–McGlamery model is shown in Figure 4.

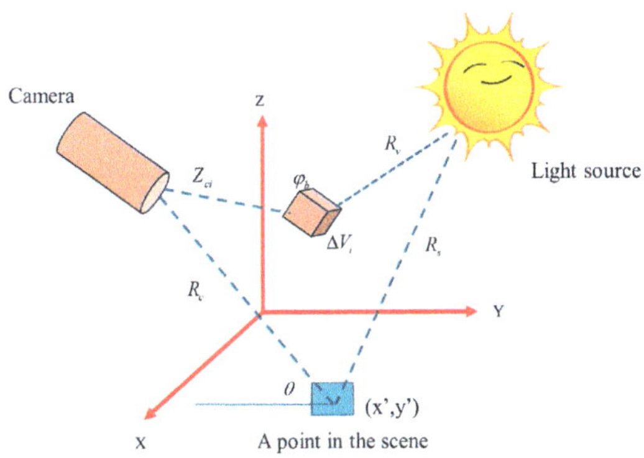

Figure 4. 3D spatial coordinates of the Jaffe–McGlamery model.

The Jaffe–McGlamery model provides a physical foundation for understanding underwater image degradation by simulating light propagation through water. It decomposes the total irradiance received by the camera into three components:

1. Direct component ($I = I_{direct} + I_{forward} + I_{back}$): light traveling from the object to the camera without scattering.

2. Forward-scattered component ($I_{forward}$): light deflected by suspended particles but still reaching the sensor.

3. Backscattered component (I_{back}): ambient light reflected by water particles towards the camera.

The original model expresses the total irradiance as follows:

$$I = I_{direct} + I_{forward} + I_{back} \tag{14}$$

However, in practical marine ranch monitoring scenarios (depth < 50 m, NTU < 20), forward scattering contributes less than 5% to the total irradiance. We thus adopt a simplified formulation:

$$I(x) = J(x)t(x) + A(1 - t(x)) \tag{15}$$

where the variables are defined as follows:

$J(x)$: scene radiance (ideal image without degradation);
$t(x)$: transmission map ($t = e^{-\beta d}$, β: attenuation coefficient);
A: background light intensity, estimated via dark channel prior [59].

This simplification aligns with field observations in aquaculture environments [60], where turbidity variations are moderate. The model's wavelength-dependent attenuation (Figure 1c) explains the dominance of blue–green hues in deep water.

2.2.2. Simplified Formulation

The complexity of the original Jaffe–McGlamery model arises from its comprehensive consideration of multiple scattering effects and spatially varying parameters. To adapt this model for practical underwater image enhancement in marine ranch monitoring, we propose three key simplifications supported by empirical and theoretical studies:

1. Neglecting Forward Scattering

Existing studies have shown that in close-range imaging scenarios (e.g., typical camera-to-target distances in aquaculture monitoring), where the optical range is short and the suspended sediment concentration is low (NTU < 20), the image blurring effect due to forward scattering tends to be reduced by the model to a minor factor, and is negligible, especially in shallow waters (<30 m).

Jaffe [61], in modeling low turbidity underwater imaging, noted that forward scattering had less than a 2% effect on image signal-to-noise ratios when the turbidity NTU < 15 and the target distance was less than 10 m.

Mobley [62] derived the radiative transfer equation to indicate that in low turbidity waters (NTU < 20), forward scattering as a proportion of the total scattered energy is typically less than 5 percent. This conclusion is supported by the measured data of Twardowski et al. [63], who found that in clear waters with NTU = 10, backward scattering accounted for more than 90 percent of the total scattering, while forward scattering contributed less than 4 percent.

Although experimental data to directly quantify the contribution of forward scattering are still scarce, theoretical derivations, numerical simulations, and model simplifications have shown that the contribution of forward scattering to the total irradiance can be reasonably neglected in typical aquaculture environments with NTU < 20 and water depths of $\leq$30 m. This assumption can be further verified by controlled experiments (e.g., using collimated light sources and high-precision irradiance sensors). Future research is needed to further validate the applicability of this assumption through controlled experiments (e.g., using collimated light sources with high-precision irradiance sensors) [64].

This simplification reduces computational complexity while maintaining fidelity, as expressed by the following:

$$I \approx I_{direct} + I_{back} \tag{16}$$

2. Spatially Uniform Background Light

Under low turbulence conditions (NTU < 20) and short imaging ranges (<10 m), the spatial heterogeneity of the backscattered light can be ignored without a significant degradation in accuracy. Experimental studies have shown that the homogeneous background assumption reduces computational complexity by 60–70% while maintaining a PSNR loss of less than 3 dB (corresponding to a visual error of ~3–5%) [65].

In turbid (NTU > 50) or deep-water environments (>30 m), spatial fluctuations in backscattered light can be as large as 20–30% due to significant spatial gradients caused by particle stratification and light attenuation, and the homogeneity assumption will lead to model failure. Neglecting these variations may lead to irradiance estimation errors of more than 15% [66,67].

Based on the above theoretical and experimental basis, it is reasonable and efficient to assume that the background light is locally constant, which is particularly suitable for real-time image enhancement tasks in stable underwater environments such as aquaculture.

In order to synthesize the practical situation, we can simplify the calculation of the background light to the following equation:

$$I = J(t) \cdot e^{-\beta d} + A \tag{17}$$

where I is the observed irradiance (image intensity). $J(t)$ is the scene radiance. β is the attenuation coefficient. d is the depth (or distance to the object). A is the spatially uniform background light intensity, which is assumed to be constant in this formulation.

This equation encapsulates the simplification that A is constant across the image, avoiding the need to model complex spatial gradients of background light [68].

The simplified transmission map $t_c(x) = e^{-\beta_c d(x)}$ assumes a constant β_c and a globally estimated $d(x)$, which is computationally efficient and suitable for shallow, stable waters (depth < 10 m, NTU < 20). However, this approach may oversimplify light propagation in complex environments, such as deep or turbid waters, where β_c varies spatially due to changes in depth, turbidity, and illumination. Factors like forward scattering and non-uniform background light, neglected here, could further impact accuracy in such scenarios.

2.3. Traditional Methodologies Example

In this section, we will perform enhancement using traditional underwater image enhancement methods; our underwater image enhancement framework uses a multi-stage iterative architecture. Color balance correction is first performed, followed by LAB spatial decomposition to separate luminance and chrominance. Adaptive histogram equalization and bilateral filtering are then applied to suppress noise while preserving edges. Finally, a multi-scale fusion strategy integrates the enhanced features through Laplace pyramid decomposition.

2.3.1. Scene-Specific Underwater Image Enhancement

Low-light underwater images often exhibit low global brightness, insufficient contrast, and a loss of local details [69]. To address these challenges, this paper employs multi-scale histogram equalization as an enhancement technique. This method effectively improves global brightness and contrast while simultaneously enhancing local details [70,71]

Multi-scale processing typically uses Gaussian pyramid decomposition or other similar methods to decompose an image into low-frequency components and multiple high-frequency components.

Cumulative distribution function:

$$P(r_k) = \frac{n_k}{MN} \tag{18}$$

$$C(r_k) = \sum_{j=0}^{K} P(r_j) \tag{19}$$

Greyscale value r_k mapped to equalized values:

$$s_k = (L-1)C(r_k) \tag{20}$$

Weight of details:

$$s_k = (L-1)C(r_k) \tag{21}$$

Multi-scale image fusion:

$$I_{enhanced} = L_K^{enhanced} + \sum_{k=1}^{K} H_k^{enhanced} \tag{22}$$

$$I_{enhanced} = \omega_0 L_K^{enhanced} + \sum_{k=1}^{K} \omega_k H_k^{enhanced} \tag{23}$$

Enhancement methods based on wavelength compensation and contrast correction are well-suited for processing color deviation images [72,73]. The wavelength compensation method compensates for each color channel by analyzing the attenuation law of light: restoring the intensity of the red channel and compensating for the long wavelength portion that is absorbed [74]. Balancing the RGB channel makes the overall color closer to the real scene. Wavelength compensation can effectively correct color bias [75,76].

Based on the properties of light attenuation in the water column, the transmittance $t_c(x)$ is estimated by Eq:

$$t_c(x) = e^{-\beta_c \cdot d(x)} \tag{24}$$

where β_c is the attenuation coefficient for each channel, reflecting the intensity of water absorption at different wavelengths (red β_r > green β_g > blue β_b). $d(x)$ is the distance from the pixel point to the camera (as appropriate).

In this study, the attenuation coefficients were set to $\beta_r = 0.1$, $\beta_g = 0.05$, and $\beta_b = 0.03$ for the red, green, and blue channels, respectively, based on typical values for clear coastal waters. These reflect the wavelength-dependent attenuation observed in shallow, low-turbidity environments (NTU < 20). The distance $d(x)$ was estimated using the dark channel prior method, adapted for underwater images, allowing scene-specific transmission maps. However, these fixed β_c values may not fully represent conditions in deeper or more turbid waters, where attenuation varies significantly.

Ambient light A_c is usually estimated from the pixel area with the highest intensity in the image:

$$A_c = \max_{x \in \Omega}(I_c(x)) \tag{25}$$

where Ω is the candidate region in the image, and usually, an area far from the camera is selected as the background.

Based on the simplified model of the underwater image (15), the formula for recovering the real image by backward derivation is as follows:

$$J_c(x) = \frac{I_c(x) - A_c}{t_c(x)} + A_c \tag{26}$$

Based on the absorption properties of water for different wavelengths of light, the attenuation of each channel is compensated for with the following commonly used formula:

$$J_c(x) = \frac{I_c(x)}{k_c} \tag{27}$$

$$k_c = e^{-\alpha_c \cdot d} \tag{28}$$

k_c is the wavelength compensation factor, usually determined by the absorption properties of water, where α_c is the absorption coefficient for the wavelength c.

Contrast improvement by adjusting pixel intensity distribution is as follows:

$$J_c{'}(x) = HE(J_c(x)) \tag{29}$$

where HE is the Histogram equalization operation.

The brightness curve is adjusted to improve details in dark areas:

$$J_c{'}(x) = J_c(x)^{\gamma} \tag{30}$$

γ is permanent, $0.4 \leq \mathrm{x} \leq 0.6$.

Enhancement is limited in areas of excessive contrast:

$$J_c{'}(x) = CLAHE(J_c(x), clipLimit) \tag{31}$$

Here, $clipLimit$ is the parameter used to limit the strength of the histogram equalization.

The combined compensated and corrected image is as follows:

$$J'(x) = Enhance(J(x)) \tag{32}$$

where $Enhance$ is the image enhancement functions (histogram equalization, gamma correction, and other combined operations).

To address the problem of blurred imaging in underwater images, we use a method based on Laplace sharpening.

The fuzzy degradation model can be expressed as follows:

$$I_{blurred}(x,y) = I_{original}(x,y) \cdot h(x,y) + n(x,y) \tag{33}$$

where $I_{blurred}$ is the blurred image. $I_{original}$ is the original clear image. $h(x,y)$ is the point spread function (PSF), which describes the blurring effect. $n(x,y)$ is the additional noise.

The calculation of the Laplace operator was mentioned earlier (5), which is based on the calculation of high-frequency details by second-order derivatives, which is simplified by changing the variables in it by a different name:

$$\nabla^2 I(x,y) = \frac{\partial I^2}{\partial x^2} + \frac{\partial I^2}{\partial y^2} \tag{34}$$

This operator highlights the regions where the brightness of the image changes drastically, i.e., the edge regions. In the discrete case, the convolution is implemented as follows:

$$\nabla^2 I(x,y) = \begin{bmatrix} 0 & -1 & 0 \\ -1 & 4 & -1 \\ 0 & -1 & 0 \end{bmatrix} \cdot I(x,y) \tag{35}$$

The core goal of sharpening is anti-blurring, i.e., the enhancement of the high-frequency component. The optimization formula is as follows:

$$I_{sharp}(x,y) = I_{blurred}(x,y) + \lambda \cdot \nabla^2 I(x,y) \tag{36}$$

λ controls the sharpening intensity, usually in the range of $0.5 \leq \lambda \leq 1.5$.

The parameter λ in Equation (36) controls the degree of sharpening applied to the image by scaling the contribution of the Laplacian term, $\nabla^2 I(x,y)$. In our implementation, we set λ = 0.5, a value determined empirically to achieve an optimal balance between enhancing fine details and preventing over-sharpening artifacts, such as ringing or noise amplification. This choice was informed by testing on a variety of images, where λ = 0.5 consistently improved sharpness while maintaining image quality, as assessed through visual inspection and quantitative metrics like PSNR, UCIQE, UIQM, RGB, and luminance. The specific image parameters are shown in Table 1.

Table 1. Picture metrics under different λ conditions.

λ	PSNR	UCIQE	UIQM	RGB	Luminance
λ = 0.5	10.528	30.866	0.249	102.9/158.8/174.3	179.569
λ = 1	10.333	30.921	0.252	100.9/158.8/175.2	178.147
λ = 1.5	10.284	31.084	0.25386	100.4/158.8/175.3	178.274

For complex scenes, the Laplace operator can also be improved using adjustable edge filters, e.g., high-pass filters:

$$H_{high-pass} = \delta \cdot \nabla^2 I(x,y) \tag{37}$$

where δ is the scale factor used to control the filter response.

Laplace sharpening amplifies not only edge information but may also enhance noise. The new method proposed in this paper further optimizes the noise suppression.

Combined with Gaussian blurring, a multi-scale image pyramid is constructed:

$$I_{smooth,s}(x,y) = I_{blurred}(x,y) \cdot G_s(x,y) \tag{38}$$

where $G_s(x,y)$ is the scale s of the Gaussian kernel. Based on the multi-scale image pyramid, the high-frequency enhancement part is selected:

$$\Delta I_{multi-scale}(x,y) = \sum_{s=1}^{N} \omega_s \cdot \nabla^2 I_s(x,y) \tag{39}$$

where ω_s is the weight of the s th layer and N is the number of pyramid layers.

Optimized sharpening is based on the traditional Laplace operator, combined with the gradient orientation:

$$I_{sharp}(x,y) = I_{blurred}(x,y) + \lambda \cdot (\nabla^2 I(x,y) \cdot cos^2(\theta)) \tag{40}$$

where $\theta = tan^{-1}(\frac{\partial I}{\partial y} / \frac{\partial I}{\partial x})$, and it indicates the direction of the gradient.

Combined with bilateral filtering, edge-hold noise smoothing is performed before sharpening:

$$I_{smoothed}(x,y) = \frac{1}{k(x,y)} \sum_{i,j\in\Omega} exp(\frac{-|I(x,y) - I(i,j)|}{2\sigma_r^2}) \cdot I(i,j) \tag{41}$$

Here, $k(x, y)$ is the normalization factor and σ_r controls the degree of the smoothing of intensity similarity. Nonlinear smoothing enhances the accuracy of edge differentiation while reducing noise accumulation after smoothing.

Parameters for Laplace sharpening λ can be further designed as an adaptive model:

$$\lambda(x,y) = \frac{1}{1 + \alpha \cdot exp(- \parallel \nabla^2 I(x,y) \parallel)} \tag{42}$$

$\parallel \nabla^2 I(x,y) \parallel$ is the magnitude of the image gradient and α is the control parameters that determine the response range.

Figure 5 shows the PSNR, UCIQE, and UIQM values of the pictures at the original stage, after one enhancement and after two enhancements.

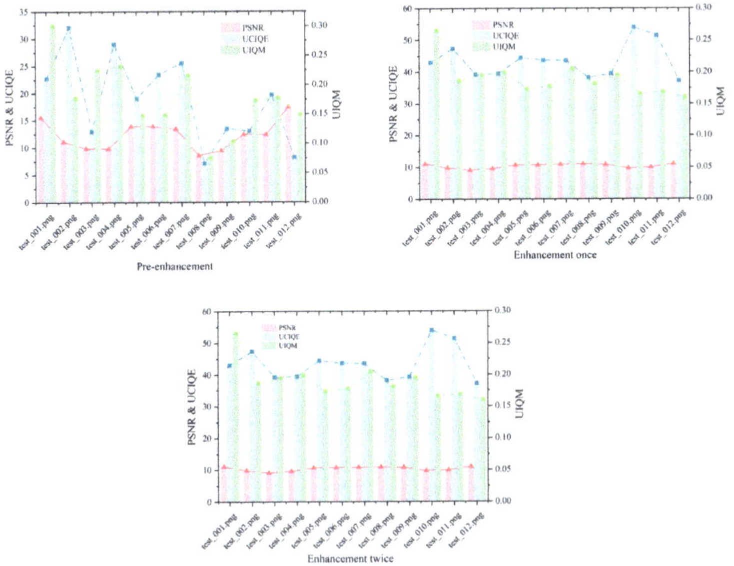

Figure 5. PSNR, UCIQE, and UIQM values for the original image, the image after one enhancement and the image after two enhancements in specific scenarios.

Figure 5 compares the image quality metrics—peak signal-to-noise ratio (PSNR), Underwater Color Image Quality Evaluation (UCIQE), and underwater image quality measure (UIQM)—for the original underwater image, after one enhancement, and after two enhancements. The PSNR values, ranging from 9 to 11, indicate that the enhancement process maintains low noise levels, though these relatively low scores suggest some residual distortion persists, an area for potential refinement. The UCIQE values, increasing from 37 to 54, reflect significant improvements in color restoration, a critical factor for enhancing the visual clarity of underwater scenes. Meanwhile, the UIQM values, stable between 0.17 and 0.27, demonstrate that the overall quality (encompassing contrast, hue, and sharpness) is preserved without substantial enhancement beyond the first iteration. This stability suggests that our method effectively targets specific distortions—such as color shifts—while maintaining the image's integrity, a balance essential for applications requiring authentic underwater visuals.

Additionally, it was found that the difference between the results of a single enhancement and two consecutive enhancements was not significant. This suggests that the first enhancement primarily addressed a single type of distortion in the image, and subsequent enhancements did not further improve the quality. In other words, after the first enhancement, the model no longer processed images with normal distortions. This indicates that the enhancement process primarily focused on addressing specific distortions rather than improving the overall image quality in multiple steps.

Figure 6 shows the mean and standard deviation of R, G, and B for the original image, the image after one enhancement, and the image after two enhancements.

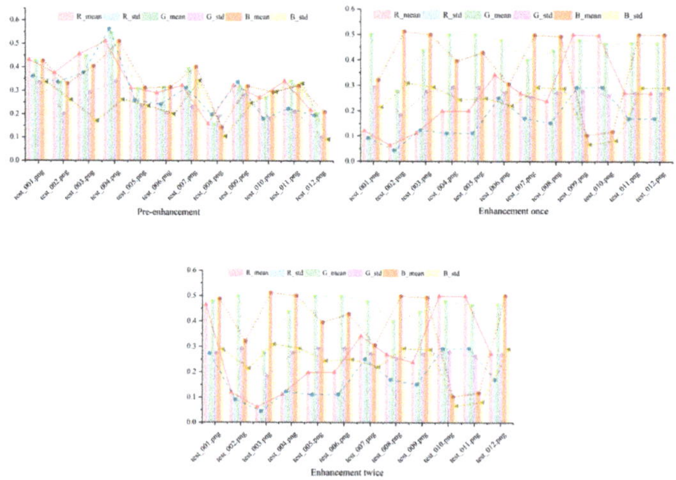

Figure 6. The mean and standard deviation of R, G, and B for the original image, the image after one enhancement, and the image after two enhancements in specific scenarios.

Figure 6 illustrates the mean and standard deviation of the red (R), green (G), and blue (B) channels across the original image, after one enhancement, and after two enhancements. The mean values shift noticeably after the first enhancement, reflecting a correction in color balance that aligns the image closer to natural underwater hues. However, between the first and second enhancements, these values remain nearly identical, suggesting that the initial enhancement sufficiently mitigates the primary color distortions—likely due to water-induced attenuation. The standard deviations, which are similarly consistent post-first enhancement, indicate that color variability across the image is stabilized, preventing over-processing artifacts.

Figure 7 shows the mean and standard deviation of brightness for the original image, after one enhancement, and after two enhancements. The mean brightness rises slightly after the first enhancement, indicating improved illumination that enhances visibility—a vital improvement for underwater environments with poor lighting. However, the lack of significant change between the first and second enhancements suggests that additional processing does not further elevate brightness, preserving the image's natural appearance. The standard deviation, which remains largely unchanged, reflects consistent brightness uniformity across all stages, implying that our enhancement avoids introducing uneven lighting effects. This outcome is advantageous for maintaining the reliability of underwater images, where uniform brightness aids in accurate object identification and analysis.

Overall, the algorithmic model for underwater image enhancement in special scenarios has a limited effect on improving image quality. While objective quality metrics such as PSNR, UCIQE, and UIQM show small improvements, these changes are insufficient to significantly enhance clarity, color recovery, contrast, and other visual aspects of the image. In particular, there is almost no noticeable difference in terms of color equalization and brightness adjustment between the enhanced image and the original. The model appears to address only a single type of distortion, and the results from the last two enhancements suggest that the first enhancement successfully addresses one specific distortion, while subsequent enhancements do not further improve the image.

These findings underscore the practical utility of our enhancement method in special underwater imaging scenarios. By correcting color distortions and enhancing visibility with minimal noise (as seen in Figures 5 and 6) and maintaining brightness uniformity (Figure 7), our approach enhances image usability for applications like marine biology research, underwater archaeology, and environmental monitoring. The efficiency of achieving substantial improvements in a single enhancement step makes it particularly suitable for real-time systems or resource-constrained settings.

The enhancement results are shown in Figure 8.

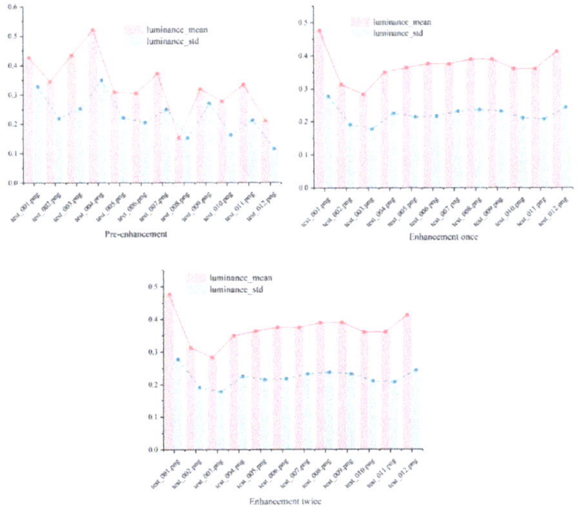

Figure 7. The mean and standard deviation of the brightness for the original image, the image after one enhancement, and the image after two enhancements in specific scenarios.

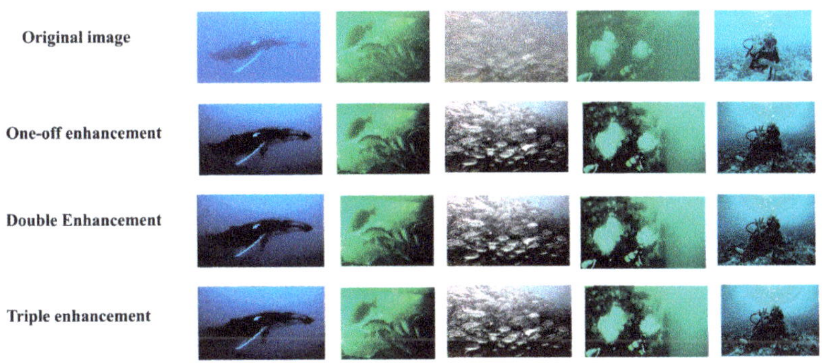

Figure 8. Enhancement results of underwater image models for specific scenarios.

2.3.2. Complex Scenarios Underwater Image Enhancement

Multi-feature fusion techniques are widely employed for underwater image enhancement, particularly in addressing the degradation challenges posed by complex underwater environments [77–79]. This advanced image processing approach integrates multiple image features to significantly improve the visual quality and informational clarity of underwater images, making them more suitable for practical applications [80,81].

The white balance method is based on the Lambertian reflection model [82], which corrects the color bias of an image according to the color temperature it presents. For a color image, the color of a point x on the surface of an object in the image can be represented by the Lambertian reflectance model.

$$I(x) = \int_{\omega} e(\lambda)S(x,\lambda)C(\lambda)d\lambda \tag{43}$$

Estimation of light source color e:

$$e = \int_{\omega} e(\lambda)C(\lambda)d\lambda \tag{44}$$

The mean value of surface reflections from objects with the same attenuation coefficient in an underwater environment is colorless.

$$\frac{\int a(x)S(x,\lambda)dx}{\int a(x)dx} \tag{45}$$

Assume that the color of the light source $a(x)$ with an attenuation factor $\hat{e}$ is a constant:

$$\hat{e}(\lambda) = a(x)e(\lambda) \tag{46}$$

Bilateral filtering is a nonlinear filtering technique that preserves edge information while smoothing an image [83,84]. It combines Gaussian filtering in both spatial and pixel-value domains so that edges are not blurred while smoothing the image [85]. In our model, we do this by mapping the image to a 3D mesh (a combination of spatial and pixel-valued domains), then applying Gaussian filtering to the mesh, and finally mapping the result back to the original image resolution by interpolation.

Gaussian Spatial Kernel:

$$W_s(p,q) = e^{-\frac{\|p-q\|^2}{2\sigma_s^2}} \tag{47}$$

Gaussian Range Kernel:

$$W_s(p,q) = e^{-\frac{\|I(p)-I(q)\|^2}{2\sigma_r^2}} \tag{48}$$

Bilateral filter formula:

$$I_{out}(p) = \frac{1}{K(p)} \sum_{q\in\Omega} W_s(p,q)\cdot W_r(p,q)\cdot I(q) \tag{49}$$

We conducted a comparison of the effect of the filter; if you do not use the parameters of the processing of the function that comes with the openCV, we found that the smoothing effect is more limited after the parameter qualification of the smoothing effect, as shown in Figure 9c.

Laplacian Contrast:

$$C_L(x,y) = \left|\nabla^2 I(x,y)\right| \tag{50}$$

Local Contrast:

$$C_{LC}(x,y) = \frac{I(x,y) - \mu(x,y)}{\sigma(x,y)} \tag{51}$$

Saliency:

$$S(x,y) = \omega_{color} S_{color}(x,y) + \omega_{texture} S_{texture}(x,y) + \omega_{edge} S_{edge}(x,y) \tag{52}$$

Exposure:

$$E(x,y) = \frac{I(x,y) - min(I)}{max(I) - min(I)} \tag{53}$$

Final fused image formula:

$$I_{fused}(x,y) = \omega'_L \cdot C_L(x,y) + \omega'_{LC} \cdot C_{LC}(x,y) + \omega'_s \cdot S(x,y) + E(x,y) \tag{54}$$

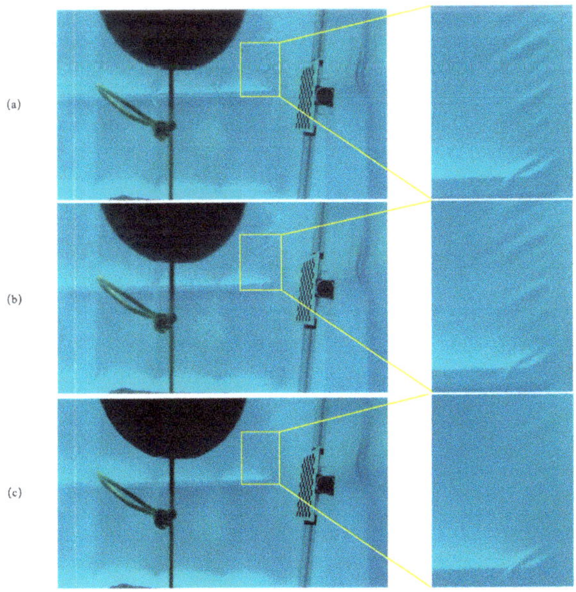

Figure 9. (**a**) Original image: the input image without any processing. (**b**) OpenCV bilateral filter result: the image after processing with cv2.bilateralFilter. (**c**) Custom bilateral filtering result: the effect of our filter function.

Figure 10 presents the PSNR, UCIQE, and UIQM values of the images after one and two enhancements using the fusion enhancement model. The PSNR values indicate that both the first and second enhancements result in some improvement in image quality. The UCIQE shows a significant enhancement after the first application, with the algorithm performing well and achieving natural, consistent color correction. However, after the second enhancement, issues such as oversaturation or color deviations (e.g., an overpowering blue channel) may arise, leading to a decrease in overall color quality. In contrast, the UIQM suggests that a single enhancement is better aligned with typical underwater image enhancement requirements in terms of color quality. The second enhancement, however, increases overall detail.

Figure 11 shows the R, G, and B mean and standard deviation of the image after one enhancement and the image after two enhancements of the enhancement fusion algorithm. The analysis of the mean and standard deviation reveals that the enhancement algorithm significantly impacts the recovery of the color channels. The standard deviation of the red channel is generally higher than that of the green and blue channels, indicating a stronger hue variation during the enhancement process. In contrast, the standard deviation of the green and blue channels is slightly lower, suggesting these channels exhibit less volatility, with the enhancement algorithm having a more stable effect on them. This observation aligns with the light absorption characteristics of water.

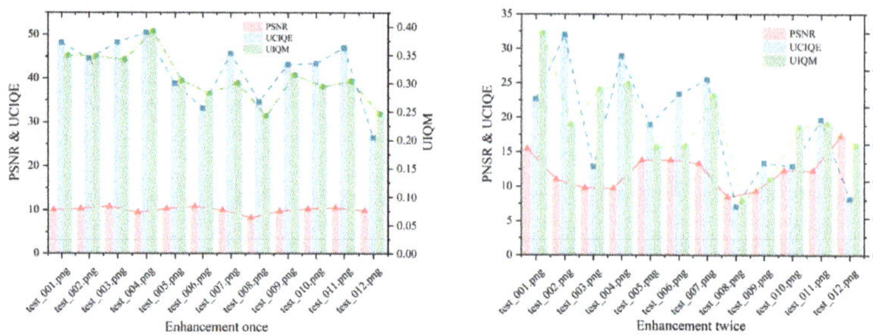

Figure 10. PSNR, UCIQE, and UIQM values for the original image, the image after one enhancement, and the image after two enhancements in complex scenarios.

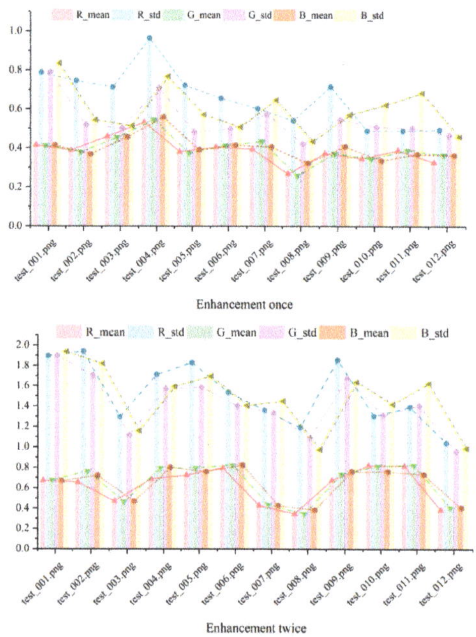

Figure 11. The mean and standard deviation of R, G, and B for the original image, the image after one enhancement, and the image after two enhancements in complex scenarios.

Figure 12 illustrates the mean and standard deviation of the image brightness after one and two enhancements using the enhancement fusion method. After the first enhancement, the image brightness is moderate and uniform, with a natural contrast. Following the second enhancement, the average brightness increases, potentially revealing more details with greater clarity.

In summary, the underwater image enhancement method proposed in this paper not only offers significant improvements in image quality, noise suppression, and detail retention, but also demonstrates strong adaptability, making it effective for various underwater environments. In shallow-water areas with strong light, the degradation is primarily caused by bluish tones and slight blurring [86]. In such cases, the enhancement delivers superior performance, with natural color correction and no noticeable distortion. In medium and deep water, where light gradually diminishes, degradation is characterized by a low contrast and a slight increase in noise [87]. A second enhancement offers some advantages in terms of brightness improvement, but care should be taken to avoid over-enhancement in detailed areas. For deep-water environments, where light is almost nonexistent and degradation is dominated by strong blur and noise, a second enhancement provides a significant improvement in visibility through increased brightness. The enhancement results are shown in Figure 13.

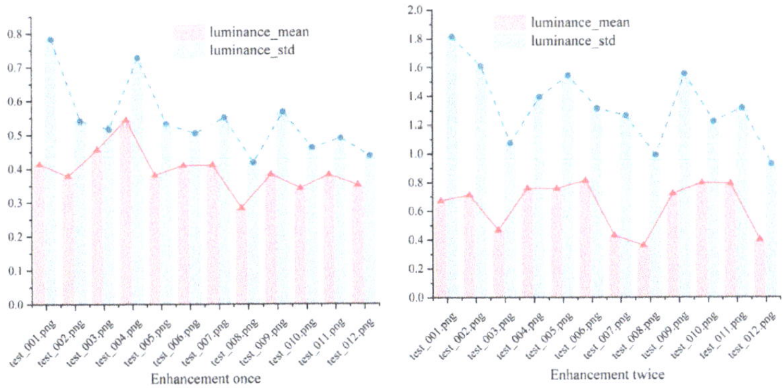

Figure 12. The mean and standard deviation of the brightness for the original image, the image after one enhancement, and the image after two enhancements in complex scenarios.

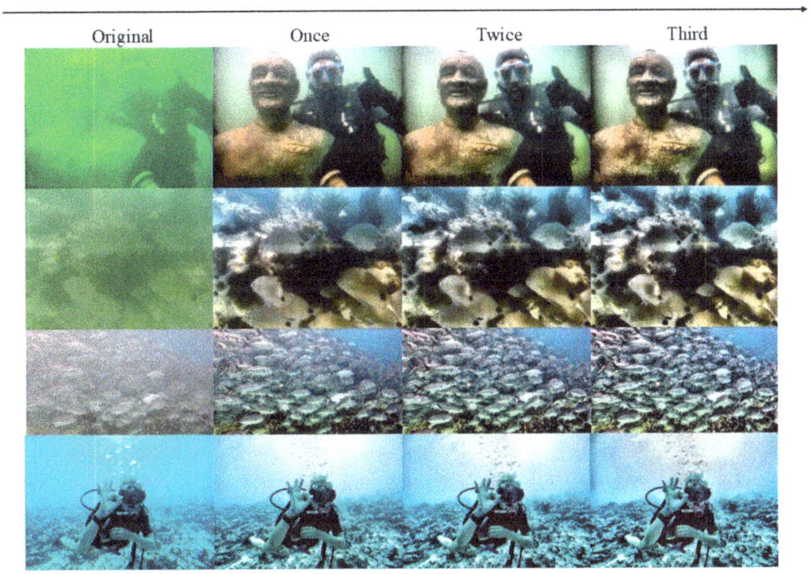

Figure 13. Enhancement results of underwater image model under enhanced fusion model.

2.3.3. Laplace Pyramid Decomposition

By implementing the aforementioned multi-feature fusion method for image enhancement, it was observed that directly applying the method can introduce negative effects, such as artifacts [88]. To address these issues, this paper incorporates the Laplacian pyramid method into the process [89,90]. This approach is built upon the Gaussian pyramid, which progressively down samples the image, creating a multi-scale representation through resolution reduction. The Laplacian pyramid further enhances this structure by generating layers that capture details between each scale, effectively representing information at different frequency levels within the image.

Similar to the input image, each weight map is processed into a multi-scale version through Gaussian pyramid decomposition. This decomposition smooths the weight maps and mitigates sharp transitions at the boundaries [91], effectively reducing the risk of introducing artifacts during the fusion process. The input image and the weight map are fused at each layer of the Laplace pyramid and Gaussian pyramid, respectively [92].

Finally, the fusion results of all layers are reconstructed by progressively combining them from the bottom up to obtain the final enhanced image. This approach ensures that the fused image preserves high-frequency details while maintaining a natural global distribution of brightness and contrast [93]. Laplace decomposition is shown in Figure 14.

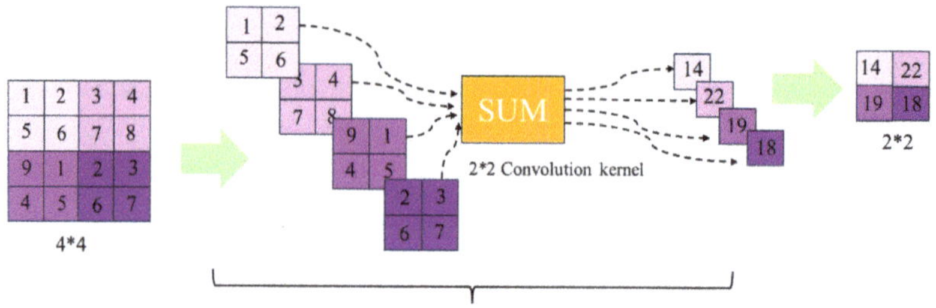

Figure 14. Laplace decomposition.

In this subsection, experiments are conducted using the constructed physical model to generate degraded images influenced by three key variables: depth, turbidity, and background light. To simulate various environmental conditions effectively, the experiment was designed with different experimental groups. Water depth was categorized into three classes: the shallow-water group (depths of 2 m and 5 m), the medium-depth group (10 m and 15 m), and the deep-water group (20 m and 25 m). Turbidity levels were set at 0, 5, 10, 15, 20, 25, 30, 40, 50, 70, and 100. Additionally, ambient light conditions were divided into two categories: low light (50) and high light (255).

The experimental results are shown in Figure 15. Observing the hotspot map reveals a significant decline in image sharpness as depth increases. This phenomenon can be attributed to light attenuation, which reduces the contrast of image details. Additionally, increased turbidity has a pronounced negative impact on sharpness, likely due to the scattering and absorption of light by suspended particles. Within a certain range, enhancing background light intensity improves clarity, though the effect tends to plateau over time.

The mean brightness value decreases progressively with increasing depth and turbidity, indicating overall light attenuation. However, brightness improves with enhanced ambient light. Red light diminishes rapidly with depth, while blue light, characterized by shorter wavelengths, penetrates more effectively underwater. Green light exhibits a pattern of attenuation between that of red and blue light. A strong negative correlation exists between depth and clarity, highlighting depth as the primary factor influencing clarity. Similarly, turbidity shows a strong negative correlation with brightness, reflecting its impact on light scattering intensity. On the other hand, background light demonstrates a strong positive correlation with brightness, suggesting that increasing background light can effectively enhance image brightness. The experimental results are in high agreement with the underwater optical properties, verifying the reasonableness of the degradation model.

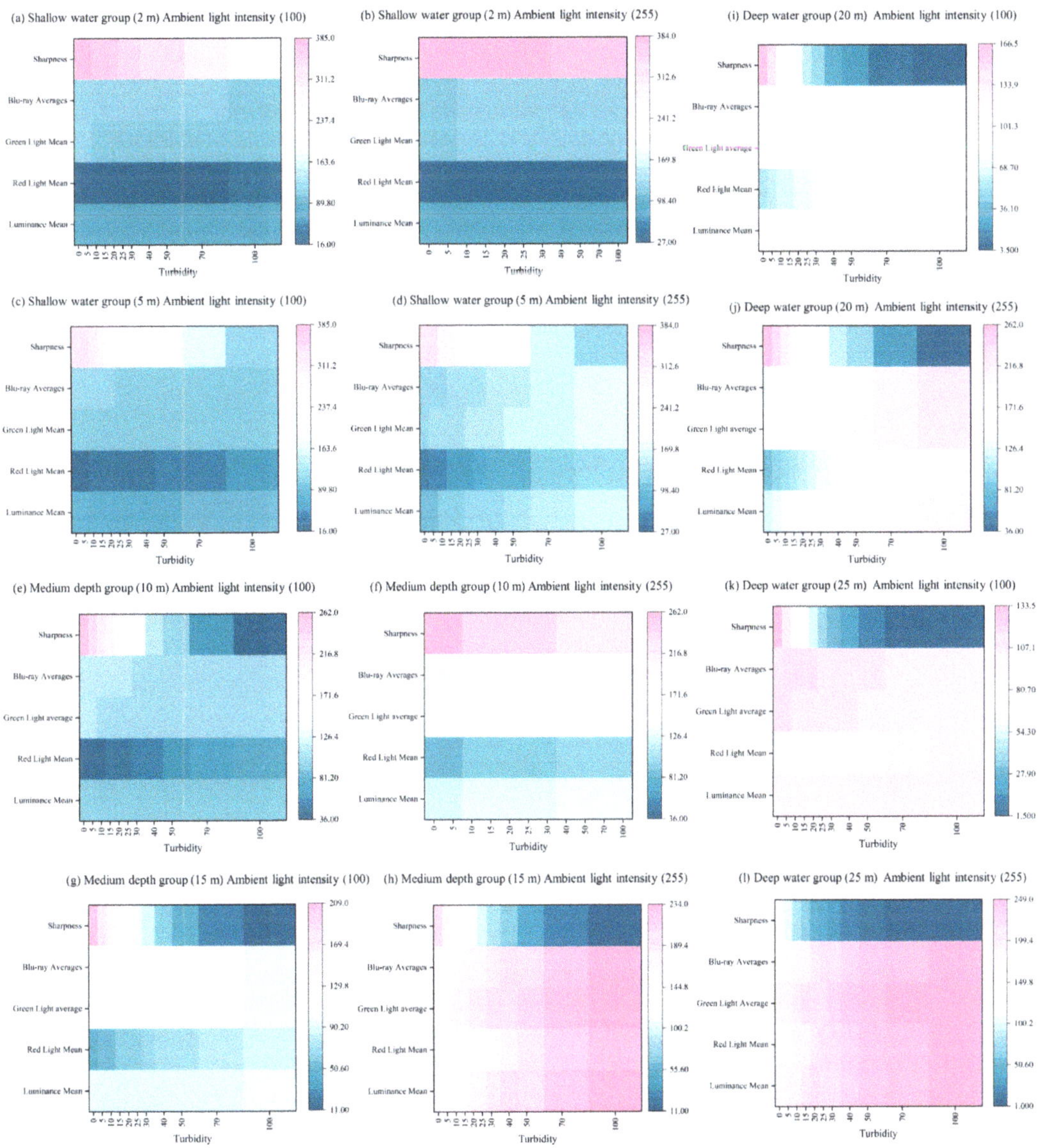

Figure 15. Results of the experiment.

3. Cluster Analysis

This study adopts a bibliometric methodology to investigate the historical progression of underwater image enhancement. Data for the analysis, including journals, research fields, national contributions, and institutional involvements, were systematically extracted from the Web of Science (WOS) database. Recognized as a leading citation index resource, WOS provides access to nearly 100 years of multidisciplinary scholarly works, representing foundational research across diverse domains. The database is distinguished by its inclusion of high-impact journals, making it an indispensable tool for researchers seeking credible academic references. Its prominence and reliability within the scholarly community are further corroborated by its authoritative standing.

The following are the key logical formulae that we used for the search: TS = ("underwater") and TS = ("image*" OR "picture*") and TS = ("enhancement" OR "restoration" OR "processing" OR "intensification"). The initial default screening searched all years on Web of Science, and approximately 2332 pieces of the relevant literature were screened. The earliest documentation in WOS dates back to 2002; we searched all years for articles published in English. For the article type, we chose "Article". In the end, we retrieved 2244 articles from Web of Science after deduplication using "CiteSpace". After manual screening, we selected 2057 references as a sample for the bibliometric data analysis. The specific search process is shown in Figure 16.

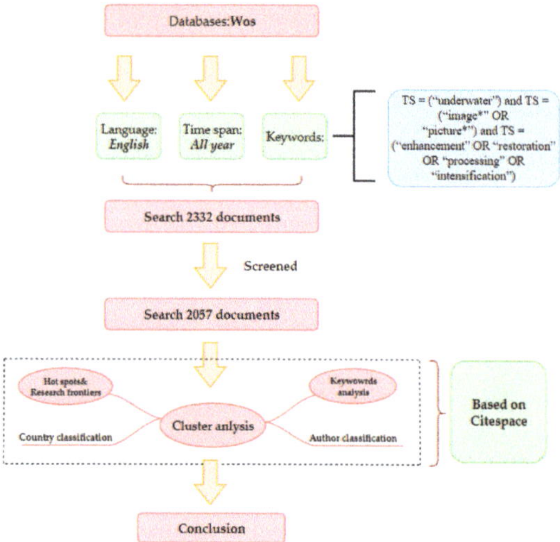

Figure 16. Schematic diagram of the search process.

The referenced papers were selected based on their relevance to underwater image enhancement, prioritizing peer-reviewed publications that address traditional physics-based methods and deep learning approaches. Seminal works laying the methodological foundation (e.g., Ref. [24]) and recent studies introducing novel techniques or datasets (e.g., Ref. [17]) were included to ensure a comprehensive and credible review. The selection also aimed to cover diverse methodologies—such as CNNs, GANs, and Transformers—to contextualize our comparative analysis.

3.1. Country Analysis

In the CiteSpace user interface, time slicing selects January 2002 to December 2025, time slicing is per year, node types selects country, and selection criteria selects the g-index, k = 25. Pruning selects Pathfinder and prunes the merged network; a picture of the finished process is shown in Figure 17.

Figure 17 shows that as many as 74 countries and regions have published papers related to underwater image enhancement. We can see that almost the whole world is focusing on the field of underwater image enhancement and contributing to the marine field as much as possible. The size of each node represents the number of posts. Among them, China, the United States, India, and Australia are extremely prominent in contributing to the field of underwater image enhancement.

The color of the node represents the year; the closer the color is to red, the closer the publication time is to the present. As can be seen in Figure 17, the node for China changes from purple to red, which means that China's image enhancement research started earlier. The larger nodes in the figure are colored red, indicating that underwater image enhancement is a focus that these countries are paying close attention to.

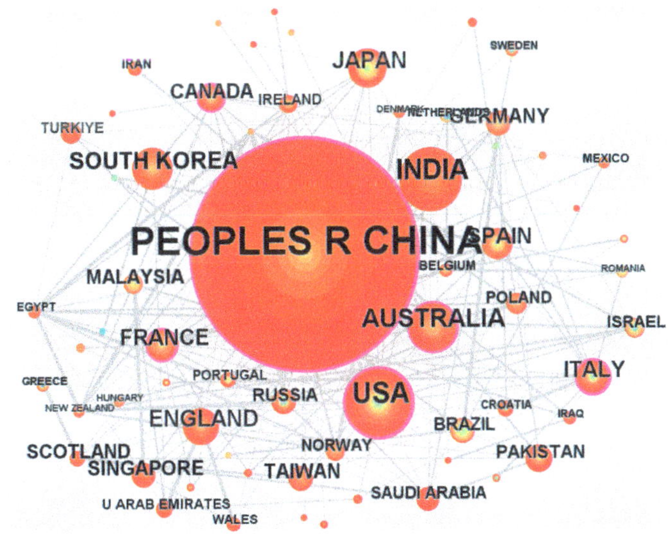

Figure 17. Country cluster analysis based on CiteSpace.

China dominates the number of publications in this area, with a count of 1361, which is almost two-thirds of the total number of publications, followed by the United States, with a count of 162 (7.88%), India, with a count of 133 (6.47%), and Australia, with a count of 83 (4.04%) (Table 2).

Table 2. Top 10 underwater image enhancement posts.

Rank	Country	Frequency	Percentage	Centrality
1	China	1361	66.2%	0.09
2	USA	162	7.88%	0.00
3	India	133	6.47%	0.00
4	Australia	83	4.0%	0.18
5	South Korea	61	2.97%	0.05
6	Japan	54	2.63%	0.05
7	England	50	2.43%	0.28
8	Italy	48	2.33%	0.44
9	France	47	2.28%	0.25
10	Spain	43	2.09%	0.22

We visualized the country data we had on a world map and the results obtained are shown in Figure 18.

As shown in Figure 18, most of the regions involved in underwater image enhancement are coastal countries, which have an inherent geographic location. The top ten countries in terms of the number of publications are all coastal regions, which have greater technological needs in this area than inland regions.

We have made a chord diagram of the cooperation between countries, as shown in Figure 19.

In Figure 19 and the chord diagram, the length of the colored band represents the number of articles issued by the country; because China issued a large number of articles, we have taken the logarithmic operation of the value. The relationship of the connecting line represents the cooperation, and the depth of the connecting line represents the amount of cooperation. In the figure, we find that the countries with a medium level of publications are closely related to each other, for example, USA and South Korea and Turkey and Saudi Arabia have darker-colored ties; on the contrary, China, which is the country with the highest number of publications, has less cooperation with other countries, with lighter-colored ties and a smaller number of ties.

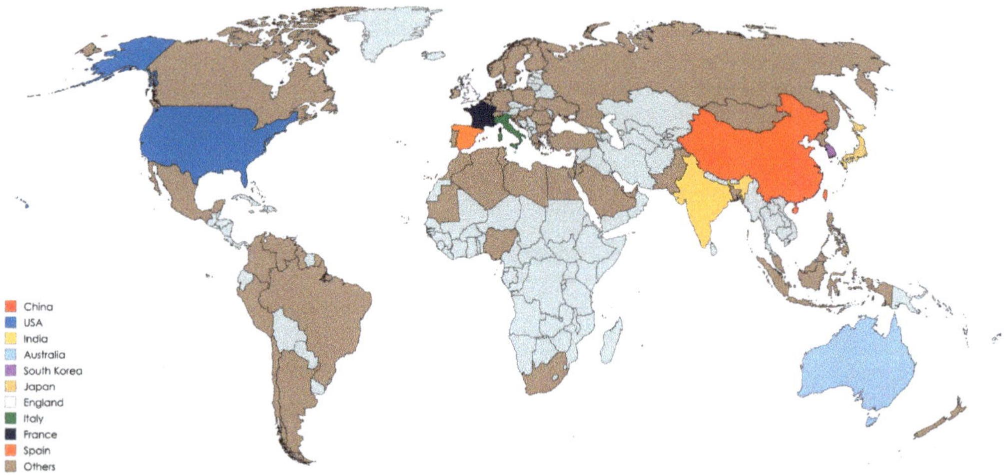

Figure 18. Distribution of countries and regions with published papers.

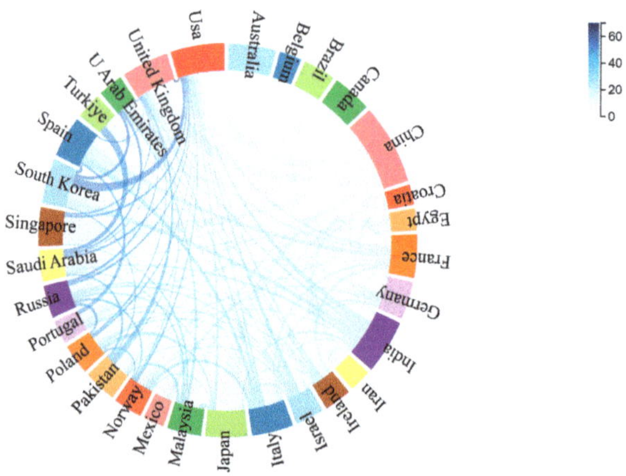

Figure 19. Chord charts of cooperation relations among countries.

3.2. Institution Analysis

In this section, we present the institutional analysis. Figure 20 is the clustering image we made using "VOSviewer"; again, the node size represents the number of postings and the connecting line represents the partnership. From the figure, we can see that the highest number of postings is Dalian Maritime University with 120 postings, followed closely by Chinese Academy of Sciences with 116 articles, followed by Harbin Engineering University with 84 articles. Table 3 shows the ranking of the top 15 institutions in terms of the number of articles or centrality.

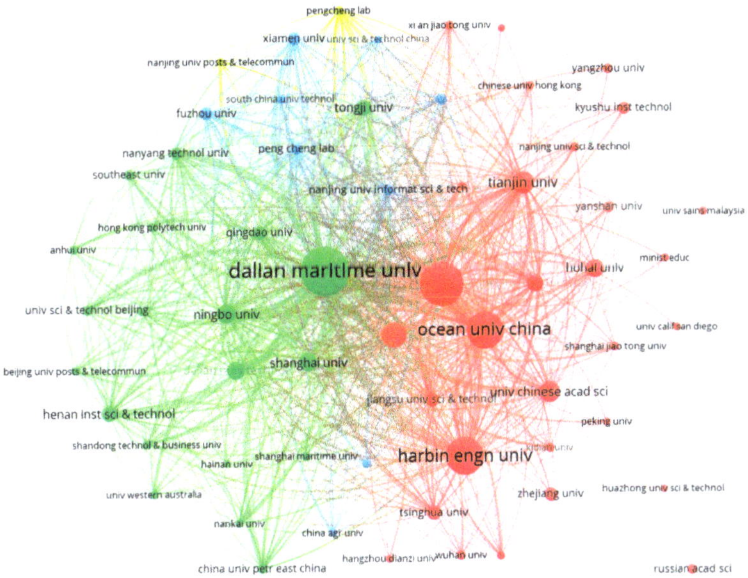

Figure 20. Institutional cluster analysis based on VOSviewer.

Table 3. Ranking of institutions by number of articles.

Rank	Count	Institution	Year	Centrality	Institution	Year
1	120	Dalian Maritime University	2015	0.23	Chinese Academy of Sciences	2013
2	116	Chinese Academy of Sciences	2013	0.20	Institute of Deep-Sea Science & Engineering	2023
3	84	Harbin Engineering University	2006	0.17	University of California System	2003
4	79	Ocean University of China	2014	0.16	Henan Institute of Science & Technology	2022
5	49	Northwestern Polytechnical University	2019	0.15	Laoshan Laboratory	2020
6	43	Tianjin University	2016	0.15	Chongqing Jiaotong University	2023
7	35	University of Chinese Academy of Sciences	2018	0.14	Centre National de la Recherche Scientifique (CNRS)	2007
8	35	Ningbo University	2021	0.14	Nanjing University of Information Science & Technology	2020
9	32	Dalian University of Technology	2020	0.13	Shanghai Maritime University	2023
10	31	Hohai University	2014	0.12	Anhui University	2021
11	30	Shanghai University	2019	0.11	Sun Yat Sen University	2024
12	29	Beijing Institute of Technology	2014	0.11	Institute of Oceanology	2020
13	27	Henan Institute of Science & Technology	2022	0.10	Tongji University	2016
14	24	Centre National de la Recherche Scientifique (CNRS)	2007	0.09	Harbin Engineering University	2006
15	24	Tongji University	2016	0.06	Dalian Maritime University	2015

From Table 3, we can see that among the top 15 organizations in terms of the number of publications, Chinese research organizations occupy 14 of them, and China's contribution to the number of publications in underwater image enhancement is the largest, which is in line with our previous conclusion. On the right side of the table, the ranking of the institutions is performed with reference to centrality, which, in "Citespace", refers to the Betweenness Centrality: this measures how often a node appears on the shortest path between other nodes. Higher values indicate that the node is more likely to be a 'bridge' between different clusters, playing a key role in the dissemination of knowledge or the evolution of the field. Nodes with a high centrality are likely to be the ones that drive important documents in the field (e.g., papers proposing new theories, methods, or techniques). The Chinese Academy of Sciences has the highest centrality value, suggesting that it has contributed significantly to the field, followed closely by the Institute of Deep-Sea Science & Engineering; we find that the number of publications is not directly correlated with centrality, and that a higher number of publications does not necessarily mean that it has contributed the most. We found that there is no direct correlation between the number of publications and centrality. Among the top 15 institutions ranked by centrality, there are 13 Chinese institutions, which shows that Chinese authors are very popular in this field.

3.3. Keywords Analysis

Keywords encapsulate the central themes of scholarly works, and their co-occurrence patterns serve as a methodological tool for mapping disciplinary foci. In constructing visual knowledge networks, the frequency with which terms appear together and their centrality metrics—indicators of conceptual influence—are pivotal analytical parameters. Temporal evaluations of these metrics, tracking shifts in prominence and interconnectivity over time, further refine insights into evolving research priorities. In this paragraph, we will use "CiteSpace" and "VOSviewer" to analyze and re-search keywords in underwater image enhancement in an attempt to find hot changes in research in the historical development of it. Figure 21 shows the keyword clusters we generated using VOSviewer.

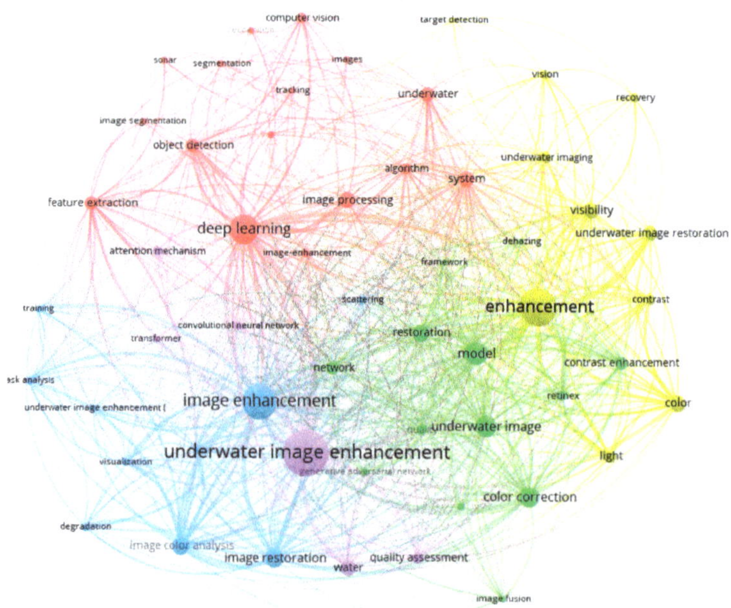

Figure 21. Cluster analysis diagram of underwater image enhancement keywords.

Figure 21 shows the map of underwater image enhancement keywords and the relationship between them; from the figure, we can see that the keyword 'underwater image enhancement' has the most occurrences, with a count of 417, followed closely by 'image enhancement' and "enhancement". These three keywords belong to the same field, and if we merge them, the count is 1036. The second ranked keyword is 'deep learning', with a count of 226, followed by model, with a count of 168.

Table 4 shows the rankings of the top 15 keywords in terms of the number of occurrences or centrality.

Table 4. Keyword rankings based on number of count or centrality.

Rank	Count	Keywords	Year	Centrality	Keywords	Year
1	1036	Underwater image enhancement	2011	0.63	Feature extraction	2006
2	226	Deep learning	2020	0.58	Shape	2006
3	168	Model	2007	0.55	Segmentation	2007
4	149	Underwater image	2013	0.52	Vision	2010
5	137	Color correction	2017	0.45	Visibility	2012
6	134	Image restoration	2006	0.39	Underwater imaging	2005
7	120	Image color analysis	2020	0.38	Underwater image enhancement	2016
8	109	Color	2017	0.34	Image restoration	2006
9	106	Visibility	2012	0.27	Color correction	2017
10	105	Image processing	2007	0.26	Navigation	2006
11	101	Restoration	2011	0.23	Model	2007
12	99	System	2008	0.23	System	2008
13	98	Water	2007	0.20	Images	2008
14	92	Underwater image restoration	2019	0.19	Image processing	2007
15	83	Feature extraction	2006	0.15	Reconstruction	2008

In the list of keywords ranked by quantity, the keywords 'Color correction', 'Feature extraction', etc., are listed as professional terms in the field of underwater images, which appeared earlier and belonged to the early traditional methods of enhancement. The deep learning keyword attracts the most attention, which appears in the latest year but has a high keyword frequency and indicates that this represents a new type of research trend. We further narrowed down the scope by limiting the keywords to 2020 and beyond, and the results are obtained in Table 4.

In Table 5, we find that in addition to the keywords of deep learning, attentional mechanism, convolutional neural network, generative adversarial network, unsupervised learning, and other keywords about deep learning are frequently occurring keywords; deep learning becomes the main force of underwater image enhancement after 2020, and the research direction gradually changes from traditional physical methods to deep learning methods.

Figure 22, produced via CiteSpace's keyword mutation function, illustrates a visualization designed to examine the thematic evolution within a research domain. This analytical tool identifies shifts in core topics and emerging trends over time by evaluating changes in terminology across academic publications. Such evaluations enable scholars to detect critical transitions in a field's focus, offering predictive insights into its developmental trajectory. By mapping temporal variations in keyword usage, this method not only highlights conceptual transformations but also aids in forecasting research priorities, thereby informing strategic directions for future scholarly inquiry.

Burst detection analysis offers a powerful approach for monitoring disciplinary trends, as it can effectively pinpoint sudden shifts in keyword activity. This method surpasses traditional bibliometric measures, such as citation counts or publication numbers, in its ability to capture the evolution of emerging research frontiers.

In the keyword mutation table in Figure 22, we find that the keywords algorithm, classification, navigation, etc., have a long time span and appear early, which suggests that researchers focused on some traditional physical algorithms for image restoration before 2020, and these algorithms have been a research hotspot. However, the keyword unsupervised learning has a high mutation intensity in recent years, and the intensity of other directions is weakening, which indicates that underwater image enhancement is shifting in the direction of unsupervised learning, which also confirms our previous conclusions that the field of underwater image enhancement is moving towards the development of the field in the deep learning direction.

Table 5. Keyword rankings based on number of count or centrality after 2020.

Rank	Count	Keywords	Year
1	226	deep learning	2020
2	120	image color analysis	2020
3	78	network	2022
4	56	quality	2020
5	53	attention mechanism	2022
6	43	convolutional neural network	2021
7	43	underwater images	2020
8	39	task analysis	2022
9	37	generative adversarial network	2021
10	34	retinex	2020
11	27	underwater image enhancement (uie)	2024
12	23	underwater object detection	2023
13	20	histogram	2022
14	19	image fusion	2023
15	18	unsupervised learning	2023

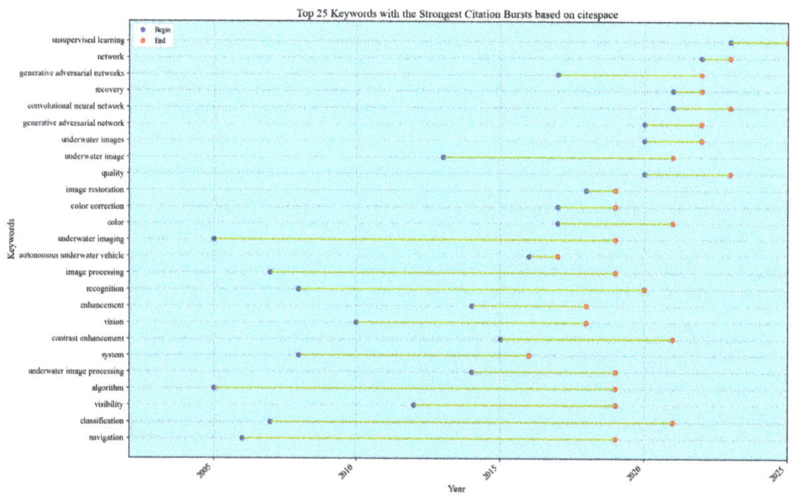

Figure 22. Top 25 keywords with the strongest citation bursts based on CiteSpace.

4. Deep Learning Theory

In recent years, deep learning technology has brought breakthroughs in the field of underwater image enhancement through its unique paradigm innovation. Compared with traditional methods that rely on physical modeling, the core advantages of deep learning are reflected in three aspects: first, based on the end-to-end learning architecture [94], the deep network is able to automatically learn the complex mapping relationship from degraded images to clear images, getting rid of the strong dependence on physical models such as the Jaffe–McGlamery model, and effectively solving the problem of error accumulation in traditional methods due to simplifying the optical transmission equation, which leads to the problem of error accumulation [95]. Secondly, the multi-scale feature fusion mechanism enables the network to simultaneously handle multimodal tasks such as color offset correction [81,96], local contrast enhancement, and scattering noise suppression [97], breaking through the efficiency bottleneck of the traditional method that needs to deal with different types of degradation in tandem at different stages [98]. More importantly, the deep neural network can accurately portray the coupling relationship between non-uniform light distribution and spatially varying scattering effects in underwater environments through the stacking of nonlinear activation functions [99], which is significantly better than linear methods, such as histogram equalization, in complex scenarios, such as coral reef-shaded areas and turbid waters with sudden changes in visibility [100]. Together, these features drive the paradigm shift from 'physics-driven' to 'data-driven' underwater visual enhancement techniques [101], validated by benchmarks (UIEB [17], ImageNet [102]) and perceptual metrics (SSIM [103], LPIPS [104]).

4.1. Evolution: From CNNs to Transformers

Underwater image enhancement methodologies in the deep learning era have predominantly revolved around three core architectural frameworks. CNN-based approaches leverage hierarchical feature extraction capabilities [25], often enhanced by embedding physical priors into network layers [105–107]. GAN-driven solutions excel in modeling complex distribution mappings through adversarial training, particularly effective for unpaired data scenarios [108,109]. Transformer-based models, while relatively nascent in this domain, demonstrate superior performance in capturing long-range dependencies via self-attention mechanisms [110,111]. In Section 5, we present a comparative analysis of the operational characteristics of these models, highlighting their respective advantages in dealing with chromatic aberration, scattering noise, and non-uniform illumination.

4.1.1. CNN

In recent years, the rapid development of deep learning technology has driven a fundamental shift in underwater image enhancement from the traditional physical model-driven to data-driven paradigm. With its powerful feature extraction capability and end-to-end learning mechanism, convolutional neural networks (CNNs) gradually overcome the limitations of traditional methods relying on simplifying assumptions, and become a core tool for solving degradation problems such as color distortion, contrast degradation, and the blurring of details in underwater images. Figure 23 illustrates the structure of the CNN.

a. Breakthrough in end-to-end multi-task learning frameworks

The UIE-Net proposed by Wang et al. [105] pioneered the construction of an end-to-end multi-task learning framework for the collaborative modeling of degraded features by jointly optimizing the color correction and defogging tasks. The network employs a pixel-shuffle strategy to enhance local feature extraction and synthesizes 200,000 training data based on a physical imaging model. Experiments show that its PSNR in the cross-scene test is improved by 23.5% compared with the traditional method, which verifies the effectiveness of multi-task learning.

b. Normalization of benchmark datasets and baseline models

To overcome the limitation of data scarcity on algorithm evaluation, Li et al. further constructed the Underwater Image Enhancement Benchmark (UIEB) dataset, which contains 950 real underwater images (including 890 reference-degradation data pairs). The proposed water-net model based on this dataset adopts a progressive enhancement architecture and achieves a PSNR of 27.8 dB on UIEB, which is a 14.2% improvement over the earlier CNN model, through a three-stage feature fusion (degradation perception $\rightarrow$ adaptive enhancement $\rightarrow$ global optimization). This work provides a standardized benchmark for algorithm performance evaluation and model generalization capability [17].

c. Multi-color spatial fusion and neural profile optimization

UIEC^2-Net, proposed by Zhang et al., innovatively integrates the dual-color spaces of RGB and HSV to break through the representation limitations of a single-color space. Its architecture contains three core modules:

(1) RGB pixel-level module: denoising and color bias correction through residual linkage;

(2) HSV global adjustment module: introducing a neural curve layer to dynamically adjust brightness and saturation;

(3) Attention fusion module: weighted fusion of bispace outputs to suppress cross-modal conflicts.

Experiments show that the method achieves 3.85 on the UCIQE metric, which is a 12.7% improvement over the single-space model, verifying the superiority of multimodal feature fusion [106].

d. Co-optimization of lightweight design and attention mechanism

Aiming at the balance between computational efficiency and detail retention, Zheng et al. proposed an improved CNN defogging network, the innovations of which include depth-separable convolution, with a 58% reduction in the amount of parameters (Equations (25)–(28)); basic attention module (BAM), focusing on key regions through channel-space dual paths; and cross-layer connection and pooling pyramid, enhancing multi-scale feature extraction.

The model reduces latency to 85 ms in 1080p image processing while maintaining UIQM 4.02, providing a viable solution for real-time underwater enhancement [107].

e. Scene a priori-driven and video enhancement extensions

The UWCNN developed by Wang et al. for the first time embeds an underwater scene prior to the CNN training process to synthesize diverse degraded data (covering five types of water quality conditions) through a physical model. Their lightweight network (only 2.1 M parameters) employs a codec structure that incorporates jump connections to retain low-frequency information. Experiments show that the model achieves a PSNR of 26.5 dB in turbid waters with NTU > 30 and can be extended to video frame-by-frame enhancement with a stable frame rate of 25 fps [112].

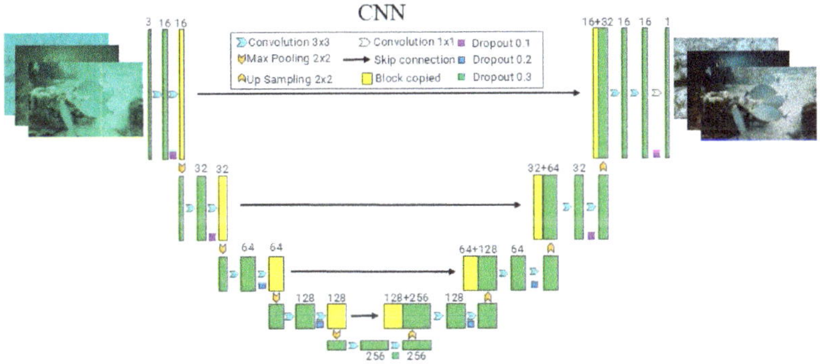

Figure 23. Fundamental principles of CNN construction.

4.1.2. GAN

Generative adversarial networks (GANs) have demonstrated significant advantages in the field of underwater image enhancement, gradually overcoming the limitations of traditional methods relying on simplifying assumptions through the deep integration of their adversarial learning mechanisms with physical models. GAN architectures (e.g., PUGAN, UW-CycleGAN) are able to simultaneously model complex light absorption-scattering effects through the dynamic game of the generator discriminator, and produce visually realistic enhancement results, which significantly improves the physical plausibility of color correction and the robustness of detail recovery. Figure 24 illustrates the structure of the GAN.

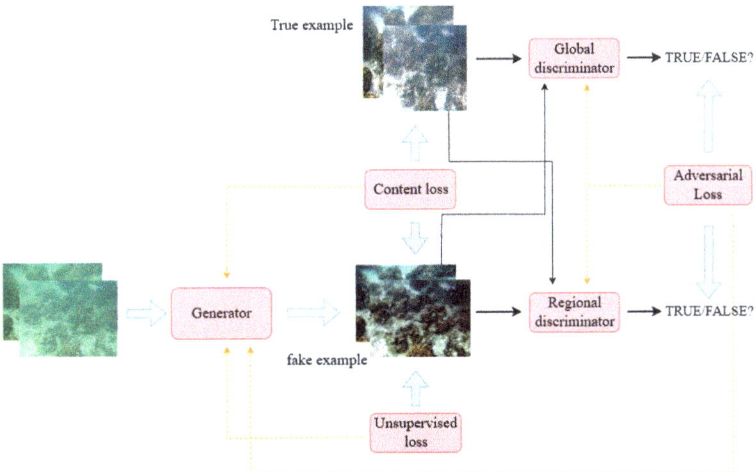

Figure 24. Fundamental principles of GAN construction.

Cong et al. proposed an underwater image enhancement method based on generative adversarial networks (GANs) and physical models called the physical model-guided underwater image enhancement using GAN with dual discriminators. Underwater images usually suffer from low contrast, color distortion, and the blurring of details due to light absorption and scattering effects of the water medium, which increase the difficulty of underwater enhancement tasks. To solve these problems, PUGAN combines the visual aesthetics advantage of GANs and the scene adaptation advantage of physical models; consequently, an architecture (TSIE-subnet) and parameter estimation subnetwork (Par-subnet) are proposed. Par-subnet is used to learn the parameters of the inversion of the physical model and generate color-enhanced images as TSIE-subnet's auxiliary information. The Degradation Quantization (DQ) module in the TSIE-subnet is used to quantify the scene degradation to enable the enhancement of critical regions. In addition, PUGAN is designed with dual discriminators for style–content adversarial constraints to improve the realism and visual aesthetics of the results [109].

Panetta et al. focused on the wide range of current target tracking methods on publicly available benchmark datasets and pointed out that these methods mainly focus on open space image data, while less attention has been paid to underwater visual data. The inherent distortion problems of color loss, poor contrast, and underexposure in underwater images due to light attenuation, refraction, and scattering in underwater environments greatly affect the visual quality of underwater data, making existing open-space trackers perform poorly on such data.

He presents the first comprehensive underwater target tracking benchmark dataset (UOT100), which aims to facilitate the development of tracking algorithms suitable for underwater environments. The dataset contains 104 underwater video sequences and over 74,000 annotated frames derived from natural and artificial underwater videos, covering a wide range of distortion types. The article also evaluates the performance of 20 state-of-the-art target tracking algorithms on this dataset and introduces a cascaded residual network for underwater image enhancement modeling to improve the accuracy and success of the tracker. The experimental results show that existing tracking algorithms have significant shortcomings on underwater data, while the generative adversarial network (GAN)-based augmentation model can significantly improve tracking performance [113].

Chen et al. explored the problem of low visual quality in underwater robots, an issue that limits their widespread application. Although several algorithms have been developed, real-time and adaptive approaches are still insufficient for practical tasks. To this end, the authors proposed a generative adversarial network (GAN)-based restoration scheme (GAN-RS), aiming to simultaneously preserve image content and remove underwater noise by means of a multi-branching discriminator (including adversarial and critic branches).

The authors not only employ adversarial learning, but also introduce a novel dark-channel a priori loss to facilitate the generator to produce more realistic visual effects. The authors also investigated the underwater index to characterize underwater features and designed an underwater index-based loss function to train the critical branch to suppress underwater noise [114].

Yan et al. proposed a model-driven cyclic coherent generative adversarial network (CycleGAN)-based model, called UW-CycleGAN, for underwater image restoration. The model is inspired by underwater image formation models and is capable of directly estimating the background light, transmission map, scene depth, and attenuation coefficient. Through comprehensive experiments, the authors demonstrate that the method outperforms other underwater image restoration methods both qualitatively and quantitatively, and is able to provide restored images with satisfactory color saturation and brightness [115].

Hambarde et al. proposed an end-to-end generative adversarial network called UW-GAN for depth estimation and image enhancement from a single underwater image. According to the literature, the coarse-grained depth map is firstly estimated by the underwater coarse-grained generative network (UWC-Net) and then the fine-grained depth map is computed by the underwater fine-grained network (UWF-Net), which splices the estimated coarse-grained depth map with the input image as an input. The UWF-Net consists of compression and excitation modules at both spatial and channel levesl for fine-grained depth estimation. In addition, the performance of the network proposed in the literature is analyzed on both real-world and synthetic underwater datasets and thoroughly evaluated on underwater images under different color dominance, contrast, and illumination conditions [116].

4.1.3. Transformer

The Transformer architecture is becoming an important technology paradigm in the field of underwater image enhancement due to its global modeling capability and self-attention mechanism. Compared with traditional convolutional neural networks (CNNs), Transformer can effectively capture the scattering effects of non-uniform illumination and spatial variations in underwater images through long-range dependent modeling (e.g., U-shape Transformer's multi-scale windowed attention mechanism), which significantly improves the enhancement effect of complex scenes (e.g., shaded coral reefs and turbid waters). In addition, by combining physical a priori (e.g., Jaffe–McGlamery model) and frequency-domain guided optimization, Transformer shows stronger robustness in color correction and detail recovery. However, high computational complexity and high training data requirements are still the main challenges. Future research can explore a lightweight design (e.g., knowledge distillation) and multimodal fusion (e.g., sonar-optical cross-modal alignment) to further promote the practical application of Transformer in underwater enhancement. Figure 25 illustrates the structure of the Transformer.

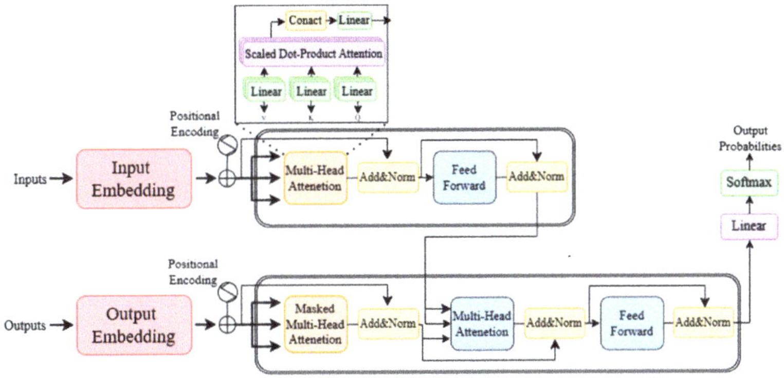

Figure 25. Fundamental principles of Transformer construction.

Gao et al. proposed a path-enhanced Transformer framework called PE-Transformer, which aims to improve the performance of underwater object detection in complex backgrounds. The authors designed a scheme for embedding local path detection information that facilitates the interaction between high-level features and low-level features, thus enhancing the semantic representation of small-scale underwater targets. Within the CSWin-Transformer framework, rich dependencies are established between high-level and low-level features, further enhancing the semantic representation in the encoding stage. A flexible and adaptive point representation detection module is designed, which is capable of covering underwater targets from any direction. Through feature selection between salient point samples and points in classification and localization, the module achieves the optimization of feature selection while improving the detection accuracy of underwater objects [117].

Due to the absorption and scattering of underwater impurities, existing data-driven methods perform poorly in the absence of large-scale datasets and high-fidelity reference images. Peng et al. constructed a large-scale underwater image dataset (LSUI), which contains 4279 sets of real-world underwater images, each accompanied by clear reference images, semantic segmentation maps, and media transport maps.

On this basis, the authors introduced the Transformer model into the UIE task for the first time and proposed a U-shaped Transformer network. The network integrates two specially designed modules: the channel-level multi-scale feature fusion Transformer (CMSFFT) and the spatial-level global feature modeling Transformer (SGFMT). These modules enhance the network's attention to heavily attenuated color channels and spatial regions.

In addition, to further enhance the contrast and saturation of the images, the authors designed a novel loss function combining RGB, LAB, and LCH color spaces, which follows the principles of human vision [110].

Shen et al. proposed an underwater image enhancement method based on a Dual Attention Transformer Block (DATB) called the UDAformer. Considering the inhomogeneity of underwater image degradation and the loss of color channels, UDAformer combines the Channel Self-Attention Transformer (CST) and the LCH loss function. The Self-Attention Transformer (CSAT) and Shifted Window Pixel Self-Attention Transformer (SW-PSAT) are also proposed in the literature; in these approaches, a new fusion method is proposed, combining channel and pixel self-attention for the efficient encoding and decoding of underwater image features. In addition, in order to improve computational efficiency, the literature also proposes a shift window method for pixel self-attention. Further, the self-attention weight matrix is computed by constructing a convolution, which enables the UDAformer to flexibly handle input images of various resolutions and reduce network parameters. Finally, underwater images are recovered by jump connections based on the design of an underwater imaging model [111].

Ummar et al. pointed out that the estimation of high-quality underwater images is an important step in the development of computer vision systems for marine environments, a step that encompasses many computer vision and robotics applications, such as ocean exploration, robotic manipulation, navigation, object detection, tracking, and marine life monitoring. To this end, Ummar et al. proposed a novel end-to-end underwater window transform generative adversarial network (UwTGAN). The algorithm consists of two main components: a transform generator for generating recovered underwater images and a transform discriminator for classifying the generated underwater images. Both components are equipped with window-based self-attention blocks (WSABs), which maximize efficiency and provide a relatively low computational cost by restricting the self-attention computation to non-overlapping local windows. The WSAB-based transform generator and discriminator are trained end-to-end, and the authors also formulate an efficient loss function to ensure that the variables are tightly integrated [118].

As a matter of fact, underwater image enhancement has evolved from relying on single-architecture deep learning models, such as CNNs for feature extraction, GANs for image generation, and Transformers for contextual understanding, to embracing fusion models that synergize these approaches. By integrating multiple architectures, these hybrid models mitigate the shortcomings of individual methods—such as CNNs' limited global awareness, GANs' training challenges, and Transformers' computational costs—thereby achieving superior performance in enhancing underwater images.

Wang et al. proposed an underwater image enhancement method (UIE-Convformer) based on a convolutional neural network (CNN) with a feature fusion Transformer, which efficiently extracts the local texture features by the ConvBlock module and models the long range dependency by combining the cross-channel self-attention mechanism of the Feaformer module, which solves the problem of traditional CNNs concerning the difficulty in dealing with underwater wide-range blurring and scattering due to the restricted receptive field [119].

Zheng et al. proposed a dual generative adversarial network (LFT-DGAN) based on reversible convolutional decomposition with a full-frequency Transformer, which introduces reversible neural networks into underwater image processing for the first time, and separates the image into low, medium, and high-frequency components by the decomposition technique without information loss, to alleviate the problem of random information loss in the traditional mathematical transforms (such as wavelet transform). Experiments demonstrate that the method significantly outperforms existing methods in complex underwater scenarios such as UCCS, UIQS, etc., and shows strong generalization ability in extended tasks such as de-fogging and de-raining [120].

5. Comparison of Results

In this section, we perform a comparative analysis of traditional physical models, water-net, UWCNN, UWCycleGAN, U-shape, and so on. To further quantitatively assess the quality of the enhanced images, we calculated the peak signal-to-noise ratio (PSNR), underwater image quality measurement (UIQM), underwater contrast, and color enhancement quality evaluation (UCIQE, RGB statistics, and luminance metrics). The PSNR quantifies the signal-to-noise ratio of the image, with higher values indicating that the enhanced image has less distortion compared to the ideal image. the UCIQE mainly evaluates the color uniformity, contrast, and saturation of underwater images. UCIQE evaluates color uniformity, contrast, and saturation of underwater images, while UIQM is a comprehensive underwater image quality metric that combines contrast, hue, and sharpness.

Our experimental data come from the UIEB dataset, and in order to compare the performance of each deep learning model in different underwater scenarios, we selected three groups of experimental subjects from the dataset, namely, the rock group, the underwater portrait group, and the marine life group. Due to some hardware constraints, the training models we use are from pre-trained models without any processing. Since some models are only trained on 256 × 256 resolution, we choose the reference image as a medium grey image of the same size, and the PSNR values are for reference only.

The physical method (b) can more directly simulate the physical conditions underwater, and in some areas, especially those with moderate water turbidity or light, clearer images can be obtained. However, in cases of severe distortion, such as deep water with low visibility, enhancement may be less effective. Color correction may be too simple or impractical, as it may not capture the full effect of light in the water (Figure 26).

The UWCNN (c) aims to address underwater image degradation more effectively. It generally improves visibility and restores more accurate colors, especially in shallow waters. However, in highly turbid waters or very deep seabed images, it may still have difficulty in restoring color and contrast. The color processing may be too biased towards certain shades and appear unnatural.

Water-net (d) ranks last in terms of brightness performance but improves contrast and reduces noise. It may work well in certain situations where contrast enhancement is needed without overdoing it.

UWCycleGAN (e) produces a natural, visually appealing image with more balanced colors and clearer details, and it ranks first among all methods in terms of brightness. However, it produces subtle artifacts with some color distortion (Table 6).

The U-shape method (f) is very effective in restoring structural details and reducing noise. This structure helps to maintain the sharpness of the image, making it an excellent choice for fine detail. It does introduce some unnatural artefacts, such as a slight halo effect around the edges of Picture 3 and the over-correction of colors, resulting in slightly 'artificial' or over-sharp results.

The physical model improves visibility and reproduces details extremely well for the first, third, and seventh images, close to the reference image (GT). However, the fourth images show severe overexposure, the colors look unnatural, the overcorrected appearance distorts the appearance of the scene, and its inability to deal with complex textures and fine details effectively produces an image that is inferior to the original image (Figure 27).

UWCNN performs mediocrely in the human group; its ability to enhance brightness is still ranked at the bottom of the list, and UWCNN colors are prone to oversaturation or artifacts. The enhancement of Picture 3 is poor, its luminance, UCIQE, and UIQM are far below average, and there is almost no color representation. These problems are due to the limited ability of the model to handle complex underwater lighting conditions (Table 7).

The water network has improved in areas such as color correction and object visibility. The model displayed a more natural fourth image with improved color accuracy. However, because of its average brightness performance, the model was sometimes unable to recover finer details, resulting in slight blurring in some areas.

UWCycleGAN, the luminance performance of which is at the top of the list, has visually enhanced images with realistic colors and details. This method performs well in maintaining a natural underwater environment. However, the first image is a bit overcorrected with the color of the third image, and some unnatural artifacts appear, especially at the boundaries of the objects.

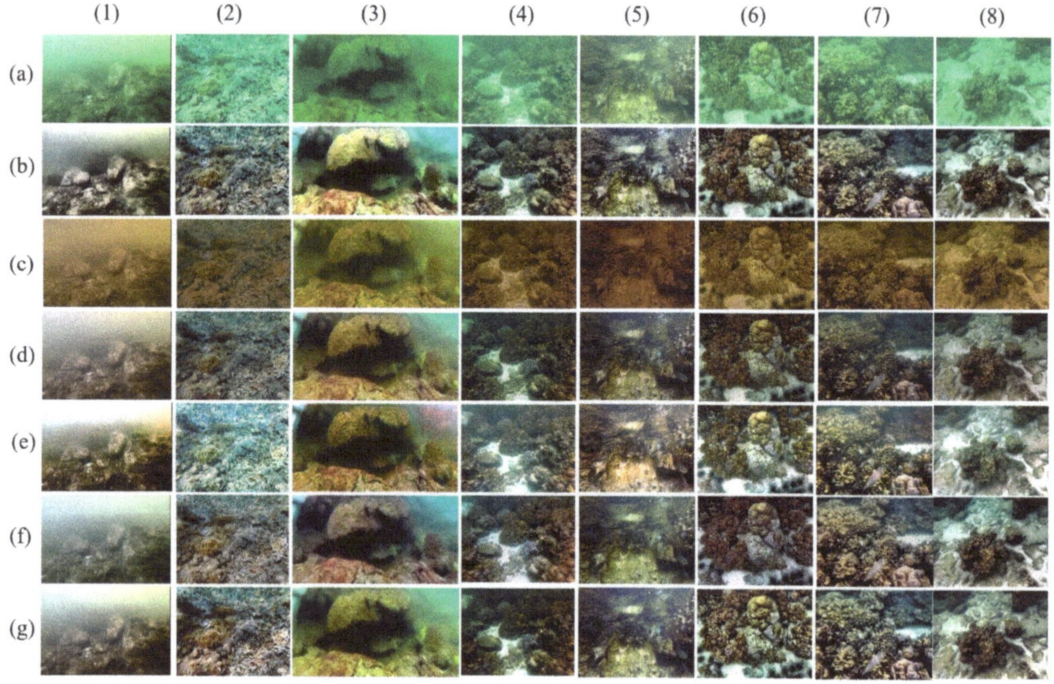

Figure 26. Enhanced image of the rock group. (**a**) Original image; (**b**) physical; (**c**) UWCNN; (**d**) water-net; (**e**) UWCycleGAN; (**f**) U-shape; and (**g**) reference images (recognized as ground truth (GT).

Table 6. Average parameters for each type of rock group.

Method/Average	PSNR	UCIQE	UIQM	R	G	B	Luminance
input	15.562	16.988	0.152	90.436	149.377	110.477	149.385
Physical	12.207	27.976	0.286	109.621	119.795	104.432	124.442
UWCNN	13.511	14.786	0.134	115.071	99.928	55.546	116.546
Water-net	13.183	22.094	0.205	100.624	105.537	88.949	109.989
UWCycleGAN	12.224	25.090	0.301	124.132	132.200	113.943	139.895
U-Shape	13.526	22.918	0.241	110.643	115.628	105.674	121.626

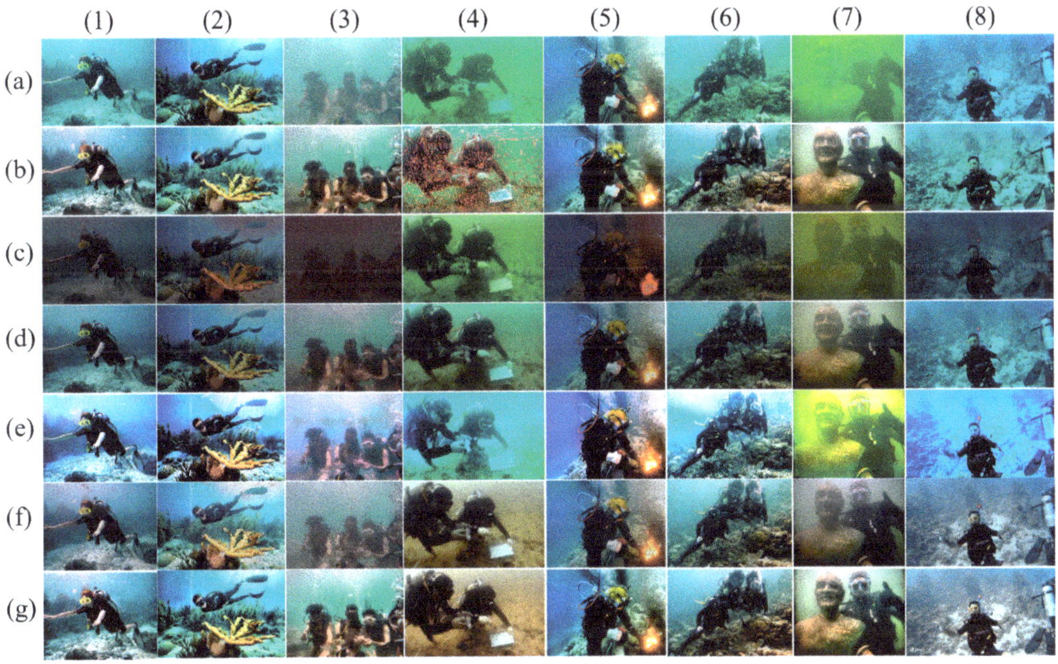

Figure 27. Enhanced image of the human group. (**a**) Original image; (**b**) physical; (**c**) UWCNN; (**d**) water-net; (**e**) UWCycleGAN; (**f**) U-shape; and (**g**) GT.

Table 7. Average parameters for each type of human group.

Method/Average	PSNR	UCIQE	UIQM	R	G	B	Luminance
input	11.411	21.648	0.147	44.670	143.691	134.548	158.017
Physical	11.276	27.888	0.223	81.421	127.246	124.898	145.738
UWCNN	12.214	15.245	0.091	67.934	84.011	69.039	91.477
Water-net	12.694	22.736	0.146	68.673	103.128	104.302	115.178
UWCycleGAN	9.413	30.565	0.268	72.859	145.800	161.942	182.180
U-Shape	13.598	23.827	0.189	95.807	119:455	121.908	131.719

The U-shape method performs best in improving the sharpness and contrast of the fourth image, which looks sharper and more detailed. However, the model sometimes has problems with tonality, resulting in unnatural gradients or overly bright areas, especially in the background, which can distract from the scene, as in images 3 and 7, which show an image that is too dark overall.

The physical models generally improve the visibility of fish and coral structures, with Figure 4 showing the best color reproduction of these models. The contrast between the fish and the background is better than in the original image. Colors are more vibrant, though not perfect, and there is a slight improvement in the clarity of the fish scales and water in particular (Figure 28).

However, in Figure 1, the model gives the water a yellow–green tint that looks unnatural. Figure 3 shows some strange darkening in the background areas, especially around the coral structures, giving it an unrealistic look.

The texture of the UWCNN fish scales and corals is better preserved than the solid model, retaining detail, but the overall performance in the benthic organism group is also unsatisfactory, with the worst recovery of brightness, and a color treatment that may be too biased towards certain shades, which looks unnatural (Table 8).

Figure 28. Enhanced image of the underwater portrait group. (**a**) Original image; (**b**) physical; (**c**) UWCNN; (**d**) water-net; (**e**) UWCycleGAN; (**f**) U-shape; and (**g**) GT.

Table 8. Average parameters for each type of fish group.

Method/Average	PSNR	UCIQE	UIQM	R	G	B	Luminance
input	13.468	19.123	0.152	81.898	135.992	126.946	152.768
Physical	12.517	25.389	0.224	94.670	121.617	124.365	139.514
UWCNN	12.732	12.237	0.086	82.135	78.048	65.592	94.324
Water-net	13.687	20.049	0.157	81.290	103.434	105.927	118.358
UWCycleGAN	10.593	24.277	0.279	111.597	121.791	129.019	167.182
U-Shape	14.314	19.309	0.203	103.049	120.037	118.490	133.613

The water-net method strikes a good balance between enhancing the image without oversaturating the colors. The fish and coral colors in columns two and seven look more natural. The coral areas and fish in images 4 and 7 look more like the ground truth, with more natural transitions between colors. Unlike the physical model, water-net seems to avoid the overexposure or darkening of certain areas, maintaining good overall brightness and visibility.

However, subtle artefacts can be seen around the fish in image 4, especially at the edges, which may be overcorrected by this method. The model sometimes struggles to maintain the finesse of the darker parts of the image (e.g., the background) and therefore can appear somewhat blurred.

UWCycleGAN shows significant improvement in color reproduction. Underwater visibility is much improved in images 3 and 7, and the background is much sharper compared to the original image. The water in the second column looks very close to the ground truth, with a natural blue color and no visible distortion.

Although the color balance has improved, the lighting in some areas (image 1) looks strange. There is an unnatural halo or strong contrast that reduces realism, and the overall photo has a strange pink color with some over-enhancement.

The U-shape method excels in just the right amount of color enhancement. Fish and corals, especially in pictures 2 and 3, appear more vibrant and detailed. The mode enhances contrast without creating noticeable artefacts. The fish are clearly defined, and the background appears brighter. There is a balance between natural and enhanced light: the lighting is more natural than other methods and helps to enhance realism. The fish in the first column is affected by the artificial lighting effect, where the light source does not match the distribution of natural light underwater, resulting in a slightly unrealistic feel.

6. Conclusions

Underwater image enhancement plays a pivotal role in marine resource exploration, ecological monitoring, and infrastructure inspection. This study systematically evaluates traditional physics-driven approaches and state-of-the-art deep learning models, revealing their distinct strengths, limitations, and applicability across diverse underwater environments. The key findings and implications are summarized as follows:

6.1. Comparative Performance of Physical and Deep Learning Models

(a) Traditional Physical Models (e.g., simplified Jaffe–McGlamery)

Strengths: High interpretability, computational efficiency, and robustness in stable, shallow environments (depth < 10 m, NTU < 20). For instance, wavelength compensation and multi-scale histogram equalization effectively restore color balance and contrast (PSNR = 12.207 in rock group).

Limitations: performance degrades in complex scenarios (e.g., deep-sea or turbid waters) due to oversimplified assumptions (e.g., uniform background light) and an inability to model nonlinear degradation interactions (e.g., Figure 27b overexposure). Physical methods (e.g., wavelength compensation, Laplace sharpening) have difficulty in capturing the coupling effect between color deviation, scattered noise, and non-uniform illumination, resulting in overexposure (Figure 27b) or artifacts (Figure 28f) in complex scenes.

The fixed $\beta_r = 0.1$, $\beta_g = 0.05$, and $\beta_b = 0.03$ used in our physical model are grounded in clear-water scenarios but lack justification for diverse depths and illumination conditions. This may contribute to overexposure or color distortion in complex scenes (e.g., Figure 27b).

(b) Deep Learning Models

The water-net and UWCNN models excel in adaptive enhancement through supervised learning (UIEB dataset), balancing noise suppression and detail preservation. However, UWCNN struggles with luminance recovery (human group: luminance = 91.477), while water-net achieves moderate contrast improvement (UIQM = 0.205).

The UWCycleGAN and U-shape Transformer methods demonstrate superior capability in extreme conditions. UWCycleGAN's adversarial training produces visually natural results (UCIQE = 30.565) but risks color oversaturation. The U-shape model, leveraging self-attention mechanisms, achieves state-of-the-art sharpness (rock group: luminance = 121.626), yet requires careful parameter tuning to avoid artifacts. Because of the hardware limitation, the training data are selected as pre-training data, while most deep learning models (e.g., UWCycleGAN, U-shape) are only trained at a 256 × 256 resolution, while real underwater devices (e.g., AUV, ROV) may acquire higher-resolution images (e.g., 1080p). Resolution normalization processing (e.g., downsampling) may lead to the loss of details, affecting the recovery of fine features such as coral texture, fish scales, etc. (Figures 26–28).

6.2. Methodological Trade-Offs

(a) Implementation Complexity

Physical models rely on explicit optical principles and minimal data, enabling real-time deployment on low-power devices (e.g., 15–20 MB memory for histogram equalization).

Deep learning methods demand substantial computational resources: water-net and UWCNN require mid-tier GPUs (25–30 fps), while Transformer-based models (U-shape) necessitate $\geq$8 GB VRAM, limiting edge-device applicability.

(b) Interpretability

Physical models provide transparent enhancement steps (e.g., LAB color space analysis), aligning with optical principles.

They also lack transparency due to their black-box nature. While attention maps (e.g., U-shape's multi-scale windows) partially reveal feature prioritization, the underlying decision-making process remains opaque. Hybrid approaches, such as embedding physical priors into network layers (e.g., UWCycleGAN's dark-channel loss), improve interpretability but require extensive validation.

6.3. Practical Implications for Marine Applications

Shallow Waters: UWCycleGAN's natural color rendering (Figure 26e) suits ecological monitoring in aquaculture ranches.

Deep-Sea Exploration: U-shape's detail recovery (Figure 28f) aids structural inspection of subsea pipelines and cables.

Real-Time Systems: Lightweight CNNs (e.g., UWCNN) or physical models are preferable for continuous monitoring on underwater drones. While lightweight CNNs (e.g., UWCNN) and physical models are suggested as preferable for real-time monitoring on underwater drones, these recommendations are based on their computational efficiency in prior studies rather than direct measurements from our experiments, due to hardware limitations. As such, claims regarding real-time applicability are derived from theoretical analysis and the literature benchmarks.

6.4. Future Directions

Hybrid Frameworks: Integrating physical priors (e.g., transmission maps) into deep architectures (e.g., CNN preprocessing layers) could enhance robustness in turbid waters.

Lightweight Designs: Techniques like depth-wise separable convolutions or knowledge distillation (Section 4.1.1) should be prioritized to reduce latency (e.g., <50 ms for 1080p images). Although some real-time underwater image enhancement techniques are now available in academia [121–124], the trade-off between image quality and processing time remains a daunting task. Future research will focus on testing these models on more powerful hardware (e.g., mid-tier GPUs with $\geq$8 GB VRAM) to quantify real-time performance metrics such as FPS and latency. Collaborations with institutions possessing advanced computational resources are also planned to validate these capabilities.

Unsupervised Learning: Expanding synthetic datasets (e.g., LSUI) and leveraging contrastive learning frameworks can address data scarcity while improving cross-environment generalization.

Adaptive Enhancement: Dynamic strategy selection based on depth, turbidity, and hardware constraints will optimize resource utilization in marine ranching.

This study underscores that no single method universally outperforms others. The choice of enhancement technique must align with environmental conditions, task requirements, and hardware capabilities. Future innovations should focus on bridging the gap between physics-driven interpretability and data-driven adaptability, fostering sustainable advancements in underwater vision technologies. By addressing these challenges, we can unlock the full potential of underwater imaging for marine resource management, ecological conservation, and global blue economy development.

Author Contributions: Conceptualization, Y.M. and D.Z.; methodology, Y.M.; software, Y.M.; validation, Y.M. and Y.C.; formal analysis, Y.M.; investigation, Y.C.; resources, D.Z. and Y.M.; data curation, Y.M.; writing—review and editing, D.Z.; visualization, Y.M.; supervision, D.Z.; project administration, Y.M.; funding acquisition, Y.M. All authors have read and agreed to the published version of the manuscript.

Funding: This research received no external funding.

Data Availability Statement: All data generated and analyzed during this study are included in this published article. The datasets used and analyzed during the current study are available from the corresponding author on reasonable request.

Conflicts of Interest: The authors declare no conflicts of interest.

References

1. Zhang, D.; Ma, Y.; Zhang, H.; Zhang, Y. Marine Equipment Siting Using Machine-Learning-Based Ocean Remote Sensing Data: Current Status and Future Prospects. *Sustainability* **2024**, *16*, 8889. [CrossRef]
2. Bax, N.; Novaglio, C.; Maxwell, K.H.; Meyers, K.; McCann, J.; Jennings, S.; Frusher, S.; Fulton, E.A.; Nursey-Bray, M.; Fischer, M.; et al. Ocean resource use: Building the coastal blue economy. *Rev. Fish Biol. Fish.* **2021**, *32*, 189–207. [CrossRef]
3. Zhang, D. Engineering Solutions to Mechanics, Marine Structures and Infrastructures. *Eng. Solut. Mech. Mar. Struct. Infrastruct.* **2024**, *1*. [CrossRef]
4. Zhang, D.; Zhang, Y.; Zhao, B.; Ma, Y.; Si, K. Exploring subsea dynamics: A comprehensive review of underwater pipelines and cables. *Phys. Fluids* **2024**, *36*, 101304. [CrossRef]
5. Levin, L.A.; Bett, B.J.; Gates, A.R.; Heimbach, P.; Howe, B.M.; Janssen, F.; McCurdy, A.; Ruhl, H.A.; Snelgrove, P.; Stocks, K.I.; et al. Global observing needs in the deep ocean. *Front. Mar. Sci.* **2019**, *6*, 241. [CrossRef]
6. Yuan, B.; Cui, Y.; An, D.; Jia, Z.; Ding, W.; Yang, L. Marine environmental pollution and offshore aquaculture structure: Evidence from China. *Front. Mar. Sci.* **2023**, *9*, 979003. [CrossRef]
7. Long, L.; Liu, H.; Cui, M.; Zhang, C.; Liu, C. Offshore aquaculture in China. *Rev. Aquac.* **2024**, *16*, 254–270. [CrossRef]
8. Zhang, Y.; Li, T.; Lin, J.; Zhang, D. The Characteristics and Advantages of Deepwater Ultra-Large Marine Ranches. *Eng. Solut. Mech. Mar. Struct. Infrastruct.* **2024**, *20*, 20.
9. Ma, Y.; Si, K.; Xie, Y.; Liang, Z.; Wu, J.; Zhang, D.; Zhang, Y.; Cai, R. Global Marine Ranching Research: Progress and Trends through Bibliometric Analysis. *Eng. Solut. Mech. Mar. Struct. Infrastruct.* **2024**, *1*, 1–23. [CrossRef]
10. Holmer, M. Environmental issues of fish farming in offshore waters: Perspectives, concerns and research needs. *Aquac. Environ. Interact.* **2010**, *1*, 57–70. [CrossRef]
11. Ubina, N.A.; Cheng, S.-C. A review of unmanned system technologies with its application to aquaculture farm monitoring and management. *Drones* **2022**, *6*, 12. [CrossRef]
12. Tan, Y.; Lou, S. Research and development of a large-scale modern recreational fishery marine ranch System ☆. *Ocean Eng.* **2021**, *233*, 108610. [CrossRef]
13. Han, M.; Lyu, Z.; Qiu, T.; Xu, M. A review on intelligence dehazing and color restoration for underwater images. *IEEE Trans. Syst. Man Cybern. Syst.* **2018**, *50*, 1820–1832. [CrossRef]
14. Zhou, J.-C.; Zhang, D.-H.; Zhang, W.-S. Classical and state-of-the-art approaches for underwater image defogging: A comprehensive survey. *Front. Inf. Technol. Electron. Eng.* **2020**, *21*, 1745–1769. [CrossRef]
15. Porto Marques, T.; Branzan Albu, A.; Hoeberechts, M. A contrast-guided approach for the enhancement of low-lighting underwater images. *J. Imaging* **2019**, *5*, 79. [CrossRef]
16. Zhou, J.; Zhang, D.; Zhang, W. A multifeature fusion method for the color distortion and low contrast of underwater images. *Multimed. Tools Appl.* **2021**, *80*, 17515–17541. [CrossRef]
17. Li, C.; Guo, C.; Ren, W.; Cong, R.; Hou, J.; Kwong, S.; Tao, D. An underwater image enhancement benchmark dataset and beyond. *IEEE Trans. Image Process.* **2019**, *29*, 4376–4389. [CrossRef]
18. Ancuti, C.O.; Ancuti, C.; De Vleeschouwer, C.; Bekaert, P. Color balance and fusion for underwater image enhancement. *IEEE Trans. Image Process.* **2017**, *27*, 379–393. [CrossRef]
19. Garg, D.; Garg, N.K.; Kumar, M. Underwater image enhancement using blending of CLAHE and percentile methodologies. *Multimed. Tools Appl.* **2018**, *77*, 26545–26561. [CrossRef]

20. Yang, M.; Hu, J.; Li, C.; Rohde, G.; Du, Y.; Hu, K. An in-depth survey of underwater image enhancement and restoration. *IEEE Access* **2019**, *7*, 123638–123657. [CrossRef]
21. Hou, G.; Zhao, X.; Pan, Z.; Yang, H.; Tan, L.; Li, J. Benchmarking underwater image enhancement and restoration, and beyond. *IEEE Access* **2020**, *8*, 122078–122091. [CrossRef]
22. Zhang, W.; Wang, Y.; Li, C. Underwater image enhancement by attenuated color channel correction and detail preserved contrast enhancement. *IEEE J. Ocean. Eng.* **2022**, *47*, 718–735. [CrossRef]
23. Guo, Y.; Li, H.; Zhuang, P. Underwater image enhancement using a multiscale dense generative adversarial network. *IEEE J. Ocean. Eng.* **2019**, *45*, 862–870. [CrossRef]
24. Schettini, R.; Corchs, S. Underwater image processing: State of the art of restoration and image enhancement methods. *EURASIP J. Adv. Signal Process.* **2010**, *2010*, 746052. [CrossRef]
25. He, K.; Zhang, X.; Ren, S.; Sun, J. Deep residual learning for image recognition. In Proceedings of the IEEE Conference on Computer Vision and Pattern Recognition, Las Vegas, NV, USA, 27–30 June 2016.
26. Li, J.; Skinner, K.A.; Eustice, R.M.; Johnson-Roberson, M. WaterGAN: Unsupervised generative network to enable real-time color correction of monocular underwater images. *IEEE Robot. Autom. Lett.* **2017**, *3*, 387–394. [CrossRef]
27. Talaat, F.M.; El-Sappagh, S.; Alnowaiser, K.; Hassan, E. Improved prostate cancer diagnosis using a modified ResNet50-based deep learning architecture. *BMC Med. Inform. Decis. Mak.* **2024**, *24*, 23. [CrossRef]
28. Yuan, X.; Guo, L.; Luo, C.; Zhou, X.; Yu, C. A survey of target detection and recognition methods in underwater turbid areas. *Appl. Sci.* **2022**, *12*, 4898. [CrossRef]
29. Han, Y.; Huang, L.; Hong, Z.; Cao, S.; Zhang, Y.; Wang, J. Deep supervised residual dense network for underwater image enhancement. *Sensors* **2021**, *21*, 3289. [CrossRef]
30. Zhang, Z.; Yan, H.; Tang, K.; Duan, Y. MetaUE: Model-based meta-learning for underwater image enhancement. *arXiv* **2023**, arXiv:2303.06543.
31. Zhang, S.; Zhao, S.; An, D.; Li, D.; Zhao, R. MDNet: A fusion generative adversarial network for underwater image enhancement. *J. Mar. Sci. Eng.* **2023**, *11*, 1183. [CrossRef]
32. Zhu, D. Underwater image enhancement based on the improved algorithm of dark channel. *Mathematics* **2023**, *11*, 1382. [CrossRef]
33. Han, J.; Shoeiby, M.; Malthus, T.; Botha, E.; Anstee, J.; Anwar, S.; Wei, R.; Armin, M.A.; Li, H.; Petersson, L. Underwater image restoration via contrastive learning and a real-world dataset. *Remote Sens.* **2022**, *14*, 4297. [CrossRef]
34. Li, C.-Y.; Guo, J.-C.; Cong, R.-M.; Pang, Y.-W.; Wang, B. Underwater image enhancement by dehazing with minimum information loss and histogram distribution prior. *IEEE Trans. Image Process.* **2016**, *25*, 5664–5677. [CrossRef]
35. Guo, C.; Li, C.; Guo, J.; Loy, C.C.; Hou, J.; Kwong, S.; Cong, R. Zero-reference deep curve estimation for low-light image enhancement. In Proceedings of the IEEE/CVF Conference on Computer Vision and Pattern Recognition, Seattle, WA, USA, 13–19 June 2020.
36. Wei, K.; Fu, Y.; Zheng, Y.; Yang, J. Physics-based noise modeling for extreme low-light photography. *IEEE Trans. Pattern Anal. Mach. Intell.* **2021**, *44*, 8520–8537. [CrossRef]
37. Lee, S.-W.; Maik, V.; Jang, J.; Shin, J.; Paik, J. Noise-adaptive spatio-temporal filter for real-time noise removal in low light level images. *IEEE Trans. Consum. Electron.* **2005**, *51*, 648–653. [CrossRef]
38. Li, C.; Tang, S.; Kwan, H.K.; Yan, J.; Zhou, T. Color correction based on cfa and enhancement based on retinex with dense pixels for underwater images. *IEEE Access* **2020**, *8*, 155732–155741. [CrossRef]
39. Cheng, F.-H.; Hsu, W.-H.; Chen, T.-W. Recovering colors in an image with chromatic illuminant. *IEEE Trans. Image Process.* **1998**, *7*, 1524–1533. [CrossRef]
40. Wang, B.; Wei, B.; Kang, Z.; Hu, L.; Li, C. Fast color balance and multi-path fusion for sandstorm image enhancement. *Signal Image Video Process.* **2021**, *15*, 637–644. [CrossRef]
41. Guo, Z.; Wang, B.; Li, C. CAT: A lightweight Color-aware transformer for sandstorm image enhancement. *Displays* **2024**, *83*, 102714. [CrossRef]
42. Yang, M.; Hu, K.; Du, Y.; Wei, Z.; Sheng, Z.; Hu, J. Underwater image enhancement based on conditional generative adversarial network. *Signal Process. Image Commun.* **2020**, *81*, 115723. [CrossRef]
43. Susstrunk, S.E.; Winkler, S. Color image quality on the internet. *Internet Imaging V* **2003**, *5304*, 118–131.
44. Yuan, W.; Poosa, S.R.P.; Dirks, R.F. Comparative analysis of color space and channel, detector, and descriptor for feature-based image registration. *J. Imaging* **2024**, *10*, 105. [CrossRef]
45. Kanan, C.; Cottrell, G.W. Color-to-grayscale: Does the method matter in image recognition? *PLoS ONE* **2012**, *7*, e29740. [CrossRef]
46. Güneş, A.; Kalkan, H.; Durmuş, E. Optimizing the color-to-grayscale conversion for image classification. *Signal Image Video Process.* **2016**, *10*, 853–860. [CrossRef]
47. Sowmya, V.; Govind, D.; Soman, K.P. Significance of incorporating chrominance information for effective color-to-grayscale image conversion. *Signal Image Video Process.* **2017**, *11*, 129–136. [CrossRef]
48. Gonzalez, R.C.; Woods, R.E. *Digital Image Processing*, 4th ed.; Pearson Education: Rotherham, UK, 2002; ISBN 978-0-13-335672-4.

49. Wang, G.; Li, W.; Gao, X.; Xiao, B.; Du, J. Functional and anatomical image fusion based on gradient enhanced decomposition model. *IEEE Trans. Instrum. Meas.* **2022**, *71*, 2508714. [CrossRef]
50. Yang, Y.; Park, D.S.; Huang, S.; Rao, N. Medical image fusion via an effective wavelet-based approach. *EURASIP J. Adv. Signal Process.* **2010**, *2010*, 579341. [CrossRef]
51. Suresha, R.; Jayanth, R.; Shriharikoushik, M.A. Computer vision approach for motion blur image restoration system. In Proceedings of the 2023 14th International Conference on Computing Communication and Networking Technologies (ICCCNT), Delhi, India, 6–8 July 2023.
52. Wang, W.; Li, Z.; Wu, S.; Zeng, L. Hazy image decolorization with color contrast restoration. *IEEE Trans. Image Process.* **2019**, *29*, 1776–1787. [CrossRef]
53. Dong, L.; Zhang, W.; Xu, W. Underwater image enhancement via integrated RGB and LAB color models. *Signal Process. Image Commun.* **2022**, *104*, 116684. [CrossRef]
54. Cernadas, E.; Fernández-Delgado, M.; González-Rufino, E.; Carrión, P. Influence of normalization and color space to color texture classification. *Pattern Recognit.* **2017**, *61*, 120–138. [CrossRef]
55. Burambekova, A.; Pakizar, S. Comparative analysis of color models for human perception and visual color difference. *arXiv* **2024**, arXiv:2406.19520.
56. Reinhard, E.; Adhikhmin, M.; Gooch, B.; Shirley, P. Color transfer between images. *IEEE Comput. Graph. Appl.* **2001**, *21*, 34–41. [CrossRef]
57. Song, Y.; Nakath, D.; She, M.; Köser, K. Optical imaging and image restoration techniques for deep ocean mapping: A comprehensive survey. *PFG–J. Photogramm. Remote Sens. Geoinf. Sci.* **2022**, *90*, 243–267. [CrossRef]
58. Lin, S.; Ning, Z.; Zhang, R. Modified optical model and optimized contrast for underwater image restoration. *Opt. Commun.* **2025**, *574*, 130942. [CrossRef]
59. He, K.; Jian, S.; Xiaoou, T. Single image haze removal using dark channel prior. *IEEE Trans. Pattern Anal. Mach. Intell.* **2010**, *33*, 2341–2353.
60. Jansen, H.; Bogaart, L.v.D.; Hommersom, A.; Capelle, J. Spatio-temporal analysis of sediment plumes formed by mussel fisheries and aquaculture in the western Wadden Sea. *Aquac. Environ. Interact.* **2023**, *15*, 145–159. [CrossRef]
61. Jaffe, J. Computer modeling and the design of optimal underwater imaging systems. *IEEE J. Ocean. Eng.* **1990**, *15*, 101–111. [CrossRef]
62. Mobley, C.D. *Light and Water: Radiative Transfer in Natural Waters*; Academic Press: San Diego, CA, USA, 1994; ISBN 978-0125027502.
63. Twardowski, M.S.; Boss, E.; Macdonald, J.B.; Pegau, W.S.; Barnard, A.H.; Zaneveld, J.R.V. A model for estimating bulk refractive index from the optical backscattering ratio and the implications for understanding particle composition in case I and case II waters. *J. Geophys. Res. Oceans* **2001**, *106*, 14129–14142. [CrossRef]
64. Kirk, J.T. *Light and Photosynthesis in Aquatic Ecosystems*; Cambridge University press: Cambridge, UK, 1994.
65. Shafuda, F.; Kondo, H. A simple method for backscattered light estimation and image restoration in turbid water. In Proceedings of the OCEANS 2021: San Diego–Porto, San Diego, CA, USA, 20–23 September 2021; IEEE: Piscataway, NJ, USA, 2021; pp. 1–6.
66. Piskozub, J.; Stramski, D.; Terrill, E.; Melville, W.K. Influence of forward and multiple light scatter on the measurement of beam attenuation in highly scattering marine environments. *Appl. Opt.* **2004**, *43*, 4723–4731. [CrossRef]
67. Garaba, S.P.; Voß, D.; Zielinski, O. Physical, bio-optical state and correlations in North–Western European Shelf Seas. *Remote Sens.* **2014**, *6*, 5042–5066. [CrossRef]
68. Trucco, E.; Olmos-Antillon, A. Self-tuning underwater image restoration. *IEEE J. Ocean. Eng.* **2006**, *31*, 511–519. [CrossRef]
69. Ji, K.; Lei, W.; Zhang, W. A deep Retinex network for underwater low-light image enhancement. *Mach. Vis. Appl.* **2023**, *34*, 122. [CrossRef]
70. Jin, Y.; Fayad, L.M.; Laine, A.F. Contrast enhancement by multiscale adaptive histogram equalization. *Wavelets Appl. Signal Image Process. IX* **2001**, *4478*, 206–214.
71. Huang, S.; Li, D.; Zhao, W.; Liu, Y. Haze removal algorithm for optical remote sensing image based on multi-scale model and histogram characteristic. *IEEE Access* **2019**, *7*, 104179–104196. [CrossRef]
72. Zhang, W.; Li, X.; Xu, S.; Li, X.; Yang, Y.; Xu, D.; Liu, T.; Hu, H. Underwater image restoration via adaptive color correction and contrast enhancement fusion. *Remote Sens.* **2023**, *15*, 4699. [CrossRef]
73. Wang, H.; Sun, S.; Ren, P. Underwater color disparities: Cues for enhancing underwater images toward natural color consistencies. *IEEE Trans. Circuits Syst. Video Technol.* **2023**, *34*, 738–753. [CrossRef]
74. Chiang, J.Y.; Chen, Y.-C. Underwater image enhancement by wavelength compensation and dehazing. *IEEE Trans. Image Process.* **2011**, *21*, 1756–1769. [CrossRef]
75. Zhou, J.; Zhang, D.; Ren, W.; Zhang, W. Auto color correction of underwater images utilizing depth information. *IEEE Geosci. Remote Sens. Lett.* **2022**, *19*, 1504805. [CrossRef]
76. Zhou, J.; Zhang, D.; Zhang, W. Underwater image enhancement method via multi-feature prior fusion. *Appl. Intell.* **2022**, *52*, 16435–16457. [CrossRef]

77. Ke, K.; Zhang, B.; Zhang, C.; Yao, B.; Guo, S.; Tang, F. Underwater image enhancement via color correction and multi-feature image fusion. *Meas. Sci. Technol.* **2024**, *35*, 096123. [CrossRef]
78. Shang, J.; Li, Y.; Xing, H.; Yuan, J. LGT: Luminance-guided transformer-based multi-feature fusion network for underwater image enhancement. *Inf. Fusion* **2025**, *118*, 102977. [CrossRef]
79. Zhou, J.; Sun, J.; Zhang, W.; Lin, Z. Multi-view underwater image enhancement method via embedded fusion mechanism. *Eng. Appl. Artif. Intell.* **2023**, *121*, 105946. [CrossRef]
80. Yin, M.; Du, X.; Liu, W.; Yu, L.; Xing, Y. Multiscale fusion algorithm for underwater image enhancement based on color preservation. *IEEE Sens. J.* **2023**, *23*, 7728–7740. [CrossRef]
81. Chen, R.; Cai, Z.; Cao, W. MFFN: An underwater sensing scene image enhancement method based on multiscale feature fusion network. *IEEE Trans. Geosci. Remote Sens.* **2021**, *60*, 4205612. [CrossRef]
82. Oren, M.; Nayar, S.K. Generalization of the Lambertian model and implications for machine vision. *Int. J. Comput. Vis.* **1995**, *14*, 227–251. [CrossRef]
83. Tomasi, C.; Manduchi, R. Bilateral filtering for gray and color images. In Proceedings of the Sixth International Conference on Computer Vision, Bombay, India, 7 January 1998.
84. Gavaskar, R.G.; Chaudhury, K.N. Fast adaptive bilateral filtering. *IEEE Trans. Image Process.* **2018**, *28*, 779–790. [CrossRef]
85. Chen, B.H.; Tseng, Y.S.; Yin, J.L. Gaussian-adaptive bilateral filter. *IEEE Signal Process. Lett.* **2020**, *27*, 1670–1674. [CrossRef]
86. Bianco, G.; Muzzupappa, M.; Bruno, F.; Garcia, R.; Neumann, L. A new color correction method for underwater imaging. *Int. Arch. Photogramm. Remote Sens. Spat. Inf. Sci.* **2015**, *40*, 25–32. [CrossRef]
87. Song, W.; Wang, Y.; Huang, D.; Liotta, A.; Perra, C. Enhancement of underwater images with statistical model of background light and optimization of transmission map. *IEEE Trans. Broadcast.* **2020**, *66*, 153–169. [CrossRef]
88. Gan, W.; Wu, X.; Wu, W.; Yang, X.; Ren, C.; He, X.; Liu, K. Infrared and visible image fusion with the use of multi-scale edge-preserving decomposition and guided image filter. *Infrared Phys. Technol.* **2015**, *72*, 37–51. [CrossRef]
89. Burt, P.J.; Adelson, E.H. The Laplacian pyramid as a compact image code. In *Readings in Computer Vision*; Morgan Kaufmann: Burlington, MA, USA, 1987; pp. 671–679.
90. Pajares, G.; De La Cruz, J.M. A wavelet-based image fusion tutorial. *Pattern Recognit.* **2004**, *37*, 1855–1872. [CrossRef]
91. Li, S.; Kang, X.; Hu, J. Image fusion with guided filtering. *IEEE Trans. Image Process.* **2013**, *22*, 2864–2875. [PubMed]
92. Mertens, T.; Kautz, J.; Van Reeth, F. Exposure fusion: A simple and practical alternative to high dynamic range photography. In *Computer Graphics Forum*; Blackwell Publishing Ltd.: Oxford, UK, 2009; Volume 28, pp. 161–171.
93. Agarwala, A.; Dontcheva, M.; Agrawala, M.; Drucker, S.; Colburn, A.; Curless, B.; Salesin, D.; Cohen, M. Interactive digital photomontage. In *ACM SIGGRAPH 2004 Papers*; University of Washington: Seattle, WA, USA, 2004; pp. 294–302.
94. Ronneberger, O.; Fischer, P.; Brox, T. U-net: Convolutional networks for biomedical image segmentation. In *Medical Image Computing and Computer-Assisted Intervention–MICCAI 2015: 18th International Conference, Munich, Germany, 5–9 October 2015, Proceedings, Part III 18*; Springer International Publishing: Cham, Switzerland, 2015.
95. Anwar, S.; Li, C.; Porikli, F. Deep underwater image enhancement. *arXiv* **2018**, arXiv:1807.03528.
96. Qi, Q.; Li, K.; Zheng, H.; Gao, X.; Hou, G.; Sun, K. SGUIE-Net: Semantic attention guided underwater image enhancement with multi-scale perception. *IEEE Trans. Image Process.* **2022**, *31*, 6816–6830. [CrossRef]
97. Li, C.; Anwar, S.; Hou, J.; Cong, R.; Guo, C.; Ren, W. Underwater image enhancement via medium transmission-guided multi-color space embedding. *IEEE Trans. Image Process.* **2021**, *30*, 4985–5000. [CrossRef]
98. Qiao, N.; Dong, L.; Sun, C. Adaptive deep learning network with multi-scale and multi-dimensional features for underwater image enhancement. *IEEE Trans. Broadcast.* **2022**, *69*, 482–494. [CrossRef]
99. Fabbri, C.; Islam, M.J.; Sattar, J. Enhancing underwater imagery using generative adversarial networks. In Proceedings of the 2018 IEEE International Conference on Robotics and Automation (ICRA), Brisbane, QLD, Australia, 21–25 May 2018.
100. Candela, A.; Edelson, K.; Gierach, M.M.; Thompson, D.R.; Woodward, G.; Wettergreen, D. Using remote sensing and in situ measurements for efficient mapping and optimal sampling of coral reefs. *Front. Mar. Sci.* **2021**, *8*, 689489. [CrossRef]
101. Zhu, J.-Y.; Park, T.; Isola, P.; Efros, A.A. Unpaired image-to-image translation using cycle-consistent adversarial networks. In Proceedings of the IEEE International Conference on Computer Vision, Venice, Italy, 22–29 October 2017; pp. 2223–2232.
102. Deng, J.; Dong, W.; Socher, R.; Li, L.J.; Li, K.; Li, F. Imagenet: A large-scale hierarchical image database. In Proceedings of the 2009 IEEE Conference on Computer Vision and Pattern Recognition, Miami, FL, USA, 20–25 June 2009.
103. Wang, Z.; Bovik, A.C.; Sheikh, H.R.; Simoncelli, E.P. Image quality assessment: From error visibility to structural similarity. *IEEE Trans. Image Process.* **2004**, *13*, 600–612. [CrossRef]
104. Zhang, R.; Isola, P.; Efros, A.A.; Shechtman, E.; Wang, O. The unreasonable effectiveness of deep features as a perceptual metric. In Proceedings of the IEEE Conference on Computer Vision and Pattern Recognition, Salt Lake City, UT, USA, 18–23 June 2018.
105. Wang, Y.; Zhang, J.; Cao, Y.; Wang, Z. A deep CNN method for underwater image enhancement. In Proceedings of the 2017 IEEE International Conference on Image Processing (ICIP), Beijing, China, 17–20 September 2017.

106. Wang, Y.; Guo, J.; Gao, H.; Yue, H. UIEC^2-Net: CNN-based underwater image enhancement using two color space. *Signal Process. Image Commun.* **2021**, *96*, 116250. [CrossRef]
107. Zheng, M.; Luo, W. Underwater image enhancement using improved CNN based defogging. *Electronics* **2022**, *11*, 150. [CrossRef]
108. Wang, J.; Li, P.; Deng, J.; Du, Y.; Zhuang, J.; Liang, P.; Liu, P. CA-GAN: Class-condition attention GAN for underwater image enhancement. *IEEE Access* **2020**, *8*, 130719–130728. [CrossRef]
109. Cong, R.; Yang, W.; Zhang, W.; Li, C.; Guo, C.-L.; Huang, Q.; Kwong, S. PUGAN: Physical model-guided underwater image enhancement using GAN with dual-discriminators. *IEEE Trans. Image Process.* **2023**, *32*, 4472–4485. [CrossRef] [PubMed]
110. Peng, L.; Zhu, C.; Bian, L. U-shape transformer for underwater image enhancement. *IEEE Trans. Image Process.* **2023**, *32*, 3066–3079. [CrossRef] [PubMed]
111. Shen, Z.; Xu, H.; Luo, T.; Song, Y.; He, Z. UDAformer: Underwater image enhancement based on dual attention transformer. *Comput. Graph.* **2023**, *111*, 77–88. [CrossRef]
112. Li, C.; Anwar, S.; Porikli, F. Underwater scene prior inspired deep underwater image and video enhancement. *Pattern Recognit.* **2020**, *98*, 107038. [CrossRef]
113. Panetta, K.; Kezebou, L.; Oludare, V.; Agaian, S. Comprehensive underwater object tracking benchmark dataset and underwater image enhancement with GAN. *IEEE J. Ocean. Eng.* **2021**, *47*, 59–75. [CrossRef]
114. Chen, X.; Yu, J.; Kong, S.; Wu, Z.; Fang, X.; Wen, L. Towards real-time advancement of underwater visual quality with GAN. *IEEE Trans. Ind. Electron.* **2019**, *66*, 9350–9359. [CrossRef]
115. Yan, H.; Zhang, Z.; Xu, J.; Wang, T.; An, P.; Wang, A.; Duan, Y. UW-CycleGAN: Model-driven CycleGAN for underwater image restoration. *IEEE Trans. Geosci. Remote Sens.* **2023**, *61*, 4207517. [CrossRef]
116. Hambarde, P.; Murala, S.; Dhall, A. UW-GAN: Single-image depth estimation and image enhancement for underwater images. *IEEE Trans. Instrum. Meas.* **2021**, *70*, 5018412. [CrossRef]
117. Gao, J.; Zhang, Y.; Geng, X.; Tang, H.; Bhatti, U.A. PE-Transformer: Path enhanced transformer for improving underwater object detection. *Expert Syst. Appl.* **2024**, *246*, 123253. [CrossRef]
118. Ummar, M.; Dharejo, F.A.; Alawode, B.; Mahbub, T.; Piran, J.; Javed, S. Window-based transformer generative adversarial network for autonomous underwater image enhancement. *Eng. Appl. Artif. Intell.* **2023**, *126*, 107069. [CrossRef]
119. Wang, B.; Xu, H.; Jiang, G.; Yu, M.; Ren, T.; Luo, T.; Zhu, Z. UIE-convformer: Underwater image enhancement based on convolution and feature fusion transformer. *IEEE Trans. Emerg. Top. Comput. Intell.* **2024**, *8*, 1952–1968. [CrossRef]
120. Zheng, S.; Wang, R.; Zheng, S.; Wang, L.; Liu, Z. A learnable full-frequency transformer dual generative adversarial network for underwater image enhancement. *Front. Mar. Sci.* **2024**, *11*, 1321549. [CrossRef]
121. Moghimi, M.K.; Mohanna, F. Real-time underwater image enhancement: A systematic review. *J. Real-Time Image Process.* **2021**, *18*, 1509–1525. [CrossRef]
122. Banerjee, J.; Ray, R.; Vadali, S.R.K.; Shome, S.N.; Nandy, S. Real-time underwater image enhancement: An improved approach for imaging with AUV-150. *Sadhana* **2016**, *41*, 225–238. [CrossRef]
123. Yang, H.; Xu, J.; Lin, Z.; He, J. LU2Net: A lightweight network for real-time underwater image enhancement. *arXiv* **2024**, arXiv:2406.14973.
124. Zhang, S.; Zhao, S.; An, D.; Li, D.; Zhao, R. LiteEnhanceNet: A lightweight network for real-time single underwater image enhancement. *Expert Syst. Appl.* **2024**, *240*, 122546. [CrossRef]

MDPI AG
Grosspeteranlage 5
4052 Basel
Switzerland
Tel.: +41 61 683 77 34

Journal of Marine Science and Engineering Editorial Office
E-mail: jmse@mdpi.com
www.mdpi.com/journal/jmse

www.ingramcontent.com/pod-product-compliance
Lightning Source LLC
Chambersburg PA
CBHW061453210926
51850CB00008B/1700
* 9 7 8 3 7 2 5 8 5 5 8 5 8 *